A Solar–Hydrogen Energy System

A Solar–Hydrogen Energy System

Eduard W. Justi
Late Professor Emeritus
Technical University
Braunschweig, Federal Republic of Germany

With the collaboration of
P. W. Brennecke
E. Broda
H. H. Ewe
W. Heiland
H. W. Kuhn
and H.-J. Selbach

Translated by
W. Schuh
and K. Claus
with the assistance of
J. O'M. Bockris

Plenum Press • New York and London

Library of Congress Cataloging in Publication Data

Justi, Eduard W.
A solar-hydrogen energy system.

"A revised and updated text based upon an authorized translation from German of Wasserstoff—die Energie für alle Zeiten...by John O'M. Bockris and Eduard W. Justi"—T.p. verso.
Includes bibliographies and index.
1. Solar energy. 2. Hydrogen as fuel. I. Bockris, J. O'M. (John O'M.), 1923–
Wasserstoff—die Energie für alle Zeiten. II. Title.
TJ810.J87 1987 665.8′1 87-7009
ISBN 0-306-42150-X

This volume is a revised and updated text based upon an authorized translation from German of *Wasserstoff—Die Energie für alle Zeiten: Konzept einer Sonnen-Wasserstoff-Wirtschaft* by John O'M. Bockris and Eduard W. Justi.

A Division of Plenum Publishing Corporation
233 Spring Street, New York, N.Y. 10013

Printed in the United States of America

Acknowledgment

This book owes its existence to our common interest in solar–hydrogen energy systems and the much-appreciated encouragement of Professor John O'M. Bockris, Texas A&M University, College Station, Texas.

Eduard W. Justi
Peter W. Brennecke
Henning H. Ewe

Preface

This book concerns one of the more persistent of the ideas that have been discussed in journals devoted to energy science during the last few years. It deals with the concept that hydrogen should be the medium of energy and the sun should be the source (and, in the interim, perhaps also coal, biomass, or nuclear fuel).

The translation has been carried out by Dr. W. Schuh and Mrs. K. Claus in collaboration with me. Certain difficulties confronted us at an early stage, and our resolution of them requires some explanation.

First, the chapters that we received from the original German authors were written at varying times during the 1980s. Some years later, for the anticipated publication in the United States, about half of the chapters were completely rewritten. The translation was done in 1984–1986.

Second, the original volume is a German book. Most of the examples in it refer to the Federal Republic of Germany, although some extend to Europe in general.

We found it inadvisable to attempt to alter this posture. To have brought the chapters all up to date as of, say, January 1986 would have involved many months of painstaking work with the last two or three years of journal articles in the field. The subsequent weighing of what should be incorporated into the text, discussion of the changes with our German colleagues, and actual revision would have consumed many extra months. Finally, the result would not have been a translation of the original book, but would steadily have evolved into a new book of mixed parentage.

As to the book's German character, we thought it best to let this remain unchanged, except for conversion of Deutschmarks to U.S. dollars.

Our attitude was affected largely by the considerable—and seminal—contributions that Eduard Justi and his colleagues have made to the subject of a hydrogen economy since the mid-1960s. It is certainly Justi who must be credited with many early contributions and concepts of hydrogen as a medium of energy to couple with solar sources. Indeed, his contributions follow the substantial advances made by German scientists and engineers to the development of synfuels, particularly such early work as the original studies on synthesizing gasoline from coal.

Thus, a person reading this book views the hydrogen economy through the eyes of Europeans, and in this sense there are certain nuances that differ from those current in the United States.

Other difficulties that faced us in translation were the ones that face all translators: The task of conveying sense from one language to another is that of striking a balance between exactness of translation and euphony in the phrasing in the English language. We have tried to come up with a translation that gives the sense of the former and the sound of the latter. I suppose we have made mistakes, and where they occur I take responsibility for them.

John O'M. Bockris

College Station, Texas

Contents

Chapter 1

Investigation, Evaluation, and Recovery Plan for an Ailing Energy Economy 1

1.1. Low Conversion Efficiency and the Effect of Power Plant Size on Cost 1
1.2. Cooling-Water Shortages Limit the Construction of New Power Plants near Population Centers 6
1.3. CO_2 Pollution of the Atmosphere Threatens a Harmful Change of Climate 8
1.4. Cost of Long-Distance Energy Transfer through Pressure Gas Pipelines May Be Significantly Cheaper than the Cost of Electrical Transmission 12
1.5. Recent Advances toward a Hydrogen Technology 17
1.6. The Possible Structure of a Hydrogen Economy 18
1.7. Right Time for a Transition from a Hydrogen to a Solar–Hydrogen Economy 20
1.8. Conclusions 23
References 24

Chapter 2

The Hydrogen Economy 27

2.1. Causes of Prospective Energy and Raw Materials Shortages 27
2.2. Future Energy Sources and Their Media 29
2.3. Future Energy Medium 30

2.4. Origin of the Hydrogen-Economy Concept 34
References 40

Chapter 3

Time Frame for Building a Hydrogen Technology 43

3.1. Energy Supplies and Energy Consumption 43
3.2. Energy Needs in the Federal Republic of Germany 47
3.3. Exhaustion of the Primary Energy Carriers 49
3.4. Controlled Nuclear Fusion 54
3.5. Fast Breeder Reactor 60
3.6. Hydrogen Technology 63
3.7. Time Frame for the Introduction of a Hydrogen Technology 64
References 65

Chapter 4

Direct Energy Conversion 69

4.1. Conversion Instead of Production 69
4.2. Direct and Indirect Energy Conversion—The DEC Matrix 69
4.3. Selected Examples of Direct Energy Conversion Effects 76
4.4. Conversion (and Production) of Wind Energy 76
4.5. Photovoltaic Direct Energy Conversion 80
4.6. Thermoelectric Direct Energy Conversion Using the Seebeck Effect 81
References 87

Chapter 5

The Basis for the Use of Solar Energy 89

5.1. Characteristics of Solar Radiation 89
5.2. Population and Living Standards 95
5.3. Use of Solar Energy on a Small Scale 96
5.4. Methods for the Collection and Conversion of Solar Energy 99
5.5. Use of Oceanic Thermal Gradients 108
5.6. Silicon Protective-Layer Cells 111
References 119

Chapter 6

Solar Cells and Solar Power Stations 123

6.1. Technological Problems Associated with Lowering the Costs of Terrestrial Silicon Solar Cells 123
6.2. Fundamental Considerations of Polycrystalline Silicon Solar Cells 124
6.3. Description of Silicon Manufacture 129
6.4. Long-Lived Heterojunctions from Thin-layer Solar Cells Consisting of CdS-$Cu_{2-x}S$ 133
6.5. Service Life of Cadmium Sulfide Solar Cells 135
6.6. Procedures for the Production of Thin-Layer Solar Cells from CdS–$Cu_{2-x}S$ 135
6.7. Solar Thermal Generating Stations with Optical Concentrators 139
6.8. One-Megawatt Solar Power Tower of the European Economic Community 142
6.9. 400 kW_{th} High-Temperature Solar Experimental Plant at Georgia Institute of Technology 144
6.10. Solar Power Plants in the United States 148
6.11. Solar Satellite Power Stations 151
References 154

Chapter 7

The Photolytic Production of Hydrogen 157

7.1. The Production of Hydrogen by Means of the Photochemical Decomposition of Water in Plants 157
7.2. An Introduction to the Production of Hydrogen by the Photochemical Decompostion of Water Using Monomolecular Layers 164
7.3. Photoelectrochemical Production of Hydrogen 170
References 174

Chapter 8

The Electrolytic Process for the Production of Hydrogen 175

8.1. Thermodynamics of Water Decomposition 176
8.2. Construction of Practical Water Electrolyzers 179
8.3. Future Possibilities and New-Type Electrolyzers 184

8.4. Eloflux Water Electrolysis Cell 187
8.5. Thermochemical Processes 193
8.6. Conclusions 195
References 195

Chapter 9

The Transmission of Energy over Large Distances 199

9.1. Direct Transmission of Electrical Energy 199
9.2. Transmission through Directed Microwave Radiation 200
9.3. Transmission by Means of Hydrogen 201
9.4. Differences between Pipeline Networks for Natural Gas and for Hydrogen 203
9.5. Operating Hydrogen Pipelines and Other Networks 206
9.6. Distribution of Hydrogen in Transportable Steel Cylinders 209
9.7. Energy Storage and Transport with Liquid and Slush Hydrogen 212
9.8. Conclusions 213
References 214

Chapter 10

The Transmission of Hydrogen in High-Pressure Pipelines and the Storage of Hydrogen in Pipes 217

10.1. Calculations for a 2150-Kilometer Hydrogen Pipeline and Distribution Network with a Capacity of 10^{10} Nm^3 H_2 per Year and with Three Pressure Stations Used for Transmission at 100, 60, and 40 Atmospheres 218
10.2. Thermodynamic Optimization of Hydrogen Transport and Pipeline Storage 238
10.3. Quantitative Calculation of the Recovery of Energy 238
References 240

Chapter 11

The Storage of Hydrogen 243

11.1. Thermal Energy Storage 243
11.2. Electrochemical Energy Storage 245
11.3. Superconducting Magnets 246
11.4. Energy Storage in Flywheels 246
11.5. Storage of Hydrogen 247

11.6. Conclusions 261
References 261

Chapter 12

Safety Aspects of Using Hydrogen 265

12.1. Physical Data and Safety-Engineering Quantities 265
12.2. Physical Dangers 268
12.3. Chemical Dangers 268
12.4. Safety Instructions 269
12.5. Experience in Safety Aspects of Dealing with Hydrogen 270
12.6. Conclusions 270
References 271

Chapter 13

The Conversion of Hydrogen into Electricity by Means of Fuel Cells 273

13.1. Introduction 273
13.2. Highly Reversible Production of Electrical Energy from Hydrogen by Means of Hydrogen–Oxygen Fuel Cells 273
13.3. Schematic Construction of a Hydrogen–Oxygen Fuel Cell 275
13.4. Alkaline Low-Temperature Fuel Cells with Raney Catalysts 276
13.5. Medium-Temperature Fuel Cells with Phosphoric Acid Electrolyte 279
13.6. Conclusions 285
References 285

Chapter 14

The Catalytic Combustion of Hydrogen 287

14.1. Introduction 287
14.2. Direct Combustion of Hydrogen 288
14.3. Catalytic Combustion of Hydrogen 289
14.4. Properties of Catalytic Hydrogen Burners 290
14.5. State of Development of Catalytic Hydrogen Burners 291
14.6. Safety Aspects of Catalytic Combustion 294
14.7. Prospect 294
References 295

Chapter 15

Industrial Applications of Hydrogen 297

15.1. Ammonia Synthesis 298
15.2. Synfuel Production 299
15.3. Direct Reduction of Iron Ore 304
15.4. Other Possibilities for the Industrial Use of Hydrogen 307
15.5. Prospects 307
References 308

Chapter 16

Hydrogen as a Fuel in Automotive and Air Transportation 309

16.1. Electrical Propulsion of Vehicles by Means of Fuel Cells and Electric Motors 310
16.2. Possible Use of Internal Combustion Engines Powered by Hydrogen for Automotive Transportation 311
16.3. Hydrogen as an Aircraft Fuel 317
16.4. Safety Aspects 319
16.5. Costs 320
16.6. Problems of the Transition 321
References 322

Index 325

CHAPTER 1

Investigation, Evaluation, and Recovery Plan for an Ailing Energy Economy

The first right of a reader is to learn what a book is all about. Therefore, let it be said that the term "hydrogen economy" refers to an energy system in which all energy sources are used to produce hydrogen (H_2); this hydrogen is then collected, stored, and distributed as a completely pollution-free, multipurpose fuel. In the case of a "solar–hydrogen economy," at least some of the primary energy that produces this hydrogen comes from direct sunlight. This precise though nonspecific definition will be illustrated and discussed in great detail in the following chapters.

1.1. LOW CONVERSION EFFICIENCY AND THE EFFECT OF POWER PLANT SIZE ON COST

If we criticize as too naïve those who blame all of our energy difficulties on the Arab states, we ought to be self-critical enough to consider whether we ourselves are wasting most of the energy we now obtain from oil, natural gas, coal, and nuclear power. For example, the gasoline engines that power our cars are the pride of the modern automotive industry, but they convert into motion only some 18% of the primary heat energy produced by combustion of gasoline. This means that 82% of the primary heat energy produced in the combustion engine is waste and expelled through the cooling systems in radiators and exhaust pipes; it also means that the atmosphere is polluted with waste gases and excessive heat. The most efficient coal power plants in Europe now have thermal efficiencies of 37% and therefore a 63% energy loss, almost two thirds. For light-water reactors, the thermal efficiencies are only 29%, with a corresponding loss of 71%. When energy losses are higher than those theoretically predicted it should be reduce these losses so that an increase of primary energy use becomes unnecessary and supplies of fossil fuels are preserved.

The remarkable, original work of Woodwell brought out the fact that atmospheric temperatures are increasing due to the greenhouse effect and that this can be attributed partly to an increase in CO_2 emissions into the atmosphere as a result of combustion processes and partly to the deforestation of tropical rain forests, resulting in a massive loss of the vegetation that absorbs excess CO_2 from the atmosphere.

Such matters, and basic facts about hydrogen, are discussed in this chapter. We have drawn on the advice of a noted expert in energy technology who is ideologically unbiased and not a member of any lobby. In his study "Is the development of large power units in power plants justified?," Buch[1] initially points out that reconstruction of German electricity supplies—from World War II until 1952—began with power outputs of up to 50 megawatts (MW), followed by power outputs of 100 MW until 1955; by 1964, the capacity of units rose to 150 MW and by 1971 to 300 MW. Not until 1975 did outputs of coal power plants double to about 600–700 MW. Nuclear power plants now give power outputs up to 1300 MW. Extensive practical experience has shown that individual construction costs are lower for plants with greater power outputs. Comparing costs between 1 coal power plant of 1200 MW and 12 plants of 100 MW, the larger plant functions at savings of more than 10% as a rule of thumb. According to Buch, however, greater concentration of power output in single locations will become more difficult due to:

1. Lack of available amounts of water or air as coolants needed to obtain adequate condensation temperatures, involving sufficiently low pressures of the condensing vapors.
2. Energy losses involved in long-distance transportation of high-voltage power to consumer centers.
3. Pollution of the environment by fumes and steam from cooling towers and by noise.
4. Operational availability and limited reserve storage in the system.

What effects would there be if a 1200-MW unit were divided into two 600-MW ones, or into four 300-MW plants? In earlier years, due to smaller power outputs and sufficient cooling water, it was possible to use vapor pressures of only 0.03 atm in the condensers, equivalent to condensation temperatures of 24°C. At present, at 600 MW, pressures of not less than 0.07–0.08 atm corresponding to 39–41°C can be attained, and at 1000 MW, particularly in the summer, with amounts of steam at the colder ends of steam turbines, vapor pressures of barely 0.1 atm, equivalent to condensation temperatures of 46°C, are usual.

Starting with these simple facts, Buch shows some interesting diagrams of numerical values of important parameters as a function of unit output. Figure 1.1. shows various values as functions of power output in MW; the top lines represent the heat consumption in kcal/kWh with two different condenser pressures (0.1 and 0.04 atm, respectively) as parameters. Below these, there are two curves for installation costs. The top curve shows the construction costs with

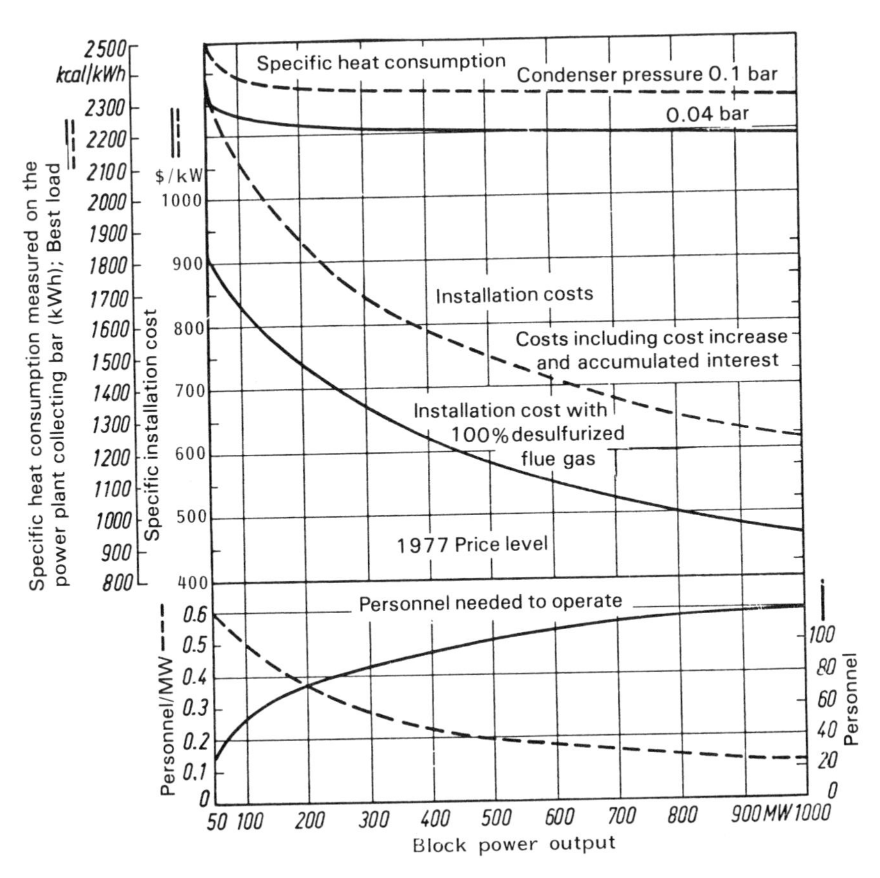

FIGURE 1.1. Modern coal power plant data as a function of block power output. *Top:* Specific heat consumption with 0.1-atm condenser pressure; *below:* with 0.04 atm. Both curves refer to the ordinate above left. *Middle:* Installation including cost increases and accumulated interest (dotted line) beneath, with 100% desulfurized flue gases; refer to the second ordinate on the left (1977 costs). *Bottom:* Personnel needed for MW block power output (left ordinate) in relation to total personnel. Live-steam condition before turbine 125–225 atm/525–530°C need of power plant (right ordinate). Heating at 27–36 atm/525–530°C, feed water temperature 230–260°C. From Buch.[1]

100% desulfurized flue gases, including cost increases and accumulated interest payments; which may extend over ten years due to lengthy authorization procedures, the center curves shows construction costs using desulfurized flue gases. The two curves at the bottom of the diagram show the total service personnel and a curve representing personnel per MW; this shows a decrease from 0.6 personnel per MW at 60 MW to 0.1 personnel per MW at 1000 MW. The total price for the plant drops from \$931/kW for the 50-MW plants to \$494/kW at 1000 MW, in which about \$61/kW arises from desulfurization of the stack gases, which has recently become a legal requirement and without which new construction of large coal-driven power plants is not permitted (however, such plants have not yet been in practical operation). The influence of condensation pressure on specific heat consumption is especially important, as shown in the uppermost pair of curves, for it rises by 5.6% with increase of condensation pressure from 0.04 to 0.1 atm (cf. Figure 1.5). It can easily be seen that these three numbers, which are for conventional coal-driven power plants, improve considerably with increasing power output. In the same way, the area needed for the construction of power plants (an important part of the economic analysis) can be reduced from 52 m^2/MW with 12 units of 100 MW each to 12 m^2/MW when the units are 1200 MW each (Figure 1.2). In Figure 1.3, production costs relevant to

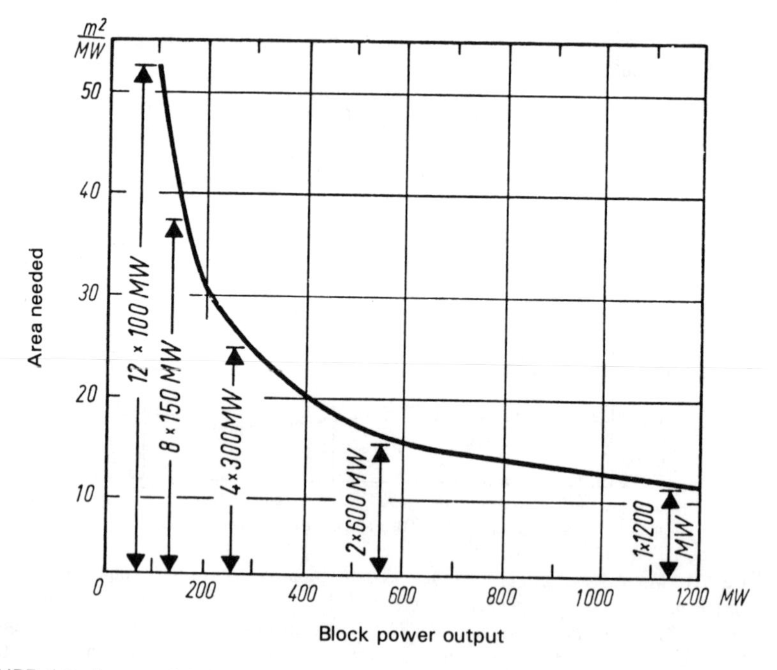

FIGURE 1.2. Area needed for coal power plants with a block power output of up to 1200 MW and divided into 1 × 1200 MW, 2 × 600 MW, 4 × 300 MW, 8 × 150 MW, and 12 × 100 MW. Cost reduction is with the larger plant size. From Buch.[1]

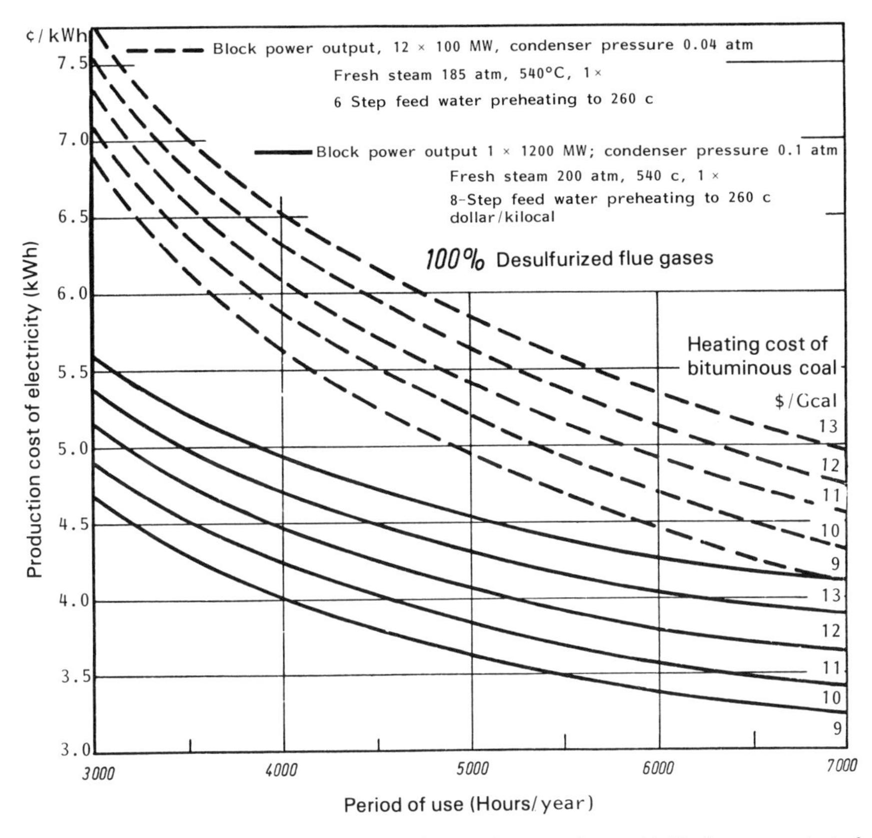

FIGURE 1.3. Production costs of electricity from coal power plants with block power output of 1 × 1200 MW, 4 × 300 MW, and 2 × 600 MW. Production data are given in the figure. From Buch.[1]

regular users are shown. The price is quoted in cents per kWh, determined by the period of use each year (hr/yr); the dashed lines show 12 units each of 100 MW at a condensation pressure of only 0.04 atm, and the continuous curves show one unit of 1200 MW at a 0.1-atm condensation pressure. The parameters of five dashed curves and those of the continuous curves indicate the calorific price of the bituminous coal, per Gcal; this figure gives fresh information because costs of the desulfurization of stack gases are taken into account. It can be seen at a glance that the price of power production for the 12 units of 100 MW is 50% more expensive than that for the use of large plants of 1200 MW, even though this already takes into account that condensation in the large unit takes place at 0.1 instead of 0.04 atm, because the total unit size does not contain as much coolant as that of several smaller, decentralized units distributed around consumption centers and providing the same total power output (Figure 1.4).

This difference in condensation pressures, which has a significant effect when comparing production costs, can be eliminated as shown in the upper pair of curves in Figure 1.1., cf. also Figure 1.5, by reducing the price of energy in the large unit by 5.6%, corresponding to the reduction in heat consumption.

Buch compares advantages and disadvantages of these building units on the basis of their cost effectiveness determined by the specific conditions at a given location. Comparing power plants on a purely thermal and financial basis shows an approximately 11% superiority of large units compared with smaller ones.

1.2. COOLING-WATER SHORTAGES LIMIT THE CONSTRUCTION OF NEW POWER PLANTS NEAR POPULATION CENTERS

Because these conclusions are independent of the type of fuel source, be it coal, natural gas, oil, or nuclear fission, we want to highlight the basic dilemma involved in choosing an appropriate site. On the one hand, continuously increasing requirements for environmental protection and project designs for pure air and water, and lack of availability of still untapped potential coolants, drive the trend to divide power plants into smaller units which are more suited to the smaller cooling capacities still available, as shown in Fig. 1.6. This strategy would mean not just a departure from building more profitable, more powerful units, but even a reversal to earlier practices, which would mean higher specific construction and power production costs. Conversely, the technical, financial, and environmental improvements in terms of free enterprise would involve transition to larger power units. Nuclear power plants in particular, with their large power outputs, strict safety precautions, and lengthy licensing procedures, have

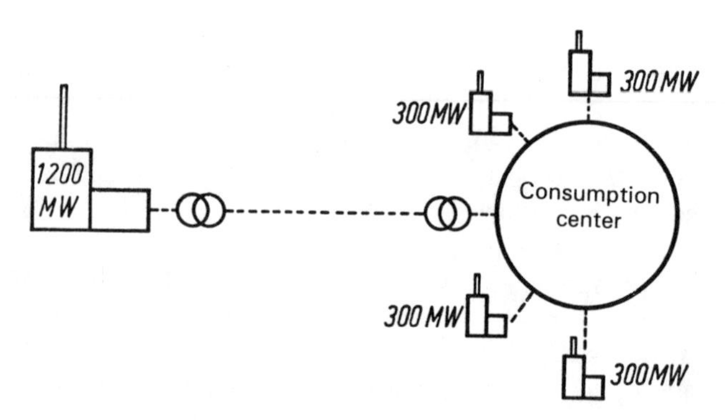

FIGURE 1.4. Diagram of a centralized and a decentralized design for block power output. Despite higher production costs, the decentralized design must be preferred because of the lack of a sufficient supply of coolant. From Buch.[1]

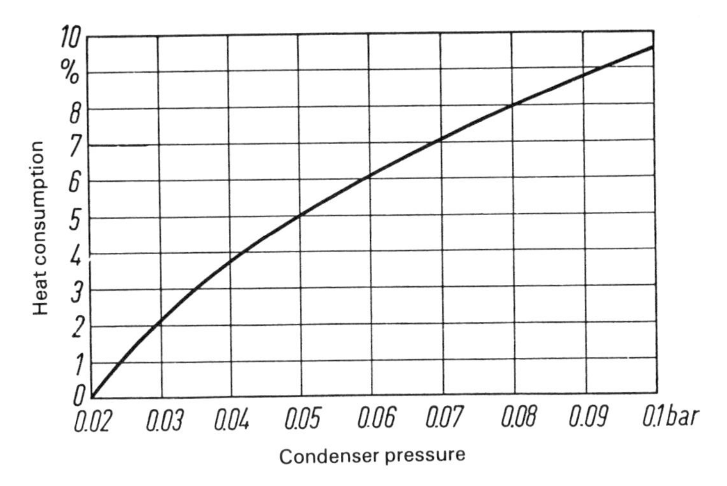

FIGURE 1.5. Influence of the vacuum (condenser pressure) on the additional specific consumption of bituminous coal power plants. From Buch.[1]

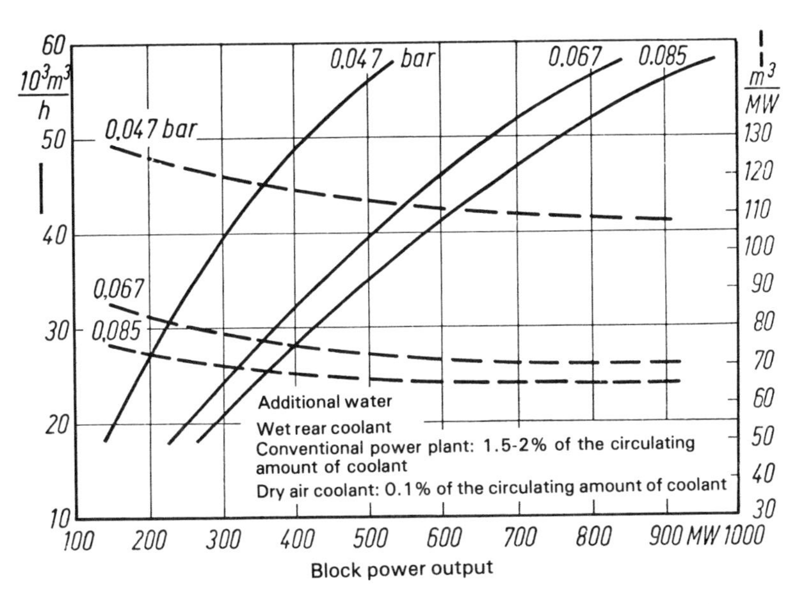

FIGURE 1.6. Dependence of the flow of coolant on the power output and degree of vacuum. *left ordinate*—solid curves: coolant consumption in 10^3 m^3/hr (from left to right) at 0.047, 0.067, and 0.085 atm as a function of block power output from 100 to 1000 MW (abscissa). *Right ordinate*—dashed curves: specific coolant consumption in m^3/MW with (top to bottom) 0.047, 0.067, and 0.085 atm. From Buch.[1]

made us aware that large power plants, contrary to present principles, should not be located near towns, but should be as far as possible from all human habitats.

In fact, United States guidelines laid down by the Nuclear Regulatory Commission specify the allowable population density within a certain radius around a reactor. The more recent German guidelines[2] contain directions less stringent than the American ones regarding population density, but more demanding with respect to safety precautions. The German guidelines require that the reactor sites be in relatively remote locations, in unpopulated isolated areas but supplied with sufficient coolants, or on islands, along coasts, or on anchored floating platforms, several hundred kilometers from the shore, with virtually unlimited potential for coolant water. These placement recommendations apply not only to nuclear power plants, but also increasingly to thermal power plants, despite the newly introduced legislation for the extensive desulfurization of stack gases (see Figure 1.1). This is due to the toxic effects of heavy-metal impurities in fly ash, which cannot be retained by even the finest of electrofilters and which enter the lungs. Furthermore, the PTB, the highest standards authority in West Germany, has determined that exhaust gases in a modern 300-MW bituminous coal power plant also contain radioactive impurities, which include 30 mCi Pb and 4 mCi Ra; these could lead to an annual bone marrow irradiation of about 14 mrem. By itself, this is harmless, but it is 100 times greater than the amount emitted by a nuclear power station of about the same size.[3]

1.3. CO_2 POLLUTION OF THE ATMOSPHERE THREATENS A HARMFUL CHANGE OF CLIMATE

Having considered the thermal, toxic, and radioactive pollution dangers of various types of power plants, and the preventive measures that can be taken, we would be remiss if we were to ignore the effects of CO_2, the raw material for plants. CO_2 causes concern due to the massive quantities of 10^{15} g/year ($\sim 5 \times 10^9$ tons/yr) emitted annually into the stratosphere as combustion products of fossil fuels. Until recently, the only data available were those shown in Figure 1.7, which shows an increase of 10–15% from 290 to 330 ppm since the beginning of industrialization, associated with an average increase in air temperature of about 0.5°C. This thermal increase has caused fears among climatologists because of possible consequent worldwide climatic changes, growth of desert areas, and the eventual melting of the arctic ice caps. Such changes would be caused by diminished infrared permeability of the atmosphere, resulting in decreased heat retention within the atmosphere of our planet ("greenhouse effect"). However, the small magnitude of the temperature increase which has occured to date, 0.5°C, the scattering of the data, and the resulting uncertainty to date, 0.5°C, the scattering of the data, and the resulting uncertainty of the

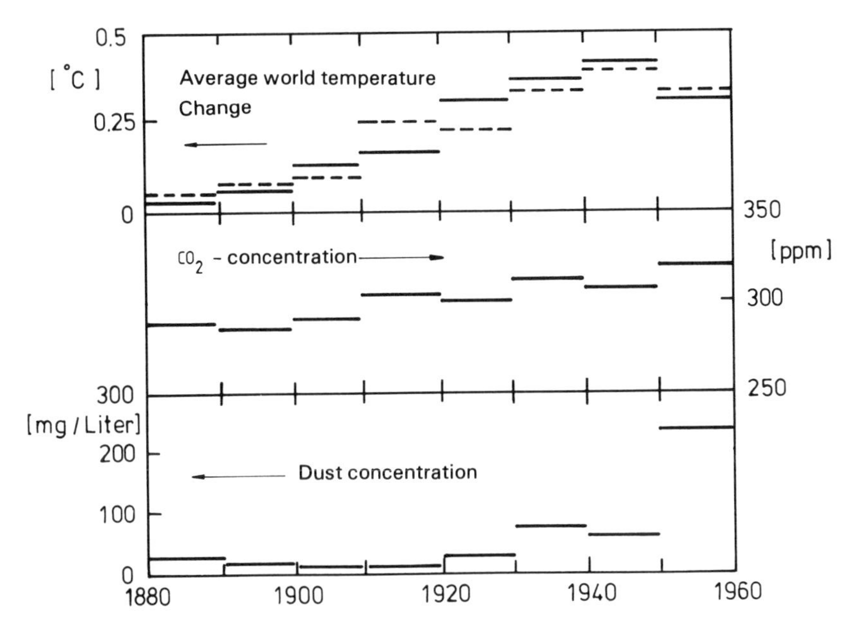

FIGURE 1.7. Changes in the earth's atmosphere since the beginning of the industrial revolution (1880) until 1960. *Top* (*left ordinate*): Increase of the average temperature of the earth's atmosphere. *middle* (*right ordinate*): Increase in the average CO_2 concentration in ppm from 1880 to 1960. These historical data are comparable to those of modern precise measurements by Keeling *et al.* (Figure 1.9). *Bottom* (*left ordinate*): Increase in the average dust concentration in mg/liter, which works against the effect of CO_2. From Bockris and Appleby.[19]

of the relationships among the three variables represented in Fig. 1.7 did not give the earlier expert opinion much support in seeing the increase in concentration of CO_2 as a definite environmental threat.

In recent times, a sharp change has been brought about by improved assessment methods and continuous recording of precision measurements in strategic locations around the planet. These indicate that the CO_2 content of the atmosphere could double by the year 2020 (Figure 1.8), due not only to the burning of fuels but also to the worldwide destruction of forests and with them their CO_2-fixing structures, trees and plants. Such destruction may occur in temperate climates, especially among the big trees in tropical rain forests in Africa and the Amazon basin.

For a convincing presentation of the increase in knowledge that has been obtained in recent times in terms of experimental and theoretical work, it is useful to look at the diagram in Figure 1.9. These data were obtained in Hawaii by Keeling *et al.*,[4] with automatic recordings of the monthly averages of the CO_2 content; according to these data, there was a 5% increase between 1958 and 1967. It is remarkable that reproducible oscillations of CO_2 content (scatter:

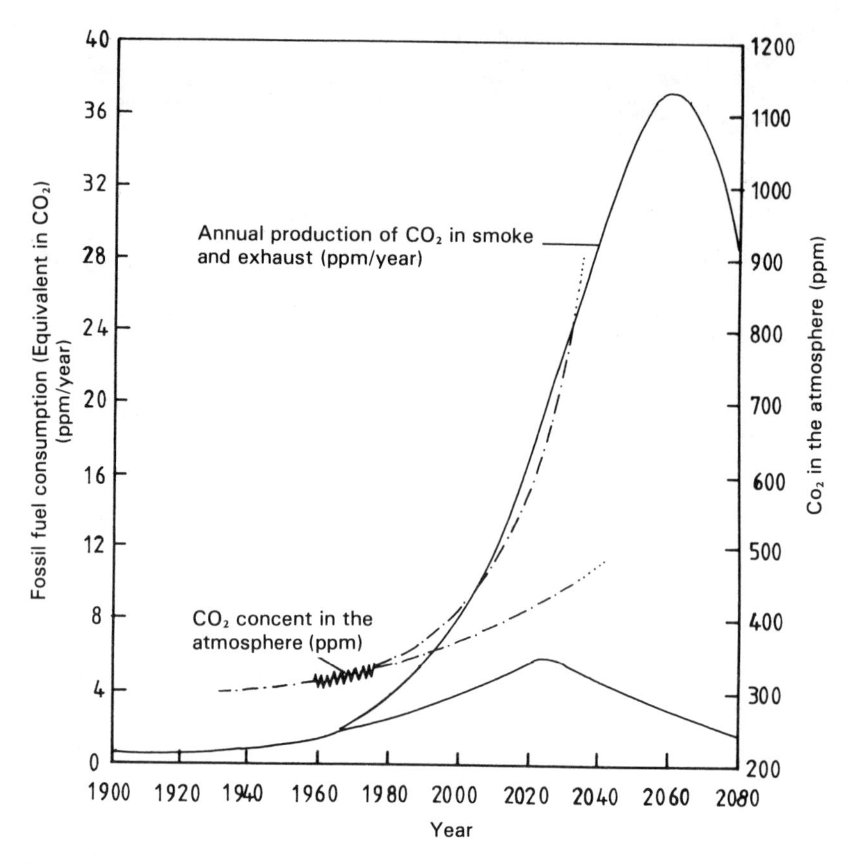

FIGURE 1.8. Present measurements and future predictions for the consumption of mineral and plant fuels, measured in CO_2 equivalents (ppm CO_2/yr) (left ordinate), as a function of the years 1900–2080 (abscissa). The bottom curve is a minimal and the top curve is a maximum estimate, both with a maximum between the years 2020 and 2060. Right ordinate: CO_2 content in the atmosphere (ppm), extrapolated under the best (bottom curve) or worst (top curve) assumptions. From Woodwell.[5]

0.5 ppm) can be detected. The period from spring to late summer shows a minimum CO_2 content due to increased photosynthesis and CO_2 absorption during the growth period, particularly in the northern hemisphere, with its extensive mainland and grassland areas. Maxima in CO_2 content during decreasing photosynthesis periods occur in late winter. Observations in carefully chosen locations showed that these oscillations are caused by the tropical rain forests with their giant trees. These trees contain 42% of the stored carbon in terrestrial vegetation, and continuously produce 32% of all vegetative substances. Altogether, the forests of the world store 90% of the vegetative carbon. The total grassland areas, however, store only 3%, and the cultivated land areas only 1%, of all combined carbon. The hope that plankton in the oceans would bind significant amounts of CO_2 has proved to be disappointing, and an international study shows that the "biota" (all living organisms) do not, in fact, reduce the CO_2, but are in fact a source thereof. The existing situation arises as a result of the breakup of the great forests of earlier times, and the subsequent decay of plants, humus, and peat, of which forest soil is composed, to form CO_2, H_2O, and heat.

To quote some numbers: The atmosphere itself contains about 700×10^{15} g carbon and its equivalents; biota $1000–3000 \times 10^{15}$; and the surface waters of the oceans (to a depth of about 100 m) 600×10^{15}. This is to be compared with only 5.10^{15} g of carbon arising from combustion processes, some 2.3×10^{15} g of which stay in the atmosphere. If, therefore, the grasslands neither gave off nor absorbed CO_2, the ocean surfaces would create a sink of 2.7×10^{15} g/yr. This seems plausible to oceanographers; the deep oceans represent the most extensive CO_2 sink (about 3500×10^{15} g CO_2), but further passage of CO_2 into the deep seas is slowed by the well-known inversion layer (the sea is warm at the top and decreases in temperature with depth).

Yet this CO_2 balance must be regarded as dubious because the latest esti-

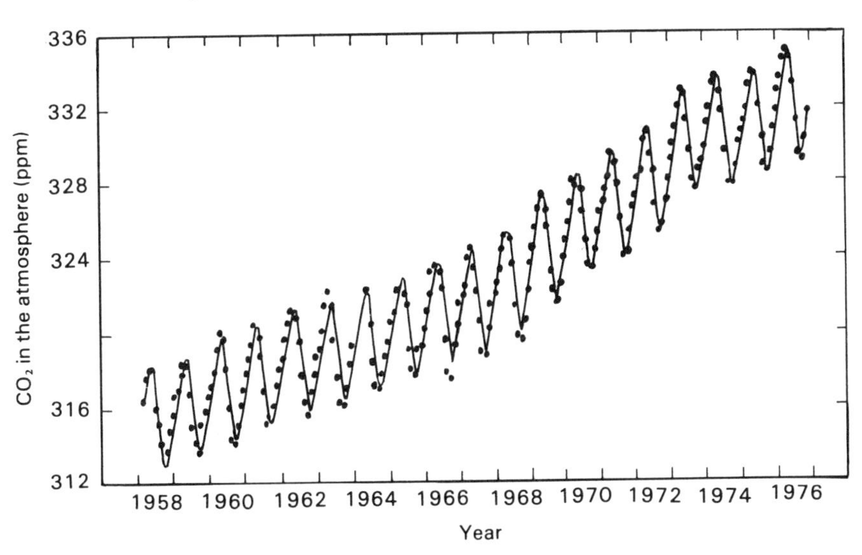

FIGURE 1.9. Precise data for the CO_2 content (ppm) of the atmosphere over Hawaii from Keeling *et al.*[4] as a function of the time of year, 1958–1976. Besides the accelerated increase in the course of the years, one recognizes the regular seasonal periods with CO_2 minimum in spring until later summer, as a consequence of the growth period of the green plants, and a maximum CO_2 consumption for photosynthesis and the corresponding CO_2 maximum as a consequence of reduced photosynthesis until late winter. Note the exact measurements within a few parts per million in the measuring periods in which the average increase from 316 to 332 ppm between 1958 and 1976 is convincingly shown. From Woodwell.[5]

mates of Woodwell[5] show that with the present rate of forest destruction of 1% per year, there is a net release of CO_2 from forests to atmosphere of about 6×10^{15} g CO_2, and further about 2×10^{15} g from oxidation of exposed humus. Thus, altogether, 8×10^{15} g CO_2, with limits between 2 and 18×10^{15} g CO_2/yr, is given off, which is far in excess of human combustion activities. These upper and lower limits are plotted in Figure 1.8 for the years between 1900 and 2080, in which the man-made emission rates are expressed in ppm/yr (lower set of curves, left ordinate) and the total CO_2 content of the atmosphere in ppm (both upper curves, right ordinate) and are explained in the caption. Thus, the CO_2 content of the atmosphere can increases from the present value of 330 to 1150 ppm before 2080—an approximate 3.5-fold increase. Even the most optimistic option (lower curve), showing a mere doubling of the CO_2 content, could give rise to very significant global climatic changes (T = 1.2–2.4°C according to Plass[6]): "There is no aspect of national or international politics which could remain unaffected by the prospect of global climatic changes." According to the most recent data, emission into the atmosphere through combustion processes is less than CO_2 emission due to forest decay. Nevertheless, our energy technology should primarily limit the burning of fossil fuels because an appeal to reason and a sense of understanding of the energy industry problem seems easier than any limiting of population growth in Africa and South America (attempts to limit population growth in India have had very limited success). A temporary solution has been suggested by Bockris,[7] who proposed that future coal power plants be built only on the shore, a solution stimulated by the diagram constructed by Hammond[8] (Figure 1.10) of a floating nuclear station. Coal could be delivered to floating islands cheaply by seaborne transports, the problem of sufficient cooling water discussed above would be solved, noise pollution would be eliminated, and the CO_2 could be discharged into the deep sea. Biologists, oceanographers, and meterologists all warn against massive flooding caused by increasing CO_2. Correspondingly, some modification has occurred in the attitude toward nuclear energy—almost totally rejected until now (compare Broda[9])—for some authors have recently recommended the development of nuclear power stations instead of thermal power stations burning fossil fuels (coal, oil, and natural gas). Woodwell concludes his report[5] with the following sentence: "CO_2, hitherto an apparently harmless trace gas in the atmosphere, can rapidly assume a central role as a significant threat to our present society."

1.4. COST OF LONG-DISTANCE ENERGY TRANSFER THROUGH PRESSURE GAS PIPELINES MAY BE SIGNIFICANTLY CHEAPER THAN THE COST OF ELECTRICAL TRANSMISSION

From our analysis of electrical power plants, we have concluded that the basic tendency is for all thermal power plants, no matter whether they burn fossil

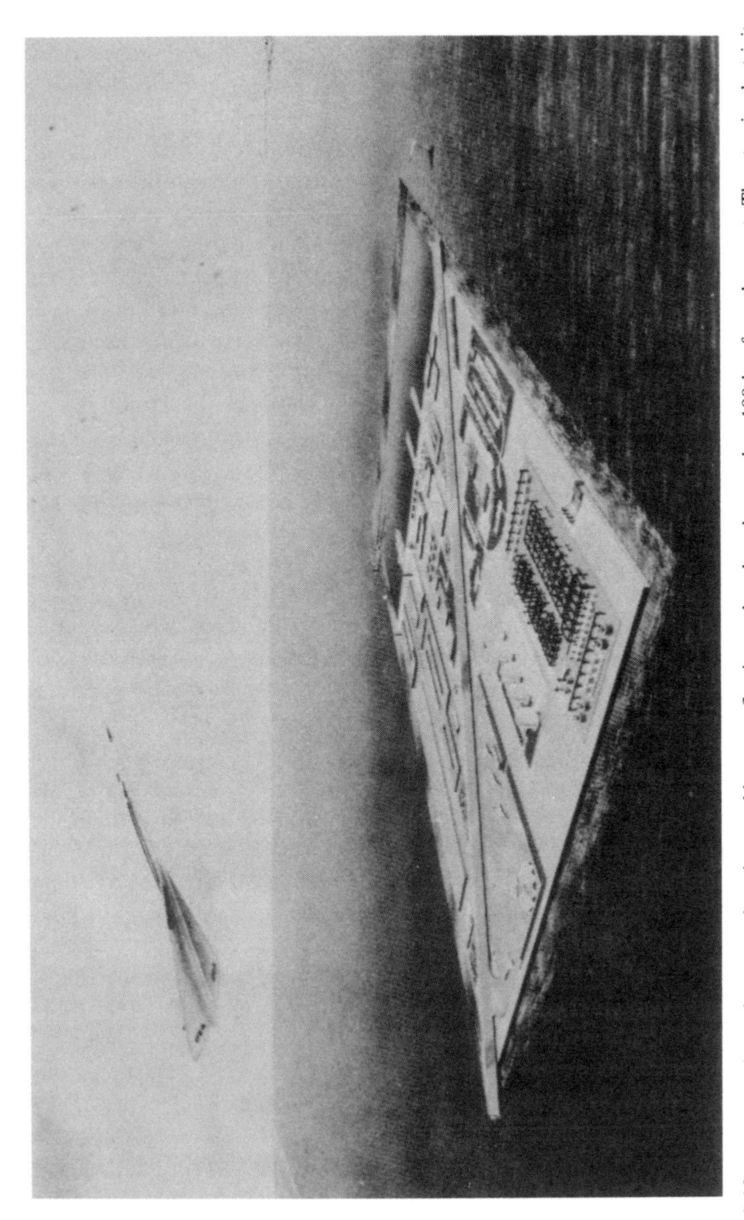

FIGURE 1.10. A proposed atomic power plant that would operate floating and anchored more than 100 km from the coast. The atomic electricity would be utilized for the production of H_2, which would be brought to land through a pipeline under the sea or by ships as frozen liquid H_2 as proposed by Bockris, who has also proposed (1961) a corresponding setup for photovoltaic or wind-powered generators. Coal power plants, floating near the coast, would have enough cold cooling water with a minimum environmental pollution and the possibility of running the exhaust fumes containing CO_2 into the sea would be an added advantage. Picture from Hammond.[8]

fuels such as coal, oil, or natural gas or run on atomic fission or fusion, to be moved from locations in overcrowded regions and settled away from population centers, e.g., in desert areas or along coastlines where coolants are plentiful and environmental problems can be minimized. The main problem arising from these considerations is the increased expense of transporting energy from peripherally located electric generators to consumption centers. Transporting electricity through high-tension cables over distances of more than about 1000 km becomes too expensive, as explained in detail in Chapter 9. Is there a cheaper alternative for energy transfer over long distances?

The decisive experiment is, of course, the largest city in the highest energy-consuming nation, New York. The most important source of primary energy for New York is natural gas from Texas, which is sent through 2000 km of steel pipes from the South; only that portion of the energy that is absolutely necessary (about 20%) is transformed into electricity at the site of consumption. There is no waste of the energy by passage through high-voltage lines, because it is sent through high-pressure steel pipes with relatively low investment and running costs.

Comparative cost figures for different electrical and gas lines as a function of distance have been prepared by the American Gas Association.[10] In Figure 1.11, we show the diagram for the investment cost per power unit, converted to SI units. The abscissa shows transportation distances from 0 to 1000 km, the right-hand ordinate shows investment costs in dollars/kilowatt, which is converted on the left-hand ordinate to dollars/kilocalorie per hour for convenience. Considerable differences are apparent. Thus, investment at 1000 km for natural gas costs only about $20/kW, compared to the $125/kW for 500 kV electrical cable, a value that can be reached by underground cables at as short a distance as 100 km. If, instead of natural gas sources, bituminous coal or nuclear power stations were used in respect to similar distances, the transmission costs of the energy they deliver could not be calculated from the calculation for natural gas because it is not possible to convert the electrical output of thermal stations into natural gas without using correspondingly large amounts of carbon. A carbon-free synthetic gas is hydrogen, which can be obtained by means of conventional electrolysis from generally available water, according to the reaction $H_2O = H_2 + \frac{1}{2}O_2$, or by any one of the ten different methods shown in Table 1.1. Table 1.2 lists ten of the important advantages of H_2 as a universal energy medium. There is, for example, 100% environmental safety because the combustion product is simply pure water vapor, which reenters the natural water cycle of the atmosphere without treatment or equipment (Figure 1.12).[12]

In the capital cost–distance diagram (Figure 1.11), the costs of the transport of H_2 obtained from electrolysis are shown in the cross-hatched region, the upper edge of which represents the current commercial price of hydrogen and the lower edge indicates the anticipated lowest cost for the year 2000. The points at which

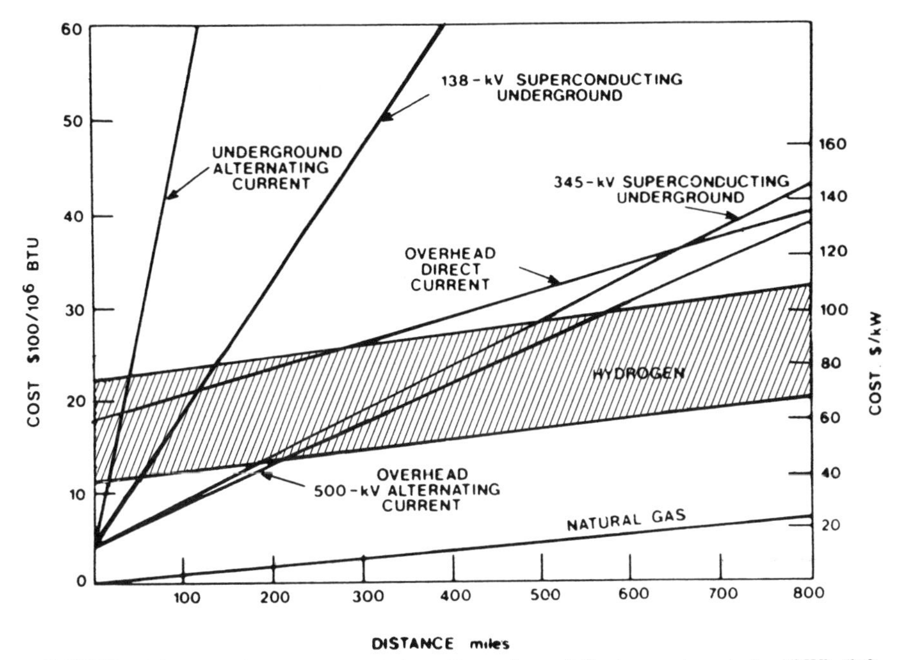

FIGURE 1.11. Cost of energy transmission. Comparison of the transport costs in ¢/kWh (left ordinate) or in dollars/kW (right ordinate) as a function of the transport distance in miles (abscissa).

the other lines intersect depend on fixed costs and the estimated cost of transmission. The line for transport of 500 kV A.C. intersects the topmost line for hydrogen transport at about 600 mi and the lowest hydrogen line at about 200 km. This means that transporting energy through 500 kV lines is cheaper when distances are less than 600 mi, but for longer distances, it is cheaper to use the

TABLE 1.1
Ten Methods of Producing Hydrogen

1. Classic electrolysis from H_2O with moderate temperatures (<100°C)
2. By-product of chlorine–alkali electrolysis
3. High-temperature semiconductor electrolysis of steam ("thermolysis")
4. Chemical cracking of carbon–hydrogen compounds in the light arc
5. Re-formation of carbon–hydrogen compounds using steam
6. Thermochemical cycles of coupled reactions ("Ispra process")
7. Water–gas process: $C + H_2O = CO + H_2$ (−28 kcal/mol from lignite)
8. Biotechnical photolysis through chloroplasts of green plants
9. Inorganic photolysis in monomolecular layers (Kuhn)
10. Photoelectrochemical cells ("Becquerel effect")

TABLE 1.2
Ten Advantages of Hydrogen as a Universal Energy Medium

1. It is 100% environmentally safe, because its combustion products are simply steam and water.
2. It is found in open circulation and fully recyclable in the biosphere without needing a return line (in contrast to alternative ideas for the exploitation of the energy of CH_4 synthesis, which uses two delivery and return pipes).
3. It is electrolytically, thermochemically, or biotechnically producible without additional raw materials.
4. It is nonpoisonous.
5. It can be ionized at ambient temperatures with inexpensive electrocatalysts, making it convertible into electricity in fuel cells, using "cold combustion" with great efficiency and avoiding the energy-wasting Carnot process.
6. It has the highest heating value per unit weight, 29 kcal/g. The corresponding value for CH_4 is 12.0 kcal/g.
7. It is a raw material for the chemical industry.
8. It can be stored in the gaseous, liquid, or absorbed state.
9. It burns in a different manner from that of carbon–hydrogen compounds, due to the absence of soot.
10. It can be transported similarly to CH_4, inexpensively and safely through high-pressure gas pipes. The compatibility between H_2 and CH_4 technology favors a gradual transition to that of H_2.

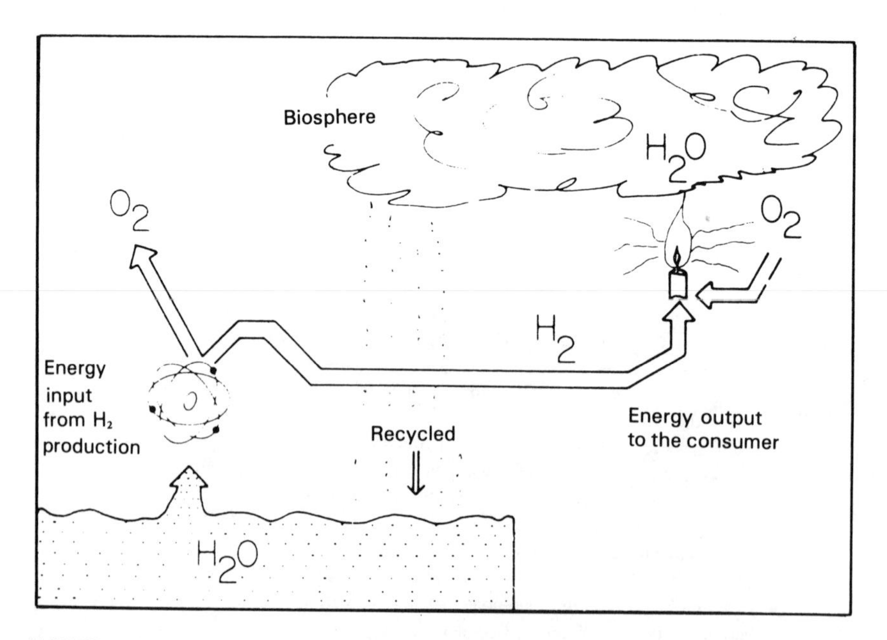

FIGURE 1.12. Production of H_2 from H_2O and recycling of the combustion product, H_2O, through the biosphere. From Seifritz.[12]

passage of hydrogen through pipes. With continued improvement of hydrogen electrolysis, this intersection point should come down to the 400 mi value. Further calculations show that transmission of energy in hydrogen is up to eight times cheaper than that through 500 kV A.C. lines at really large distances. Alternating current is not suitable for long-distance transmission (>600 mi) because at a frequency of 50 Hz, nodes and antinodes develop. Only high-voltage D.C. transmission can be considered for long-distance transmission. This method has been under investigation for decades; it is expensive and has not yet been tested over really long distances. It seems unlikely that it can compete with the simplicity and reliability of the gas pipeline. Unfortunately, high-voltage D.C. transmission lines have recently (1987) been shown to provide a substantial health hazard.

1.5. RECENT ADVANCES TOWARD A HYDROGEN TECHNOLOGY

Technical and commercial details of various kinds for electrical and gas power lines are to be found in Chapter 9. Chapter 7 gives detailed discussions of various methods of hydrogen production by photolysis, as in green plants, which produce several billion tons of hydrogen annually, exceeding the yearly production by such man-made means as industries and mines, without the aid of high temperatures, pressures, metallic components, chemical agents, or catalysts. The chapter also discusses how recent bioelectrochemical research is being used to try to produce hydrogen by means of solar energy. Correspondingly, Chapter 10 describes how conversion from electrical to hydrogen operation solves the problem of energy storage, not only for short times associated with peak loading, but also for daily and seasonal changes. Chapter 12 gives specific information about explosion hazards in hydrogen technology. In Chapter 13, there is a description of the direct conversion of hydrogen to electricity by means of fuel cells, so-called "cold combustion." Chapter 14 describes less-known applications of hydrogen, e.g., in decentralized space heating with hydrogen fuel, which avoids the formation of even traces of nitrogen-containing compounds (which, like NO, are capable of destroying the ozone layer). The direct reduction of cheap iron ores by hydrogen is described in Chapter 15.

Chapter 16 shows that air travel can look forward to improvements because it will be possible to introduce a change from the usual fuel ("JP" = jet petrol), now becoming increasingly expensive, to liquid hydrogen (LH_2) with its high specific calorific value per unit weight (Table 1.2), independently of whether the total concept of a hydrogen economy is taken up. The same chapter describes advances in automotive engines modified for use with hydrogen. Such motors carry out combustion smoothly due to superior combustion velocity—referred to the same heating value—and give 50% more driving distance per unit of energy used (when compared to gasoline). For passenger cars, low-temperature LH_2 (B.P. 20K) may be too dangerous. Intensive research on new hydrogen-type fuels is being carried out. Our preference is to include toward hydrazine (N_2H_4), outwardly similar to water, which would give only nitrogen and water vapor as waste products when used in a properly engineered way.

This book is not a compilation of practical knowledge; it is, however, a complete, clear, and realistic plan for the near and more distant future of our energy resources. Politicians seem to be becoming favorably inclined to take such challenges seriously.

1.6. THE POSSIBLE STRUCTURE OF A HYDROGEN ECONOMY

By analyzing the difficulties of the present energy technology, and by extensively reviewing the most recent specific and interdisciplinary research information contained in the following chapters, we can use a graphic approach to make a preview of the hydrogen economy (Figure 1.13). The right side of the diagram shows the various ways hydrogen is consumed in West Germany, particularly in highly populated areas. The lower portion of the diagram shows how hydrogen could be used in decentralized heating, space cooling, ferrous metallurgy, and chemical industries; the upper part of the diagram illustrates processes in which hydrogen would first have to be processed through fuel cells or turbo generators. Hydrogen would flow from the producer (left) to the consumer (right) through steel pipe approximately 1 m wide and 1000–2000 km long, at a maximum pressure of 100 atm. Such a system would have a storage capacity of about 30 hr without any additional storage devices such as underground caves. The primary energy can be of any desired type, e.g., fission and breeder reactors or coal power plants. Some of these sources, however, would need to be at remote distances, preferably offshore, to negate the difficulties described above.[14] All these sources of energy would supply electric power that would then be converted to hydrogen in high-pressure electrolyzers. According to the reaction $H_2O = H_2 + \frac{1}{2} O_2$, every unit of hydrogen produces half that volume of oxygen as well. According to a suggestion made by Justi,[13] the oxygen should also be piped and distributed over shorter distances. Low-cost oxygen is increasingly needed in industries such as the steel industry, in which labor costs are high and it is desirable that the production costs of quality steels remain competitive. Low-cost oxygen has recently played a decisive role in the construction of effective purification plants of various kinds for municipal and industrial waste waters. Following the remarkable large-scale experiments carried out in the German chemical industry, e.g., by Hoechst or BASF,[15] there seems to be a realistic possibility of cleaning up the highly polluted Rhine river so that it will be once more able to support edible fish such as trout.

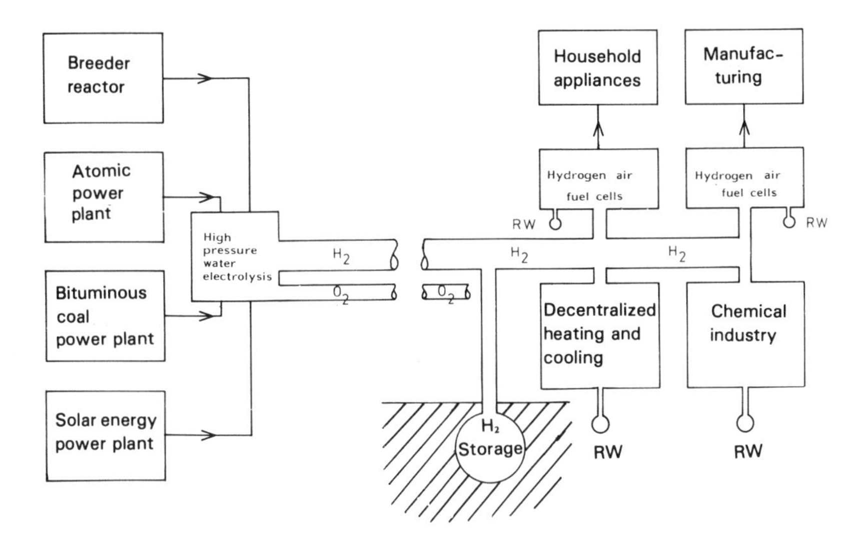

FIGURE 1.13. General block diagram of a hydrogen economy. At left are the typical primary energy sources from which thermal electricity is produced. These are used to produce high-pressure H_2 and O_2 by means of electrolysis. At right are the modes of consumption, e.g., the production of low-temperature heat. Above is the partial conversion of H_2 to electricity, preferably through cold combustion in fuel cells to satisfy the need for electrical energy such as light and power. In the middle are connections to the high-pressure gas pipes, which from 300 km deliver energy cheaper than by means of the transmission of electricity through wires. RW H_2O: = receptacles for the pure water, a product of fuel cell action.

The tendencies of nuclear and conventional thermal power plants to spread away from town centers can be seen in the different ways in which they pollute the environment. However, thermal and photovoltaic solar energy conversion plants using only nonpolluting and inexhaustible energy sources must also be planned to be in locations far from city centers, since maximal solar radiation is incompatible with the moderate climates preferred by man. This can be seen in the block diagram (Figure 1.13), in the lower left-hand corner. The best location is a hot, dry area, with the requisite cooling water for the Carnot process, such as on the southwest coast of Spain. Clearly, then, the general use of solar energy can become commercially competitive only when it is used in combination with H_2 which would serve to store energy in the diurnal variations, and allow its transport over long distances to industrial areas. During the International Solar Energy Meeting in Hamburg in 1978, it was stressed that, in the post-fossil fuel period, the massive export of solar energy by means of H_2 from the present oil states could lead to a permanent reduction in the economic inequalities between the northern and southern hemispheres.[16]

1.7. RIGHT TIME FOR A TRANSITION FROM A HYDROGEN TO A SOLAR–HYDROGEN ECONOMY

We have explained the need for a transition to a "solar–hydrogen economy" that has solar radiation as its primary source for production of the universal medium of energy. Frequently, the mistaken assumption is made that the solar energy required must be collected in a region where it is at highest concentration. Thus, the map shown in Figure 1.14 shows that equatorial regions are widely covered by rain forests, in which mist limits the annual exposure to sunlight to

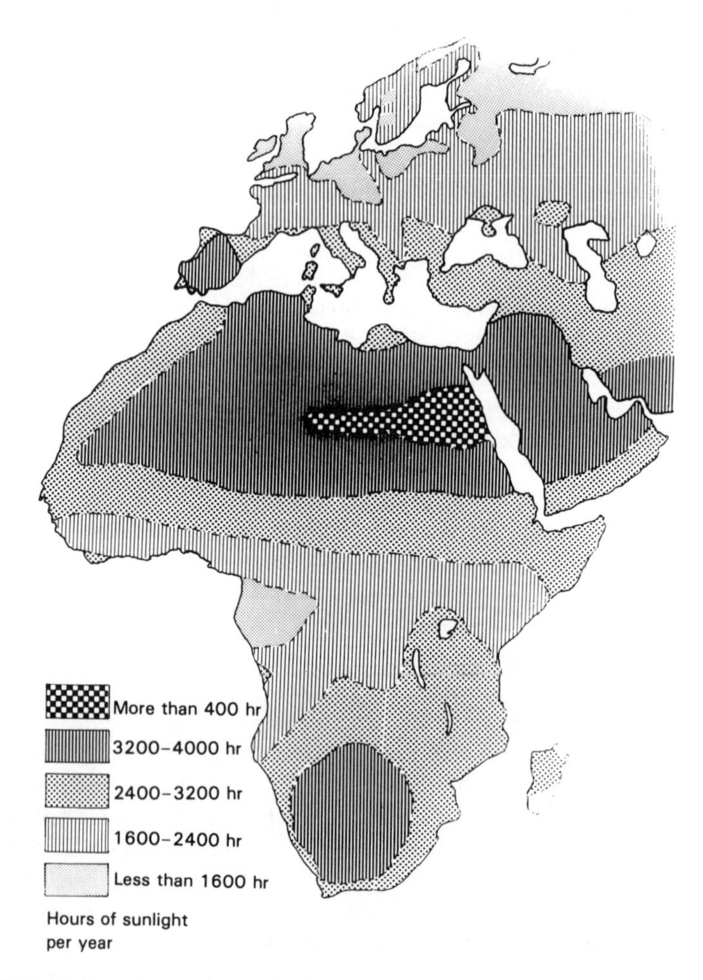

FIGURE 1.14. Map of Europe and Africa with labels for the areas with the same amount of sunlight using five different hatching patterns. It can be seen that because of the rain forests, the equatorial latitudes have less sunlight (<1600 hr/yr) than other regions (>4000 hr/yr). From Justi.[13]

some 2400–3400 hr. The planet's desert belts to the north and south of this forest zone are unsuitable, despite the associated high value of about 4000 hr of annual solar exposure, because massive water supplies are needed for cooling of steam generators and solar cells.[11] The most suitable regions, therefore, are between 20 and 40° North, as well as south of the equator, which receive 2300–4000 hr/year. Zones such as those shown on our map are available to us at fairly close range. An example of such an area is southwest Spain near the Portugese border, where there is a plentiful influx of cold cooling water from cold ocean currents. There are sufficient agriculturally worthless and arid sites of land, a good work force, and stable political conditions. There are other suitable areas in Spain, such as those in Calabria.

Instead of the original diagram (see Figure 2.3) of the solar–hydrogen economy drawn by Justi[13] in 1968, it is interesting to consider his more recent, more specific version, drawn in 1974 (Figure 1.15). It is concerned with the study of a solar farm, near Huelva, that would produce 10^{10} m^3 H_2/year (standard cubic meters measured at 1 atm and 25°C) to be sent through a 2150-km pressure pipe to Karlsruhe. Since the details of this project are explained in Chapter 10, only a general outline will be given in this introductory chapter. The solar technology consists of fixed flat-plate collectors, 137 km^2 in size, supplying steam at 120–150°C; this runs low-pressure turbo generators of 460 GW_{th} for 10 hr/day with an output of 15% of the solar energy intake. These will run pressure electrolyzers that are followed by several radial-flow compressors to 25 atm and in the second stage by piston compressors to 100 atm. At this pressure, the hydrogen enters a tripartite long-distance pipeline interrupted by two pressure regulators. This pipeline consists of welded pipes, approximately 1 m in diameter, of standard steel, in which pressures of 100, 60, and 40 atm prevail in each of their 700-km-long sections. Before entering the distribution network, the hydrogen will be passed through an expansion apparatus (cf. Justi[17]) in place of the usual baffles; thus, work will be done on an expansion apparatus to reduce pressure from 40 atm to low pressure, whereby almost half the work of compression, 575 MW, can be recovered. Thus, adiabatic work is proportional to the log of the pressure ratios, which would be log (100/1) for the compression and log (40/1) for the expansion. In this manner, the hydrogen compression costs, an important part of the transportation costs, would be reduced by about 40% as illustrated in Chapter 10.

Regarding the costs of the solar power plant, the flat-plate collectors, due to their large size, are the determining factors, rather than other machinery. Will it be possible to reduce the costs from $200/$m^2$ to an ideal of less than $50/$m^2$? Some relevant information on this matter has been obtained from one of the most experienced German firms in this field. It has offered a quotation for hydrogen transport and storage tubes, including all subsidiary costs such as buildings of $925,000,000 (as of 1973). This includes $450,000,000 for the compressor stations with their 35 compressors and a firm delivery time of 2 years

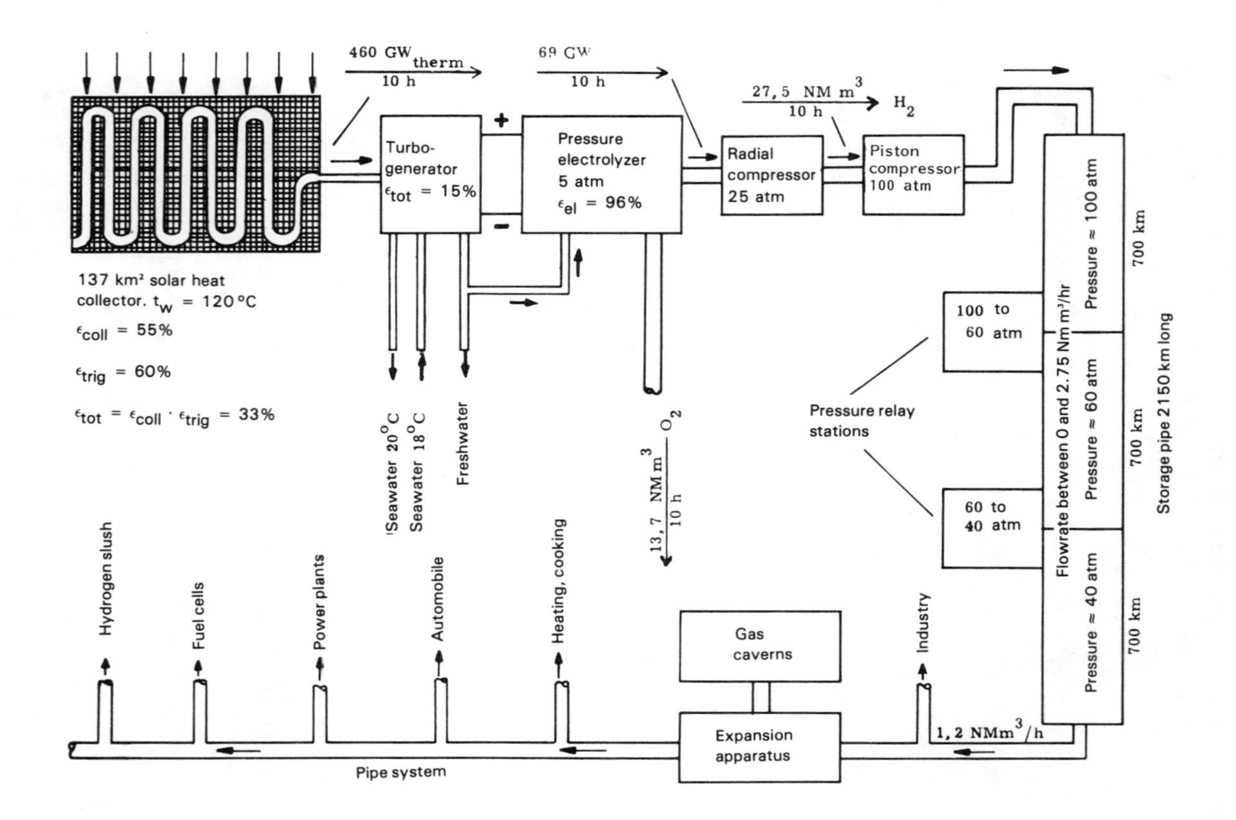
460 GW_{therm} / 10 h
69 GW / 10 h
27,5 NM m³ / 10 h H_2
Turbo-generator ϵ_{tot} = 15%
Pressure electrolyzer 5 atm ϵ_{el} = 96%
Radial compressor 25 atm
Piston compressor 100 atm
137 km² solar heat collector. t_W = 120 °C
ϵ_{coll} = 55%
ϵ_{trig} = 60%
ϵ_{tot} = $\epsilon_{coll} \cdot \epsilon_{trig}$ = 33%
Seawater 20°C
Seawater 18°C
Freshwater
13,7 NM m³ / 10 h O_2
100 to 60 atm
60 to 40 atm
Pressure relay stations
Pressure ≈ 100 atm
Pressure ≈ 60 atm
Pressure ≈ 40 atm
700 km
700 km
700 km
Flowrate between 0 and 2.75 Nm m³/hr
Storage pipe 2 150 km long
Hydrogen slush
Fuel cells
Power plants
Automobile
Heating, cooking
Gas caverns
Industry
1,2 NMm³/h
Expansion apparatus
Pipe system

from award of the contract. To convey a clear concept of its dimensions, it must be mentioned that this plant would mean a hydrogen production of 1.10^{10} m^3 H_2 per year. This is equal to 29.9×10^9 kWh (thermal), and this is 23% of the entire electricity consumption of the Federal Republic of Germany, which is 122×10^9 kWh (elec).[18] According to American experience with methane, about 80% of the hydrogen namely, 24×10^9 kWh (thermal), can be used without conversion to electricity; if, furthermore, the remaining 20%, namely, 5.9×10^9 kWh, can be reversibly converted by means of fuel cells with an average efficiency of 40%, one therefore would have 2.36×10^9 kWh (elec), a total of $(24 + 2.36) \times 10^9$, i.e., 26.4×10^9 kWh/yr. This is equivalent to a mean efficiency of $26.4/122 = 21.7\%$, and roughly one fifth of our current electricity consumption. An estimated solar collector 631 km^2 in area would be sufficient to supply all of the Federal Republic of Germany with solar energy in a usable form. This is equivalent to a square having sides about 25 km long. In terms of size and order of magnitude, this is of the same order as the area occupied by the conventional thermal power stations, with their attendant coal mines, railway tracks, canals, and harbors taken into account.

It is therefore not valid to state that solar energy would cost orders of magnitude (capital and land investment) more than present power plant systems.

1.8. CONCLUSIONS

Our theory that oil shortages are attributable more to our own wastefulness of oil, natural gas, and coal, rather than to the curtailment of oil deliveries by the Arab countries, has been tested in a detailed analysis of energy production from fossil fuels. In this analysis, the thermal efficiency of energy production in the most modern plants is only about 38%, meaning that about two thirds of the heat of combustion of these energy sources escapes unused through cooling waters and into the atmosphere. More serious is the fact that this waste pollutes

FIGURE 1.15. Concrete example of the solar–hydrogen economy (Justi, 1974) for the production and storage of 1×10^{10} m^3 H_2/yr (standard cubic meters measured at 1 atm and 25°C), equivalent to one tenth of the German consumption of electricity. The necessary area for fixed photochemical flat collectors in southern Spain is 137 km^2 with an efficiency of 15% of the electricity-producing low-pressure turbines; outdated compared to 1974 because of the introduction of collectors with selective blackening, and higher thermal efficiency of the turbines. The H_2 coming from the electrolyzers is compressed through radial and piston compressors to 100 atm and is piped through pipes of 1 m diameter and 3150 km total length to the Federal Republic of Germany. In Germany, the H_2 arrives at 40 atm and can be decompressed usefully in expansion. This reduces by more than half the energy used for compressing, and then the transportation price is reduced from 5 to 7% of the energy equivalent of H_2. The diagram visualizes the many areas of application of H_2 in households and industry. The storage capacity of the pipe system is enough for 36 hr at average demand and without separate gas underground storage.

and warms up the environment. The most wasteful process in modern technology, however, is combustion of fuel in cars, in which the average efficiency is only 18%, so that 82% escapes as waste heat and exhaust gases through radiators and exhaust systems. Nuclear plants of the low-pressure type show about 29% efficiency, 71% wasted. Recently, it has become possible to reduce the sulfur content of the exhaust gases, but a newly realized danger is the rising content of CO_2 in the atmosphere, a component that was hitherto regarded as harmless. This CO_2 content has been increasing steadily since the beginning of industrialization. It is also due to the remorseless cutting down of the large tropical rain forests in Africa and South America, so that their ability to store CO_2 in green plants, in wood, and in humus is being lost. The resultant increasing amounts of CO_2 in the atmosphere tend to absorb infrared radiation from the sun, which increases the greenhouse effect in the atmosphere and will raise its temperature by several degrees. Thus, we are threatened by worldwide and catastrophic changes of climate. The United States Government has set up a commission aimed at averting increases in atmospheric CO_2. It is little consolation that destruction of the ozone layer, at a height of about 50 km above sea level, through nitrogen oxide exhausts of cars and planes, and freon from spray cans, all give rise to possible worse threats, because it is the ozone layer alone that holds back the solar UV radiation that otherwise can cause cancer and, eventually, kill all organic life.

In this situation, there is only one solution, which is to discontinue the combustion of fuels and switch to hydrogen, in which the combustion product is pure water vapor and totally nonpolluting. Solar power plants must be located where they receive 3000–4000 hr of sunlight/year and, at the same time, where plenty of cooling water is available, e.g., Sicily or New Mexico. Basically, we are speaking about conventional thermal power plants, in which the turbines are driven, not by coal, oil, gas, or atomic fission, but by solar radiation, which is concentrated by computer-directed concave mirrors (see Figure 6.19). With the energy produced in this way, electrolytic hydrogen is produced from water, compressed to 100 atm, and routed through steel pipes to the place of consumption. For distances greater than about 500 km, the transmission of hydrogen can be significantly cheaper than transmission of electricity through wires. The transport pipelines also tend to create a solution to the storage problem simply and inexpensively. In the following chapter, the details of a possible solar–hydrogen economy will be discussed.

REFERENCES

1. A. Buch: *Energie* **29**(7):198–204 (1977).
2. W. D. Gutschmidt: Die Bevölkerungsverteilung an Standorten kerntechnischer Anlagen in der BRD [The population distribution in the surrounds of nuclear plants in West Germany], Inst. f. Reaktorsicherheit der TÜV, Cologne, IRS-W-17 (1975).

3. N. N.: Auch Kohl-Radioaktivität gering [Minor radioactivity in coal], in: *Kernenergie und Umwelt,* No. 3, pp. 1 and 3 (1978).
4. C. D. Keeling, R. B. Bacastow, A. E. Bainbridge, C. A. Ekdahl Jr., P. R. Guenther, L. S. Waterman, J. F. S. Chin: Atmospheric carbon dioxide variations: Hawaii, *Tellus* **28**(6)**:**538–551 (1976).
5. G. M. Woodwell: *Sci. Am.* **238**(1)**:**33–44 (1978).
6. G. Plass: *Tellus* **8:**140 (1956).
7. J. O'M. Bockris: Personal communication to E. Justi (6/22/1978).
8. J. O'M. Bockris, R. P. Hammond: in: *The Electrochemistry of Cleaner Environments,* Plenum Press, New York (1972).
9. E. Broda: *The Evolution of the Bioenergetic Processes,* Wiley Interscience, New York and Braunschweig, pp. 167–168 (1975).
10. D. P. Gregory, M. Gregory: *A Hydrogen Energy System,* American Gas Association, Chicago (1972).
11. E. Justi: Problematik, Stand und Aussichten der Solar-wasserstaff-Wirtschaft [Problems, situatins and future prospects of the solar-hydrogen economy], in: U. Bossel (ed.) Proceeding of the 1st German Solar Forum, Hamburg, September 26–28, 1977, Vol. 2, pp. 517–549, Verlag, DGS, Munich (1977).
12. W. Seifritz: *Chimia* **28**(7)**:**323–340 (1974).
13. E. Justi: *Leitungsmechanismus und Energieumwandlung in Festkörpern* [Mechanism of Conductivity and Energy Conversion in Solids], Vandenhoeck and Ruprecht Verlag, Göttingen, pp. 450–455 (1965).
14. W. Gries, *Blick in die Wirtschaft* of March 2, 1978, and repartee of March 30, 1978. (Remark: *Blick in die Wirtschaff* is the title of a German newspaper.)
15. H. G. Peine: *Die BASF,* No. II, pp. 41–45 (1977).
16. A. Derichsweiler: Grupswort des Präsidenten der Deutschen Gesellschaft für Sonnenenergie e.V. (DGS), an das 2. Internationale Sonnenforum und die Comples [Opening address of the president of the German Society for Solar Energy (DGS) to the participants of the 2nd International Solar Forum and the Comples, in: A. Derichsweiler, H. Krinninges (eds.), Proceddings of the German Solar Forum, July 12–14, 1978, Vol. 1, p. 21, Verlag DGS, Munich (1978).
17. E. Justi: Sonnenenergie—die dritte Energiemacht? [Solar energy—the Third World power?], Akad. d. Wiss. u. d. Literatur Mainz 1949–1974, F. Steiner Verlag, Wiesbaden (1974), pp. 41–53.
18. A. Buch: Personnal communication to E. Justi (1976).
19. J. O'M. Bockris, A. J. Appleby: *Environment This Month* **1**(1)**:**29–35 (1972).

CHAPTER 2

The Hydrogen Economy

2.1. CAUSES OF PROSPECTIVE ENERGY AND RAW MATERIALS SHORTAGES

A rather obvious but long-ignored fact is that the supplies of raw materials and fossil fuels such as coal, natural gas, and oil on our planet are limited. For decades, it was believed that these deposits were so enormous that in the foreseeable future, we would not have to fear the exhaustion of those known deposits or of the resources not yet economically exploitable. Recently, however, since the report of the Club of Rome and in particular the diagrams published by the Meadows on the limits of growth, this viewpoint is no longer defensible.[1] Because of the recent uncontrolled exponential growth of the world's population, and the additional growth of energy demands due to increasing living standards per capita of population, exhaustion of known ore deposits and fossil fuels can be expected in the foreseeable future (see Table 2.1). The scenarios of the Meadows indicate a decade of breakdown, somewhere between 2020 and 2070, in which the per capita amount of food available for the world population will undergo sharp ($>10\%$ per year) cutbacks and the production–time curve will have passed through its maximum.

The Meadows' work was widely criticized.[3,4] For example, an examination of the computer data discovered an error that changed the predicted time of collapse. Moreover, the report *Global 2000* (see Table 2.2) shows that although the present known reserves that are economically viable will be exhausted in a few years, the so-called resource potential (i.e., the profitable deposits available only at considerably increased expense) is very large. The predictions of *Global*

This chapter authored by Dr. P. W. Brennecke, Braunschweig, and Prof. H. H. Ewe, Hamburg.

TABLE 2.1
Available World Reserves (1976) of Selected Raw Materials (in Metric Tons) with Two Different Projections of Consumption[2]

Material	1976 Reserves	1976 Consumption	Predicted increase in demand	Time left (years): Consumption at the level of 1976	Time left (years): Consumption according to predicted increase
Fluorine	34×10^6	1.9×10^6	4.58%	18	13
Silver	0.19×10^6	9.5×10^3	2.33%	20	17
Zinc	151×10^6	5.8×10^6	3.05%	26	19
Lead	136×10^6	3.7×10^6	3.14%	37	25
Tungsten	1.9×10^6	0.04×10^6	3.26%	52	31
Tin	10×10^9	241×10^6	2.05%	41	31
Copper	456×10^6	7.3×10^6	2.94%	63	36
Nickel	54×10^6	0.6×10^6	2.94%	86	43
Platinum	9.2×10^3	0.08×10^3	3.75%	110	44
Phosphate	26×10^9	107×10^6	5.17%	240	51
Manganese	1.6×10^9	10×10^6	3.36%	164	56
Iron ore	93×10^9	0.5×10^9	2.95%	172	62
Aluminum	5.1×10^9	16×10^6	4.29%	312	63
Chrome	752×10^6	2×10^6	3.27%	377	80

2000 do not point to an early exhaustion of low-grade ores. As Table 2.2 indicates, the needed additional developments will require considerable capital expenditures and large energy inputs to keep the output level of certain raw materials at a desired rate.[2] From the report *Global 2000* and the Meadows' work, it follows that mankind's supply of important raw materials is going to become more and more difficult to obtain and need more and more energy; and that fossil fuels will become increasingly more expensive. Because of the exhaustion of economically exploitable fossil fuels, mainly coal, there will arise in the foreseeable future considerable problems in the production and utilization of these fuels that may be difficult to surmount. In addition, we have to take into account the growing CO_2 concentration of the atmosphere that will result from the combustion of such fuels—and/or the increase in aerosol concentration—which will certainly give rise through the greenhouse effect to considerable climatic consequences, the magnitude of which is difficult to see clearly at this time. Additionally, the pollutants NO, CO, and SO_2 pose considerable health and ecology risks.[5]

Thus, unlimited growth in the production of coal will cause a counter reaction that will damage and destroy the earth's surface and our habitable space. It will certainly be difficult to make the transition to an energy supply that is independent of fossil fuels, but it is a transition that cannot be avoided. Even efforts to put off this change for a few decades will cause considerable problems—e.g., the death of much of the tree population in the larger forests because of the increasing acidity of the forest floor owing to the excess SO_2 output by power plants, as has been dramatically illustrated in recent years. The work of the Meadows school shows that if the development of technology continues with sufficient intensity in the future, and if we are able to make the timely transition to nonfossil, clean energy sources, catastrophic breakdown can be avoided. *One absolute prerequisite for this is the limitation of the earth's population to a total that is ecologically acceptable*. At the same time, there must be an almost complete recycling of raw materials and development of heretofore uneconomic resources. A high availability of energy per capita, but without fossil fuels, and a high degree of recycling are the deciding parameters in the avoidance of an otherwise inevitable catastrophe.

2.2. FUTURE ENERGY SOURCES AND THEIR MEDIA

If we agree that it is absolutely necessary to give up burning fossil fuels, then there remain only the following permanent energy sources that are renewable

TABLE 2.2
Estimated World Production, Reserves, and Resources 1977 (in Millions of Metric Tons) for 14 Elements[2]

Material	Production	Reserves	Other resources	Potential resources (exploitable)	Total resource base (earth crust)
Aluminum	17[a]	5,200[a]	2,800[a]	3,519,000	1,999,000,000,000
Iron	495[b]	93,100	143,000[c]	2,035,000	1,392,000,000,000
Manganese	10[d]	2,200	1,100[e]	42,000	31,200,000,000
Phosphorus	14[f]	3,400[f]	12,000[f]	51,000	28,800,000,000
Fluorine	2[g]	72	270	20,000	10,800,000,000
Chrome	3[h]	780[h]	6,000[h]	3,260	2,600,000,000
Zinc	6	159	4,000	3,400	2,250,000,000
Nickel	0.7	54	103	2,590	2,130,000,000
Copper	8	456	1,770[i]	2,120	1,510,000,000
Lead	4	123	1,250	550	290,000,000
Tin	0.2	10	27	68	40,800,000
Tungsten	0.04	1	3.4	51	26,400,000
Silver	0.010	0.2	0.5	2.8	1,800,000
Platinum group (Pt, Pd, Ir, Rh, Ru)	0.0002	0.02	0.05[j]	1.2[k]	1,100,000

[a] In bauxite, dry form, with an assumed average content of 21% extractable aluminum.
[b] In ores and concentrates with an assumed average content of 50% extractable iron.
[c] In ores and concentrates with an assumed average content of 26% extractable iron.
[d] In ores and concentrates with an assumed average Mn content of 40%.
[e] Without the metal that is contained in deep-sea knolls and with nickel, the nonidentifiable resources.
[f] In phosphates and concentrates with an assumed average P content of 13%.
[g] In fluorite phosphate-containing stone, and concentrates with an assumed average F content of 44%.
[h] In ores and concentrates with an assumed average Cr content of 32%.
[i] Including 690 × 10^6 metric tons in deep sea knolls.
[j] Approximated mean value of the established resources of (0.03–0.06) × 10^6 metric tons.
[k] Platinum only.

or have long life: atomic resources and/or inexhaustible energy sources such as solar, hydroelectric, and wind. The light-water reactors that are already in use do not offer any long-term solution because they are using up exclusively ^{235}U, and in a few decades they will have used up the economically available uranium supplies. These light-water reactors will therefore be able to produce electrical energy for only a few years more.

The potentially interesting hydroelectric supplies are already to a great extent in use, so that the replacement of fossil fuels by hydroelectric plants and also by wind plants and geothermal energy can therefore take place only to a small extent at a very few places on the earth. The long-term solution of the energy problem must therefore depend only on some improved version of nuclear reactors (breeders, fusion), and upon solar power plants.

2.3. FUTURE ENERGY MEDIUM

In all the known types of nuclear reactors, the electrical power delivered is proportionally cheaper with increased power output and therefore with increasing heat dissipation. In this way, these reactors can produce relatively cheap energy at their locations, but because of their enormous heat dissipation and the nuclear waste and risk, they must be located at great distances from heavily populated areas. Thus, suitable places for them would be in distant areas that are virtually uninhabitated, e.g., northern Canada and Siberia, where the rivers can easily dissipate the heat pollution. Alternatively, they could be placed on ocean-borne floating islands at substantial distances from the shore.[6]

Correspondingly, large solar plants will have to be built at great distances from population centers. At first, there was an idea that unpopulated zones such as the Sahara Desert could be used, but such concepts did not take into account that because of the Carnot efficiency factor involved in the conversion of solar heat to electricity using heat engines, enormous quantities of cooling water or cooled air will be needed.[7] Further, the conversion of solar light to electricity with photovoltaic cells cannot be carried out without cooling. Thus, only a fraction of the incoming solar energy is converted to electricity and the rest is converted to heat, so that solar cells rapidly increase in temperature unless they are cooled with water. With increasing temperature of operation, the efficiency of the cells decreases so quickly that they must be continuously cooled for the operation to be successful (see Figure 5.19). Thus, the most suitable sites for large solar power plants are not arid zones, but seashore sites, where both abundant sunshine and water for cooling are available. A good example is the Mediterranean coast.[8]

Consequently, there will have to be very considerable distances between atomic and solar power plants and the centers of energy consumption; indeed, it is reasonable to expect that the energy will have to be transported over distances of up to 2000 km. The transport of electrical energy over such large distances is well known to be associated with considerable losses of energy (see also Chapter 9) and is therefore uneconomical. In addition, the storage of electrical energy to overcome the diurnal variation is difficult. Therefore, in numerous studies, these factors have given rise to a search for an energy medium on which a number of restrictions would have to be put:

1. The energy medium has to be efficiently and economically producible from electrical energy.
2. The necessary raw materials must be cheaply and easily available in the vicinity of large power plants.
3. Neither the production nor the use of this energy medium may give rise to pollution.
4. The material must be light, easily transported, and not difficult to store.
5. The medium must be usable in a multipurpose way both in private homes and in industry or in transportation.

These considerations indicate chemical energy media such as H_2, NH_3, and N_2H_4, of which H_2 clearly has the most favorable properties for the purpose. Studies show that it would be cheaper to convert solar or nuclear energy into electrical energy and then to H_2, transport it through pipes with relatively low pumping energy, and then store it. At the user terminal, it would be reconverted to electricity and used directly for heating or turned into mechanical power. The distance at which it becomes cheaper to transport energy in the form of hydrogen instead of electricity depends on the production costs of hydrogen and in particular on the potential at which the electrical power can be transmitted. According to Figure 2.1, the critical distance decreases rapidly with the potential for transmission, which in Europe is about 380 kV; the critical distance, then, is about 360 km.[9–11]

Hydrogen, like natural gas, is a gaseous energy medium. A comparison of these two fuels is therefore relevant. Natural gas contains about three times as much energy as hydrogen in the same volume, but only one third as much in the same weight. Methane can be more easily handled and stored. But the future belongs, despite these factors, to hydrogen, because economical supplies of methane will last only a few decades. Synthetic methane from coal[12] does not give a solution to this situation. Thus, this conversion, as recent research shows, is expensive and therefore uneconomical and requires a comparatively large amount of coal for the production of methane. Apart from this, the earlier considerations apply: Coal supplies are limited; in the combustion of CH_4, CO_2 would be produced and be added to that in the atmosphere, where it would absorb infrared radiation from reflected sunlight and thus contribute to the greenhouse effect and to a catastrophic alteration of the climate on our planet.

The foregoing considerations are based on the concept that the available energy will have to be transported over large distances and lead to hydrogen as

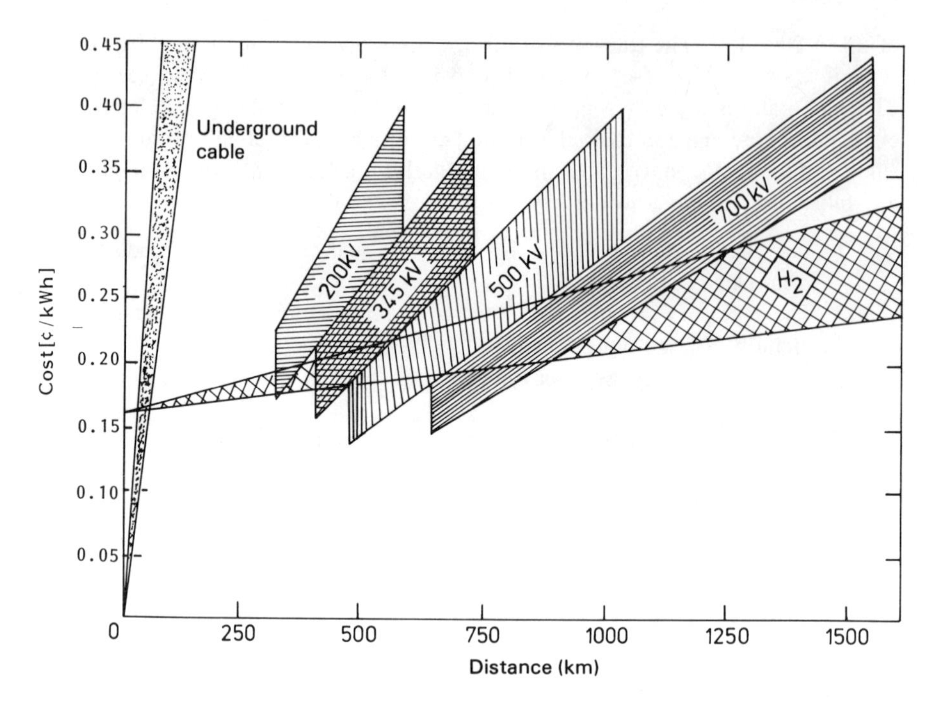

FIGURE 2.1. Comparison of transportation costs of energy through overhead electrical power lines (diagonal hatching) and underground cables (stippled) and hydrogen through high-pressure steel pipes (cross-hatching). Parameters of the variously shaped bands: potential of transmission for three-phase current. The transport of energy through H_2 can be cheaper than that using overhead power lines from a distance of 380 km. Underground cables are the most expensive according to Gregory *et al.*[9]

the energy medium of the future. The so-called hydrogen economy is, however, a concept of a wider scope. By this, we mean that it is desirable that a single energy medium answer all requirements. The future sources of clean energy should be used in a manner that is at the same time the most economical but avoids all atmospheric pollution. Correspondingly, hydrogen has specific advantages in respect to transmission and storage of energy. The hydrogen economy can therefore be described as the use of hydrogen to take energy from distant primary energy sources, as well as the extensive application of this medium as a chemical raw material in industrial chemistry and technology. It would be further used in transportation and in household energy supplies, the primary product of combustion being water, which after further use as drinking water would be recycled via the sea and the biosphere (see Figure 1.12).

Some of the advantages of the hydrogen economy can be seen by comparing it with the present energy media and the costly and varied infrastructure that is necessary for their transport, storage, and use. Between 10 and 20% of our present energy usage is in the form of electricity, which has to be delivered by means of a stationary conducting network, while a further percentage is delivered as natural gas through a pipeline network. The largest amounts of energy are to be found, however, in fossil fuels. Hydrogen would take over not only the tasks now performed by fossil fuels, but also, to a large extent, the areas at present served by other forms of energy:

1. Electrical energy. This energy can be supplied by hydrogen by means of normal combustion via heat, steam, turbines, or more directly, without loss through the Carnot factor, by means of hydrogen–oxygen fuel cells (see Chapter 13).
2. Low-temperature heat. Space heating can be provided by hydrogen just as it is by natural gas. However, with hydrogen, flame-free combustion can be carried out via catalysts, thus eliminating the need for chimneys, the only product being water (see Chapter 14).
3. Chemical technology. Sufficiently low-cost hydrogen could be used in many chemical reduction processes and thus reduce their costs while reducing atmospheric pollution to negligible proportions.
4. Metallurgy and refining. A number of processes can be carried out much more cheaply with hydrogen than with other reducing agents, e.g., carbon. Another reason for increasing the use of hydrogen would be a reduction of air pollution and, thereby, an increase in the usefulness of low-grade ores (see Chapter 15).
5. Liquid wastes. The regeneration of liquid wastes could occur much more easily by means of cheap electricity and/or the oxygen that a hydrogen economy would make readily available at low cost and in abundant amounts. Furthermore, sewage could be treated in the home by electrolytic oxidation to CO_2 or chemical treatment in molten salts at high temperatures (producing CO_2).
6. Water. Part of our drinking water supply would be readily available were hydrogen to be the universal energy medium because the final product of hydrogen combustion is pure water. If the energy use rate were 10 kW/person, one would get about 50 liters of drinking water/person per day, and at least 5 liters of this will be produced in the home.
7. Vehicular propulsion. Hydrogen can be used directly as a fuel for automobiles, and the necessary modifications are already well known (see Chapter 16). Of course, it would be expected that the final result of this transformation would be the use of hydrogen–air fuel cells, because then the efficiency of energy conversion would be more than 50% instead of the present 20%—a considerable difference. Other advantages would be that steam would be the only exhaust gas and no noise would be associated with the operation of cars. The state of development of hydrogen-powered automobiles is such that they have now achieved a weight/power ratio of 10 kg/kW (Chapter 13). Rail transportation, where the question

of weight is less important, could be powered by fuel cells that could be supplied with hydrogen fuel in the liquid form.

8. Shipping. At present, large tankers are used for the transportation of liquefied natural gas. Similarly, it would be possible to use cryogenically insulated tanks of liquid hydrogen and eventually also oxygen as fuel for large tankers. The power for the ships' engines could then be obtained either by combustion in the normal way or better, because of a higher degree of efficiency, by the use of fuel cells as electrochemical direct energy converters.
9. Aircraft. The advantages of liquid hydrogen in driving airplanes are to be noted in respect to supersonic and hypersonic aircraft (see Chapter 16). The principal fact that has to be understood here is that for aircraft, the most important factor is weight (not volume, as is the case for land vehicles). Thus, the value for hydrogen of 30.0 kWh/kg is the highest available and is *almost 3 times higher than that of gasoline*. In support of this, NASA and Lockheed have jointly made plans for a supersonic aircraft fueled by liquid hydrogen that would have a cruising speed of 6500 km/hr and a payload, range, and economic feasibility that would be superior to those of the Concorde.[12]*

2.4. ORIGIN OF THE HYDROGEN-ECONOMY CONCEPT

In explaining the concept of a hydrogen-economy, it is important to distinguish between the simple idea of the use of hydrogen and the concept of a hydrogen economy as presented herein, i.e., the all-inclusive production and use of hydrogen, taking into account the energetic, ecological, and commercial viewpoints.

The idea of a hydrogen economy seems to have been implied by Jules Verne's fictional Captain Nemo.[13] Numerous authors, both in conventional literature and in science fiction, have repeated this idea, without coming any nearer to its realization. For example, Niederreither received in 1937 a German patent for an electrochemical cell in which there was to be first electrolysis storage of the hydrogen and then recombination in the fuel-cell principle, thus storing electrical energy and being able to reobtain it when necessary; this system, however, never materialized.[14]

Serious attempts to convert electrical energy into chemical energy in the

*Translator's note: In 1986, NASA and the U. S. Air Force announced plans for the development of a "suborbital vehicle." In its civilian version, this aircraft would be able to reach Tokyo from New York in 2 hours ("The Orient Express"). This vehicle, together with a similar British concept, could only be fueled by hydrogen. After rising above the atmosphere, little energy would be needed to propel it at great speed.

form of hydrogen and vice versa became possible only after the development of the hydrogen–oxygen fuel cell, which enables one to convert hydrogen and oxygen directly into electricity without the intervening step of heat production. E. Baur and his students started off using coal as the fuel for cells of this type, but they soon came to realize that the ideal fuel would be highly reactive hydrogen together with the use of non-noble-metal electrocatalysts, and fuel cells were built on this principle.[15] However, F. T. Bacon, working in Cambridge, was the first to develop a really practical hydrogen–oxygen cell for water electrolysis, separate storage of hydrogen and oxygen, and subsequent and current-producing recombination to water in a fuel cell.[16] This hydrogen–oxygen cell was later produced in an improved version by raising the operational temperature to more than 200°C, with gas pressures up to 40 atm, thereby increasing the performance to 5 kW. This fuel cell was later developed by Pratt and Whitney and became the basis for energy production in American space vehicles and also in moon landings.

After the hydrogen–oxygen fuel cell had been thus realized and efficient electrolyzers had been developed for the commercial production of hydrogen from water via electrical energy, Lindström[17] rediscovered the suggestion made by Lawaczek in 1933[18] (Figure 2.2). According to his suggestion, hydropower from northern areas of Sweden, situated about 1000 km north of consumer centers, would be converted directly into hydrogen and transported at 500 atm through steel pipes before being reconverted to electricity again (instead of being

Backöfen, wenn deren elektrische Beheizung durchgeführt wäre. Für den Winter wenigstens kämen die Heizungen als Dauerverbraucher auch des Nachts in Frage. Es wäre denkbar, Öfen für Wohnräume zu bauen, mit solcher Wärmespeicherung, daß sie den Tag über vor= halten, wenn sie des Nachts geheizt werden. Vielleicht ließe sich des Nachts ein Wasserstofferzeuger in Gang setzen, der groß genug bemessen wird um den tagüber nötigen Brennstoff zu liefern. Diesen Anwendungs= möglichkeiten war noch bisher der gegen Kohlenwärme noch zu hohe Preis für die elektrische Wärme hindernd im Wege. Außerdem wür= den die Verteilungsnetze der Häuser unzulässig überlastet, hätten sie noch den Heizstrom am Tage zu führen. Immerhin sind in der

3*

FIGURE 2.2. Facsimile of a publication of F. Lawazcek (1921), in which he proposed for the first time the production and storage of electrolytic H_2 with surplus electricity, generated at night, and proposed utilizing it for heating purposes during the daytime. This is not yet a concept of H_2 as a universal energy medium, but effectively an improved temporary utilization of the distribution network to residential areas, which was poorly organized at the time. From Weyss.[25]

transmitted through AC electrical lines, hardly practical because of resistance losses). According to Lindström's estimate, the transport as gas amounts to only 10–20% of the cost of the corresponding transport in the form of electricity, and this saving would more than compensate for the cost of the electrolyzer. Lawatzek did not mention, however, the possibilities of energy storage or specific applications except for the propulsion of certain types of submarines.

An additional step in the direction of the hydrogen economy was made by Justi and his co-workers in their work on fuel cells and electrolyzers that worked with economical catalysts at the relatively low temperature of 80°C.[19] These electrochemical cells allowed energy storage by electrolysis and recombination with an efficiency of about 80%. This success was the origin for the extension of this type of energy storage in the form of a hydrogen economy that was represented in a diagram by Justi in a 1965 publication (Figure 2.3).[20] An essential point of this scheme is the far-reaching avoidance of Carnot processes: Low-temperature heating for household and industry is supplied by means of hydrogen combustion. Electricity is to be produced only when it is indispensable in large power plants or for low-use situations in fuel cells. Hydrogen should be a fuel in the chemical industry. According to this plan, electrolytically produced oxygen would also be collected and transported in parallel pipes to supply the chemical industry as well as to be used in such processes as welding. It was also planned to increase the storage capacity of the steel pipes by additional gasometers. In this plan, cheap energy transport was seen as taking place over distances of more than 2000 km (e.g., from the Atlas Mountains in Morrocco to the Federal Republic of Germany) with the objective of thermoelectrically converting primary solar energy from the Mediterranean area.

In 1962, Bockris suggested to the Westinghouse Company[21] that American towns be provided with solar hydrogen. The idea was to have ocean-borne floating platforms (see Figure 1.10) that would carry solar cells, resulting in the electrolysis of seawater, the hydrogen being transported back to land through pipes and reconverted to electricity by means of fuel cells. In 1969, Bockris again made efforts to introduce the idea of floating islands for nuclear or solar energy, with hydrogen as the energy medium. He attempted to interest workers at the Institute of Gas Technology in Chicago in carrying out a project in this direction; in fact, calculations originating from this initiative were published by Gregory, Ng, and Long.[9] The term "hydrogen economy" was first used in a discussion among Bockris, Triner, and others in the General Motors Technical Center in 1970.[22] They had been discussing various scenarios whereby the polluting properties of gasoline could be avoided and had come to the conclusion that most of them ended up being in some sense connected with hydrogen. It was Bockris who in 1971 published the general concept of replacing natural gas and gasoline by means of hydrogen, with its low-cost transmission properties.[23]

In its application to vehicular transportation, gaseous hydrogen must be considerably compressed or brought to a very low temperature in the liquid form.

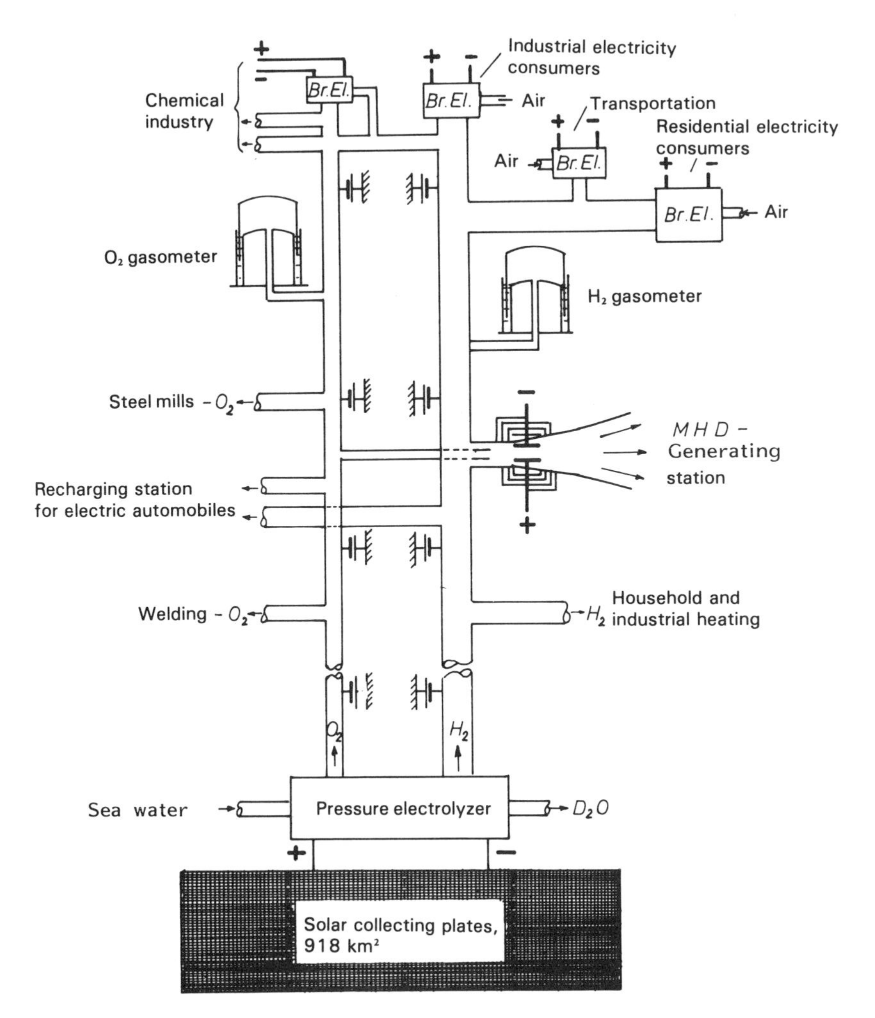

FIGURE 2.3. The first clear and quantitative block diagram of a solar–hydrogen economy, by Justi, 1964.[8] In the diagram given, 900 km^2 of heat-collector surface is supposed to give a 6% conversion efficiency of solar light to electricity with the thermo-electric Seebeck cells, developed by Justi and others. The production would have been sufficient to supply the entire electricity needs of West Germany in the 1960's. The direct currents produced were to be used to produce hydrogen in high pressure electrolyzers, also as a by-product, D_2O for use in atomic reactors. The plan was to run the hydrogen and oxygen through street tubes under pressure (and this is cheaper than passing the electricity through wires for distances of 350 km or more). The pipe network would also act as a storage system at 65% efficiency, almost as good as pumped storage. Most of the hydrogen arriving into Germany would probably be used directly, particularly for low temperature heat, thus avoiding the loss of about two-thirds of the heating value which occurs in normal fuel use. Hydrogen is the only fuel which is environmentally acceptable, and its combustion product (water) is recycled back to the biosphere (see Fig. 1.2). This diagram of Justi's is the nearest realization of the 1962 suggestion of Bockris. cf. the solar–hydrogen economy described by him in 1971.[24]

FIGURE 2.4. An automatic device for the cold combustion of N_2H_4-derived H_2 with air. The arrangement gives a high conversion efficiency and 1 kW nominal output. *Bottom:* Fuel cells in 200 W modules (the output increases exponentially with temperature). *At right:* Production of H_2 from the liquid energy carrier hydrazine by means of catalytic–thermal degradation. *Top left:* Circulation pumps, inert gas release, and controls. Photo: VARTA Batterie A.G., Kelkheim.

Both methods are very expensive, because of the heavy steel vessels required or the liquefaction process, respectively. For such purposes, liquid, C-free hydrides, such as hydrazine, N_2H_4, or ammonia, NH_3, are both suitable and obtainable using plentifully available nitrogen from the air. Figure 2.4 shows a fully automatic hydrazine fuel cell made by the Varta Corporation from a design by Justi and Winsel[24] that operates at an output of 1 kW. Inexpensive iron catalysts decompose hydrazine into N_2 and H_2, and the gaseous mixture can then be reconverted to electricity in fuel cells. The production of hydrazine, however, is at present not economically feasible.

From a historical viewpoint, the summary of the hydrogen economy by *Chemical Engineering News*[25] is quite interesting. In this summary of some of the early workers, the names of Lawaczek and Justi are mentioned, but the pioneering contributions of, e.g., Bockris, Veziroglou, Lindström, and other authors active in the founding of the concepts of a hydrogen economy are omitted. Thus:

1. DeBeni and Marchetti,[26] Ispra, conceived in 1970 the idea of thermochemical reaction cycles whereby hydrogen could be produced at less than 800°C by means of nuclear heat from helium-cooled high-temperature reactors (pebble-bed reactors).
2. R. Schulten[27] suggested that methane could be synthesized by using heat from a high-temperature reactor, passing the gas through pipes under

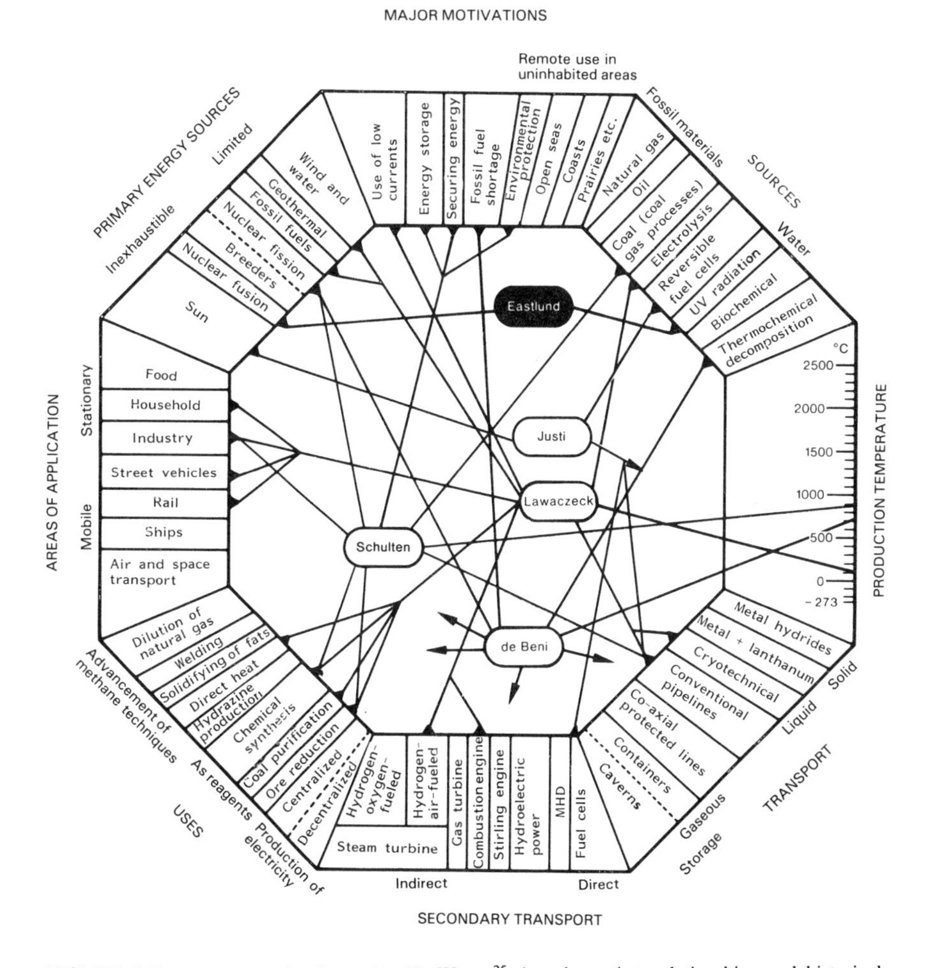

FIGURE 2.5. An octagonal scheme by N. Weyss[25] that shows interrelationships and historical development of a hydrogen economy with respect to eight divisions.

pressure to the consumer, decomposing this catalytically to yield heat by dissociation to H_2 and CO_2, and then conducting these gases through a second pipe system to be recombined in the high-temperature reactor.

3. B. Eastlund[28] of the AEC suggested that the efficiency of future fusion reactors could be increased by using UV radiation for photolysis of water, utilizing a model of photosynthesis.

In Weyss's report,[25] an octagonal diagram is used. It attempts to show some of the characteristics of the solar–hydrogen economy along with its historical development (Figure 2.5), including mutual interrelationships. Unfortunately, in this diagram no references are made to the contributions of, e. g., Haldane (1923), Stuart (1927). Bacon (1950), or Bockris (1962, 1971).

REFERENCES

1. D. H. Meadows, D. L. Meadows, J. Randers, W. W. Behrens: *The Limits of Growth*, Potomac Association, Universe Books, New York (1972).
2. *The Global 2000 Report of the President*, eds. Council on Environmental Quality and the U.S. State Department, U.S. Government Printing Office (1980).
3. J. Boyle: *Nature* **245:**172 (1973); E. Pestel, M. Mesarovic: Menschheit am Wendepunkt [Humanity at the turning point], 2. Report of the Club of Rome, Stuttgart (1974); E. Pestel: Menschheit vor neuen Imverativen [Humanity faces new initiatives], *Abhdlg. d. Braunschweig. Wiss. Ges.* **26:**107–117, Göttingen (1976).
4. D. H. Meadows, D. L. Meadows: *Nature* **247:**97 (1974).
5. G. M. Woodwell: *Sci. Am.* **238**(1)**:** 34–44 (1978).
6. J. O'M. Bockris: *Energy: The Solar–Hydrogen Alternative,* The Architectural Press, London (1975); *Energy Options*, The Architectural Press, London (1978).
7. A. Buch: *Energie* **20**(7)**:**198–204 (1977).
8. E. Justi: 2. Int. Sonnenforum [Second Solar Energy Forum], Hamburg Proc., Vol. 1, pp. 397–408, Deutsche Gesellschaft für Sonnenenergie, Munich (1978).
9. D. P. Gregory *et al., A Hydrogen Energy System,* American Gas Association, Chicago (1972).
10. On the Way to New Energy Systems: III. Hydrogen and the Nonfossil Energy Media. A project supported by the West German Ministry for Research and Technology, Bonn (1975).
11. A. Buch: *Kohle—Grundstoff der Energie* [*Coal—A Fundamental Energy Source*], Pfriemer-Verlag, Munich (1979).
12. W. Hawkins: Lockheed-California, Press release "ddp" (Nov. 7, 1977).
13. J. Verne: *L'ile mysterieuse,* Paris (1874).
14. H. Niederreither: DRP 648 941, Elektrochemische Zelle für H_2-Speicherung [An Electrochemical Cell for the Storage of Hydrogen] (1932).
15. E. Baur, E. Ehrenberg: *Z. Elektrochem.* **18:**1002 (1912); **44:**695 (1938).
16. F. T. Bacon: *BEAMA J.* (Jan. 1954), p. 2; World Sci. Rev. (April 1959), p. 21; USA Pat. 2,716,670 (1955); in: *Fuel Cells,* G. J. Young (ed.), Reinhold Publishing Corp, New York (1963), p. 51.
17. O. Lindström, *ASEA J.* **37:**(1) (1964).
18. F. Lawaczek: *Technik und Wirtschaft im Dritten Reich*, Eher Verlag, Munich (1932); *Elektrowirtschaft*. J. F. Lehmanns Verlag, Munich (1936).
19. E. Justi: Naturwissenschaften **48:**289 (1960); *ibid.* **49:**537 (1961); *Jahrb. Akad. Wiss. Lit.* (Mainz), pp. 200 and 250, Franz Steiner Verlag, Wiesbaden (1955); E. Justi, A. Winsel: *Cold Combustion,* Franz Steiner Verlag, Wiesbaden (1962).

20. E. Justi: *Leistungsmechanismus und Energieumwandlung in Festkörpern,* [*Conduction Mechanisms and Energy Transformation in Solids*], Verlag Vandenhoeck and Ruprecht, Göttingen (1965).
21. J. O'M. Bockris: Memo to Westinghouse Co. (C. Zener) (1962); *Chem. Eng. News* (Oct. 1972).
22. J. O'M. Bockris, E. Triner: *Chem. Eng. News* (Oct. 1972).
23. J. O'M. Bockris: *Environment* **13:**51 (Dec. 1971).
24. E. Justi, A. Winsel: *Physikal. Blätter* **29**(1/2)**:**20–33, 71–79 (1973).
25. N. V. Weyss: Wasserstoff-Information zum Wasserstoff-Konzept (Hydrogen economy) [Hydrogen information on the hydrogen energy concept—hydrogen energy], *Energie* **26**(1): 2–12; **26**(2): 40–50.
26. G. de Beni, C. Marchetti: Wasserstoff—Energieträger der Zukunft [Hydrogen—the energy medium of the future], *Euro Spectra* **2**(2)**:**46–50 (1970).
27. R. Schulten: Synthetisches Erdgas ermöglicht die Fernübertragung nuklearer Energie [Synthetic natural gas as a medium for nuclear energy], *Umschau* **73**(2) ("Adam und Eva") (1973).
28. B. Eastlund, W. C. Gough: Generation of Hydrogen by U. V. Light Produced by the Fusion Torch, 163. National Meeting of the American Chemical Society, Boston (1972).

CHAPTER 3

Time Frame for Building a Hydrogen Technology

Following the discussion of the need for a transition to a hydrogen technology in Chapter 2, this chapter will present material that leads to an estimate of the time needed for the introduction and buildup of a hydrogen economy. Material concerning energy supplies that are available at present and the rate of use of these supplies, together with some prognoses about the exhaustion of fossil fuels, forms the basis for this presentation.

3.1. ENERGY SUPPLIES AND ENERGY CONSUMPTION

Energy resources are linked not only with the question of the need for energy but also with important technological, economic, and sociopolitical questions. Without some consideration of the use of energy, of future needs, of reserves and resources, as well as various feasible technological developments, there can be no realistic discussion of energy problems. In the following sections, these aspects of the energy problem will be treated in some detail to form the basis for statements about future developments and the timely introduction of appropriate measures.

An overview based on present knowledge of worldwide reserves of important principal energy carriers is given in Table 3.1, in which the current economically available amounts of natural gas, oil, coal, and uranium are given, together with a statement of their energy contents. Coal, with 64% of the total,

This chapter authored by Dr. P. W. Brennecke, Braunschweig, and Prof. H. H. Ewe, Hamburg.

TABLE 3.1
Technically and Economically Exploitable World Energy Resources (1980)

Primary energy sources	Energy content (GJ)
Natural gas	3.51×10^{12}
Oil	4.18×10^{12}
Coal	2.01×10^{13}
Uranium	3.54×10^{12}

is the greatest fraction of energy reserves; i.e., for a long time to come, coal will make an important contribution to the energy supply.* To calculate the worldwide supplies of coal, we must take into account not only the reserves stated in Table 3.1 (i.e., confirmed deposits that will yield coal with present-day technology under given economic conditions) but also the resources. By resources is meant reserves and known deposits, which, because of either economic or technical limitations, cannot yet be made to yield coal, together with deposits not yet discovered. The global total coal supplies amount to about $13{,}500 \times 10^9$ tons, and these have an energy content of about 3.2×10^{14} GJ (as of 1980). Of these supplies, about 1320×10^9 tons of coal (3.1×10^{13} GJ) (or about 9.8%) have been clearly confirmed.[1] The other $12{,}180 \times 10^9$ tons of coal should be considered as raw material only with caution. The distribution of coal-bearing strata in the earth is relatively well known. It is not very likely that completely new large deposits of coal deep underground that would be exploitable with conventional mining techniques will be discovered in the near future. However, there is the possibility that a number of smaller deposits (in particular, lignite) will be found.

Considerations similar to these concerning coal may also be applied to other fossil-fuel primary energy sources such as natural gas and oil, as well as uranium.[2]

At present, the world's primary energy need is about 2.8×10^{11} GJ per year. With a world population of over 4 billion people, this corresponds to an average consumption rate of about 73.2 GJ per person per year. However, about 75% of the population gets less than the mean value; 22% of the world population uses between 58.6 and 221.2 GJ and 3% uses over 221.2 GJ per person per

* Translator's note: The role given to CO_2-producing coal in this chapter has to be reviewed in the light of the new findings concerning the relevant dates at which the Greenhouse Effect will become significant. According to work done at the Goddard Space Center (*New York Times,* June 11, 1986), significantly earlier predictions concerning the time at which a world temperature increase of 1°C will occur arise if not only CO_2 itself but also other products of combustion (NO_2, SO_2) are taken into account. The United States would be too arid for agriculture by 2030. The impliction seems to be that conversion to non-CO_2-producing energy schemes should start as soon as possible and not after 2000. Full conversion to an energy base of nuclear and/or solar energy will take 5–6 decades because of huge capital needs involved.

year. In considering the global energy-consumption rate, it must be observed that the growth rates have been exponential in recent decades. In Figure 3.1, the development of world energy needs from 1880 to 1980 is given in terms of primary energy carriers. As the diagram shows, the worldwide rise in the energy-use rate amounts to about 5% per year. Furthermore, the transition from coal (virtually the only carrier up to about 1910) to natural gas and oil can clearly be seen in this figure. The increased use of both these fossil-fuel energy sources is primarily responsible for the considerable increase in the global energy-use rate.

Concerning the future growth of energy use, there is no way to make a reliable prognosis with reliable and watertight data because of the complex relationship between technical matters in the area of energy developments and economics and energy politics. An example can be given in terms of future needs for uranium. Predictions concerning the use of uranium are more uncertain today than ever before, because of changes in nuclear-energy projects and programs as well as uncertainties in respect to future reactors with and without fast breeders and other types of reactors.

On the other hand, there are certain points that allow a rough estimate of further development in respect to the energy-consumption rate. Of special significance is the relationship between growth of the gross national product and

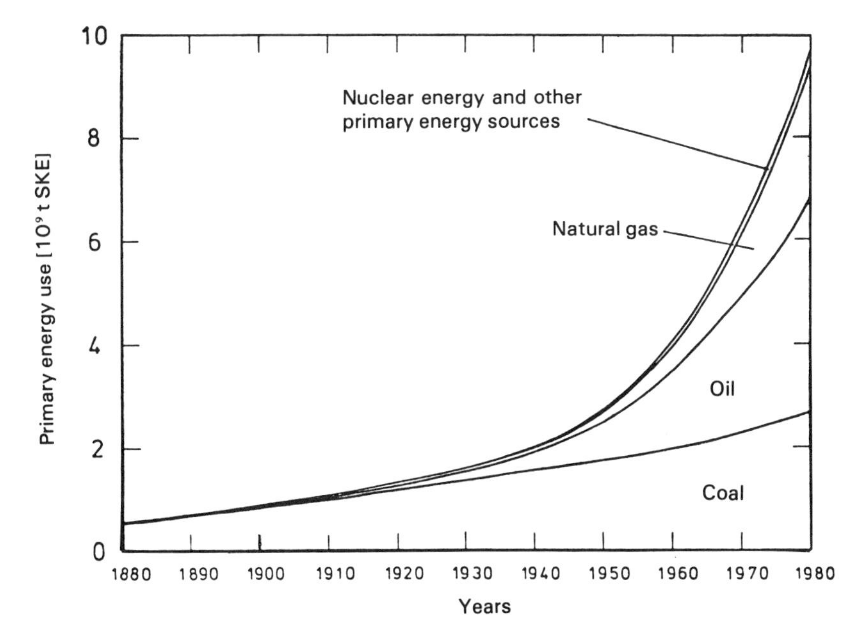

FIGURE 3.1. Schematic diagram of the development of worldwide energy consumption from 1880 to 1980 (1t SKE ≅ 29.3 GJ).

the rate of energy consumption. On average, a 1% increase in the gross national product is linked with a 0.8–1% increase in the rate of consumption of energy.[3] A corresponding comparison for different nations is given in Figure 3.2. The line drawn through the points has a slope of 1, which indicates a direct proportionality between the gross national product and the rate at which energy is consumed. In general, a high per-capita gross national product corresponds to a high need for energy per person. Corresponding to Figure 3.2, the Western industrial countries with the highest gross national products have the greatest needs for energy. This ratio of growth rates has often been discussed during the last few years. It is sometimes regarded as a law that there must always be a relatively close relationship between the two quantities of output and energy input. On the other hand, there is also the view that the gross national product and the growth of the energy-use rate can be uncoupled from one another; i.e., become independent of each other.

Some fairly concrete concepts concerning the future growth rate in world

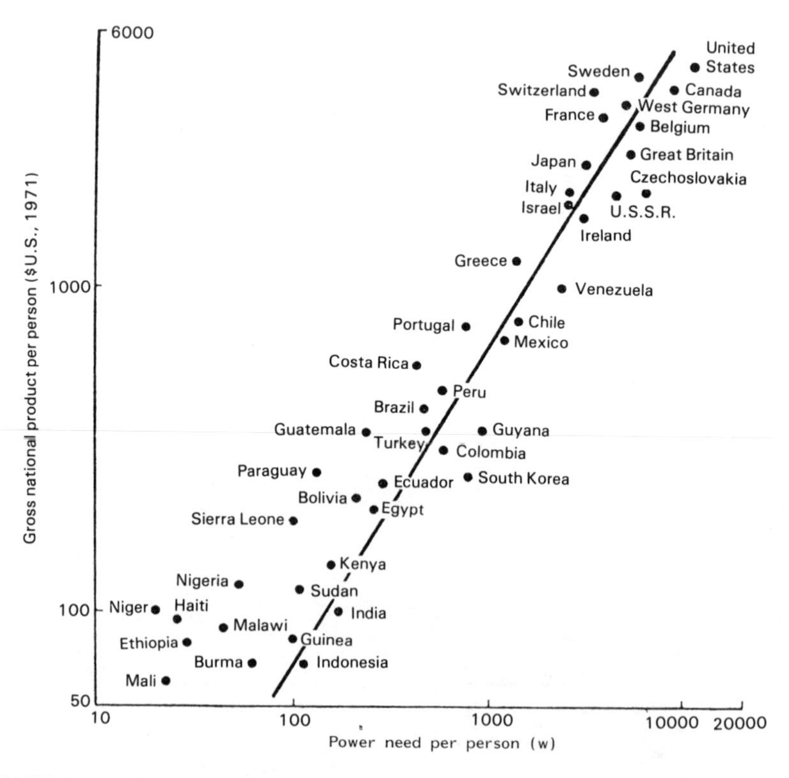

FIGURE 3.2. Relationship between gross national product per person and primary energy consumption (power need) per person for different countries.

energy demand can be obtained from the results of the 11th World Energy Conference in Munich in 1980[4] or from the published study of global energy problems by the International Institute for Applied Systems Analysis, Laxenburg (near Vienna).[5] In both cases, because of the rapidly increasing world population, the necessity for the developing countries to catch up in their needs, and further economic growth in the industrialized countries, it is predicted that there will be continuous sharp growth of the primary energy need. Thus, the present need of 2.8×10^{11} GJ will escalate to between 7.0 and 11.1×10^{11} GJ per year by 2030, and in these estimates it has already been assumed that considerable energy conservation will be made in all areas. Such an increase in need can be satisfied only if all available energy sources are developed and used and furthermore all possibilities are intensively used in the direction of rational energy applications. According to the prognoses that are developed here, it will be necessary to rely on coal and nuclear energy as the principal means of meeting the growing need for primary energy. Only a small—but important—contribution is expected at present from renewable energy sources in this picture of future energy supplies.[6]

3.2. ENERGY NEEDS IN THE FEDERAL REPUBLIC OF GERMANY

To clarify the structure of the energy-use rate discussed in the preceding section, it is useful to take an example, namely, that of the various primary energy sources being used in the Federal Republic of Germany. This subject will be treated here in somewhat greater detail.

There was roughly a 3-fold increase in primary energy use in West Germany between the years 1950 and 1980. During this time, the amounts changed from 3.9×10^9 to 1.1×10^{10} GJ. This increase was connected with a basic change in the structure of the supply network, from the dominance of coal that existed during the 1950s and the beginning of the 1960s to today's dominance of oil.[7] Thus, in 1950, the contributions of anthracite and lignite as primary energy sources were 72.8 and 15.2%, respectively, and oil had only the relatively small contribution of 4.7%. However, by 1980, the contributions of anthracite and lignite had decreased to 19.7 and 9.9%, respectively, while the contribution of oil had risen by a factor of 10 to 47.8%.

The introduction of oil into the German energy market was accompanied by a large increase in the use of natural gas. While natural gas accounted for only 0.1% of the total primary energy supply in 1950, it grew to 16.7% by 1980. Thus, today, natural gas in the Federal Republic of Germany is almost on a par, as far as an energy supply goes, with anthracite. Nuclear energy contributed nothing in 1950, but by 1980, its contribution was 3.6%. Other energy sources (e.g., hydropower) were reduced in their contributions from 7.2% in 1950 to 2.3% in 1980.

The dependence and weaknesses of this supply structure can be seen by comparing the energy production from Germany's own resources with that from imports. Here, the situation is particularly critical in the case of oil, for only 4.5% of the need (1980 figures) is obtained from fields within the Federal Republic of Germany.[8] About 75% of the supply of crude oil to Germany comes from the OPEC states, among which the Arab countries and North Africa supply the largest amount. Since the energy content of the economically available oil in Germany amounts to only 2.7×10^9 GJ, the importing of oil and dependence on the OPEC states will have to continue for the time being, though with a somewhat decreasing trend.

When it comes to the supply of natural gas, the lines of dependence are somewhat weaker. The energy content of the economically available natural gas reserves in Germany is around 5.7×10^9 GJ and hence is about double that of the oil reserves. Of the need for natural gas, 31% can be met from German supplies, while the remaining 69% can be obtained from three countries—about 37% from the Netherlands, 15% from Norway, and 17% from the U.S.S.R. (as of 1980). Diversification among countries from which imports are made looks attractive from the viewpoint of a sure supply.[3] It must be taken into account that natural gas delivery contracts are always signed for very long terms, namely, 20–25 years, and a number of the more comprehensive supply agreements extend into the next century.

Lesser problems in the primary energy supply for the Federal Republic of Germany are presented by coal. The energy content of 7.1×10^{11} and 2.9×10^{10} GJ is given by the economically available anthracite and lignite reserves, which undoubtedly are foremost with respect to security of a long-term supply. The anthracite production for 1980 was 2.5×10^9 GJ[9] while the lignite production was between 110 and 120×10^6 tons.[3] Comparison of the known reserves with the production figures shows clearly that even with some increase in the production rate, the coal supply is secure into the next century. Moreover, environmental and accessibility problems together with cost aspects would allow only a very small increase in present coal production.

Because the Federal Republic of Germany does not have appreciable uranium reserves, the entire requirement must be supplied by imports; i.e., the dependence on imports for this commodity is even greater than that for oil. The import of unrefined nuclear fuels (uranium, plutonium, thorium) was 1651 tons in 1980. The proportion of uranium enriched to the extent of 3–10% with ^{235}U amounted to 404.3 tons.[10] The most important countries that supply this uranium are France, Great Britain, the United States, and the U.S.S.R. In terms of supply, there do not seem to be any particular foreseeable problems.

Some use of what can be regarded for all practical purposes as inexhaustible energy sources (solar and wind energy) has been made since the energy crisis of 1973–1974. The long-term contributions that these sources may make is going to be influenced by numerous factors (among others, the possibility of further technical developments, penetration of the energy market, and economics). Rough estimates of the cost of these inexhaustible energy sources lead one to expect that in the near future they may make a contribution to the German primary energy supply of about 4%.[3]

Statements about the future energy supply in the Federal Republic of Germany and about possibilities of its being met until the year 1995 have been made by, for example, the recently published scenarios of three research institutes in economics.[11,12] Despite somewhat differing assumptions, the basic statements of the three scenarios agree. Until about 1995, there will be a yearly growth of gross natural product of between 2.2 and 3.4%. According to these predictions, the primary energy need will go up by only an average yearly increase of 1–1.4%. In this case, the need in the year 1995 will be 1.3—1.4 $\times$ 10^{10} GJ. The primary energy-use rate will thus grow at a significantly more slowly than the gross national product. In fact, the energy-use rate should increase between 0.4 and 0.7% per year and for the year 1995 is calculated to be 8.1—8.5 $\times$ 10^9 GJ. Thus, consumption of crude oil and lignite will be cut back by 34 and 8% respectively, while the use of anthracite will be increased to about 22%. The natural gas portion will remain at about 16%, almost unchanged. The contribution of nuclear energy to the total energy supply has been estimated to be 17% for the year 1995. The contributions of other energy sources (e.g., hydro, other fuels) is estimated to be 3%. In general, the three scenarios indicate a further increase of the dependence of the energy supply of Germany on imports. However, because of the changed structure of need, there may be a greater spread of import risk.

3.3. EXHAUSTION OF THE PRIMARY ENERGY CARRIERS

According to the foregoing presentation, the world energy supply depends on the four primary energy sources—natural gas, oil, coal, and uranium. Since the reserves and resources of these materials are not infinite, future supplies with an unaltered energy-need structure and rates of energy use cannot be assured. Thus, the finite extent of the supplies leads inevitably to an expected exhaustion of natural gas, oil, coal, and uranium.

To predict the future energy need and to devise some mid- and long-term strategies by which energy supply for the future can be assured, various energy scenarios for future energy consumption have been suggested. Using mathematical models for such prognoses, some fairly realistic scenarios can be conceptualized.[13] The input data for these situations are the known values concerning available energy supplies and data about the structure of the energy-use rate. Several limiting conditions within which a certain development can take place are usually incorporated. These energy-use predictions, the results of which differ considerably according to the model and limitations used, have been published repeatedly in recent years. In the following presentations, a few of these approaches will be discussed in an attempt to work out the time frame for the expected exhaustion of primary energy carriers—and to present the necessary timely countermeasures.

It has already been mentioned that coal is the most available of the fossil primary energy sources. Estimates of the future availability of coal are largely based on the work of Hubbert[14,15] as well as that of Elliott and Turner.[16] The world need for coal assumed by Hubbert was based on the use of coal to produce electricity and its application in various metallurgical processes. The result of these considerations is given in Figure 3.3. According to various assumed total coal resources, maximum coal production will be reached some time between 2100 and 2140. After this, the exhaustion of this primary energy source will begin, and there will have to be an increase in the use of other energy sources.

In comparison, Elliot and Turner take into consideration an increasing use of coal in the form of synthetic liquid and gaseous hydrocarbons; i.e., they take into account the liquefaction and gasification of coal as a substitute for natural gas and oil. Their estimate is given in Figure 3.4, and the maximum here for primary energy production, variable according to the various global total supplies, is placed between 2030 and 2050. The solid line corresponds to what at the time of their prognosis was a realistic estimate of the total supply of fossil primary energy, namely, 8.8×10^{13} GJ. The primary energy supply will then reach a maximum of 1.7×10^{12} GJ in 2030 and fall in a few decades. Elliot and Turner's other assumptions of the total energy supply of 1.7×10^{14} GJ and 2.9×10^{14} GJ (dashed and dotted lines, respectively) would shift the maximum value only to the year 2044 and 2050, but would correspond to a distinctly larger

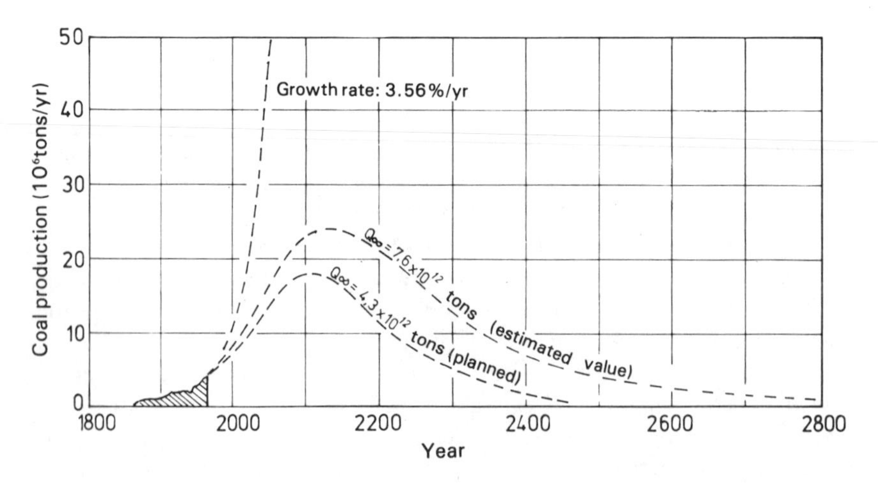

FIGURE 3.3. Prognosis for world production of coal. (Q_∞) Assumed total supply. From Hubbert.[15]

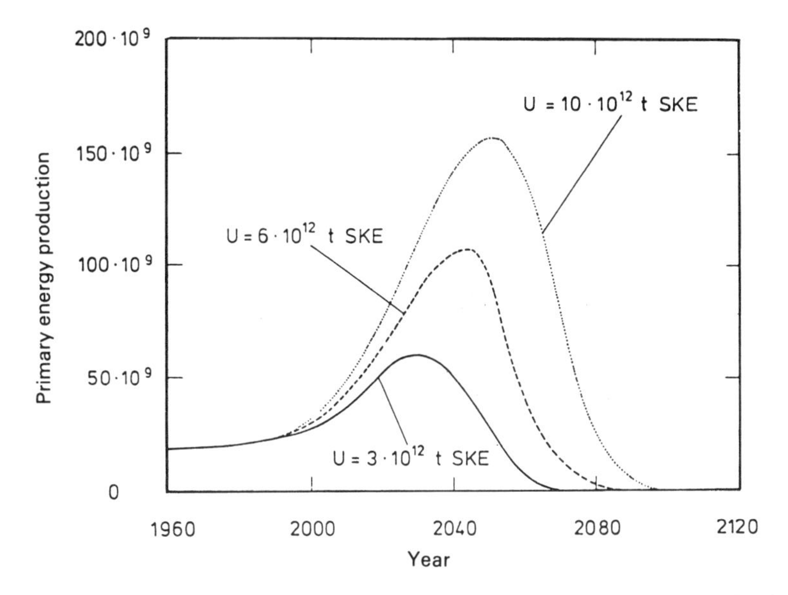

FIGURE 3.4. Prognosis for world production of fossil fuels. (U) Estimated total supply. From Elliot and Turner.[16]

maximal primary energy production of 3.1×10^{12} GJ and 4.6×10^{12} GJ, respectively.

If one compares these estimates by Elliott and Turner with what is known of the total coal reserves today which have an energy content of about 3.2×10^{14} GJ, exhaustion of the primary energy supply would not occur until the year 2050. Since only 9.8% of this total supply has actually been proved to be present, i.e., 3.1×10^{13} GJ, the conclusion is—corresponding to the preliminary calculations of Elliott and Turner—*that the proven coal deposits will be exhausted before 2030*.

Evaluations of the future availability of oil show that the maximum production rate should also be expected here. In Figure 3.5, two predictions concerning future crude oil production in OPEC countries are presented without taking into account any possible limitations on production.[17] According to this figure, maximum production will be reached around 1995. Other scenarios indicate that world oil production will reach a high point between 1990 and 2010.[2] Thus, as estimated at present, technically obtainable and economically available crude oil will be exhausted in about 20–25 years.[3,18]

Observations corresponding to those given here for coal and crude oil can also be made for the other primary energy sources, natural gas and uranium.[2,4–6,13] If one looks at the whole picture of these examinations and studies, and on the basis of their results calculates some kind of mean value concerning the subject

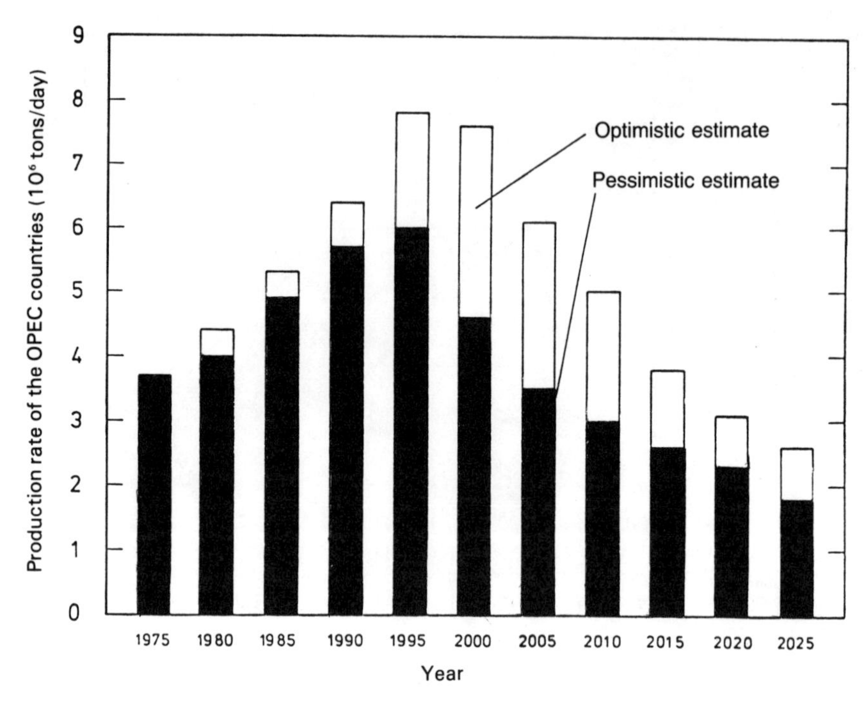

FIGURE 3.5. Prognosis for the oil production rate of the OPEC countries for the period from 1975 to 2025.

of future available energy, then we must reckon with an exhaustion of natural gas, oil, coal, and uranium around the year 2030. From this point, other energy sources will have to take up the need. Thus, we have about 40 years, of which half will have to be used for the necessary research and development work and half for the building of new energy systems.

This prediction of the year 2030 as the year in which the supply of energy from coal will begin to diminish must be evaluated in the context of the fact that there is a less than exact knowledge of both reserves and resources, the time of the possible introduction to new technologies, and the effect of energy conservation. Thus, increased activity and exploration might lead to the finding of further supplies. Further development in recovery techniques could make possible the availability of deposits that hitherto have not been counted as usable. With respect to natural gas supplies, for example, relatively little of the continental shelf and the offshore regions has been examined.[19] Among the newer techniques for mining, one must mention, among others, the development of deep mining techniques. With respect to oil, there is the possibility (in conjunction with second- and third-stage recovery from known oil fields) of exploiting deposits on the continental shelves and in polar regions as well as the development of shale oil and tar sands.[3,13] However, it is questionable whether exploitation of these deposits will develop rapidly.

Energy conservation and the possibility of more rational energy use present a further potential which is rapidly growing in importance. For example, primary energy use in the Federal Republic of Germany was 5% less in 1981 than in the previous year; i.e., it dropped to 5.6×10^8 GJ.[20] Finally, we must point out the divergence that is often observed between predictions of energy need and actual energy use. Figure 3.6, for example, shows predictions of primary energy use in the Federal Republic of Germany in the framework of the German government's energy program; it can be seen that real energy use has hitherto always been less than had been predicted.

The limited nature of the primary energy sources of natural gas, oil, coal, and uranium, and the absolute political necessity of finding substitutes for them, lead inevitably to the inexhaustible energy sources. Solar energy is the obvious energy source of the future, but breeder reactors and reactors involving fusion can also be counted as virtually inexhaustible. Thus, with regard to the bigger picture and the technical solution of the long-term energy problem, the following three possibilities seem to be in sight at present: the fusion reactor, the breeder reactor, and hydrogen technology. These energy sources will be covered in greater detail in the following sections (for solar energy, see Chapters 5 and 6).

In comparison with conventional electricity-generating stations, the various

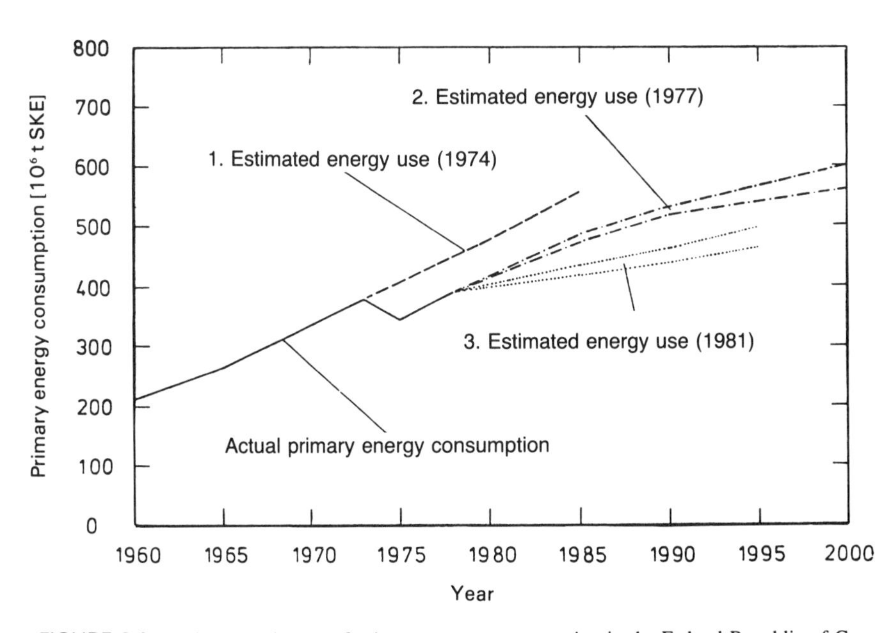

FIGURE 3.6. Projected estimates of primary energy consumption in the Federal Republic of Germany within the framework of the government's proposed energy program (1t SKE ≅ 29.3 GJ).

newer sources of energy—fusion reactor, breeder reactor, and hydrogen technology—have lower operating costs but represent a considerable capital investment. Because the fuels for these systems are virtually inexhaustible, their applications are limited only by the availability of capital for their development.[21] In such a transition, the present primary energy sources are in some sense replaced by capital, to be invested in the machinery for extracting energy from these inexhaustible energy sources. In this relationship, one can speak of a transition to an intensive use of energy resources.[5,13] Thus, it is of great importance for the energy politics of the next 50 years that this transition be started at the earliest possible moment and the construction of the new energy supply system be begun. One of the problems may be the bridging of the transition period from the use of exhaustible (natural gas, oil, coal, and uranium sources) to practically inexhaustible sources. In this transition time, it will be necessary to use both fossil and nonfossil energy sources in a rational manner.[22]

3.4. CONTROLLED NUCLEAR FUSION*

Hahn and Strassmann showed in a historic experiment that in the fission of a heavy atom into two atoms of medium atomic weight, enormous energies are released. This first example of emission of energy in nuclear fission led first to the explosion of the uranium bomb and later to its control in the atomic fission reactor. In fusion, the energy is obtained by the melting together of atoms that are very light and up to medium atomic weight.† The first time such fusion was carried out was in the hydrogen bomb, but if fusion is to be useful as an energy source, it must of course be made to occur continuously, rather than explosively.

From the fusion of atomic nucleii, energy is set free just as it is in chemical combustion reactions. The reactions that are most studied are the fusion of deuterium with tritium as well as the fusion of tritium with deuterium:

$$D + T \rightarrow He\ (3.5\ MeV) + n(14.1\ MeV)$$

The heat of reaction (given in parentheses) is given off as the kinetic energy of the reaction products. In all the reactor concepts, the idea is that the energy of the α particles (the He atoms are immediately ionized) stays in the reaction space (the plasma), while the neutrons escape and are used, e.g., to heat Li.

* Translator's note: It is important that the reader remembers that although research on fusion was commenced in 1953, no stable fusion of H atoms has yet been achieved. This section authored by Prof. W. N. Heiland, Garching.

† Translator's note: Workers in the field of fusion often present the possibilities that they have been theoretically calculated as though they were actualities. Again it must be clear to the reader that, as of 1987, these concepts have not been experimentally realized.

There can thus be made to occur a very desirable reaction with the lighter Li isotope:

$$^{6}\text{Li} + n \rightarrow \text{He} + \text{T} + (4.78 \text{ MeV})$$

This means that tritium is being bred. Thus, the amount of ^{6}Li (present to 7% in naturally occuring Li) becomes the material which determines the basic fusion reaction, and not synthetic tritium, which is very costly to produce.

The second reaction considered in fusion is one which is not useful at present, namely:

$$\text{D} + \text{D} \rightarrow \begin{cases} \text{He } (0.8 \text{ MeV}) + n \text{ } (2.5 \text{ MeV}) \\ \text{T } (1.0 \text{ MeV}) + p \text{ } (3.0 \text{ MeV}) \end{cases}$$

These latter reactions are in principle more attractive than the ones previously mentioned because deuterium is naturally available (16.7 ppm D in natural water) in sufficient amounts. No breeding reaction is necessary, and the products are relatively easy to handle (having a smaller amount of neutron energy and more charged particles). On the other hand, for the D–D reaction, an ignition temperature about 10 times higher than that for the D–T reaction, which starts up at about 4 keV $= 40 \times 10^{6}$K, is needed.

At present, there are two main reactor concepts that are being discussed, and these depend, on the one hand, on a magnetic confinement of plasma, or, on the other, on inertial confinement. For net energy to be obtained from the reactor, there must be a sufficient number of fusion reactions per unit time, which means that depending on the temperature of the reaction partners T_i (keV), the product of the number of particles per cubic centimeters, n, and the confinement time, τ_E (sec), must reach a value that is determined by the action cross section of the reaction so that the gain in energy from the fusion reactions is larger than the unavoidable radiation losses (Lawson criterion[23]) (Figure 3.7). The elements H and He ionize completely in the region of 10 keV, so that fusion reactions take place in the plasma, and this undergoes losses mainly by continuous radiation and synchrotron radiation.

The concept of magnetic confinement uses a magnetic field which is intended to hold the charged particles together so that a large enough value of $n \times \tau_E$ can be reached. Of the various configurations used, such as the mirror, the stellarator, the thyroidal theta pinch, and tokamak (cf. Post [24]), the tokamak principle is the most-discussed reactor concept (cf. Badger *et al.*[25] and Davis and Kulcinski[26]) (Figure 3.8). The idea of a tokamak involves taking a plasma in the form of a ring and then through ohmic heating raising the plasma temperature. The plasma discharge resembles the action of the secondary winding of a transformer (Figure 3.9). As the plasma temperature increases, the con-

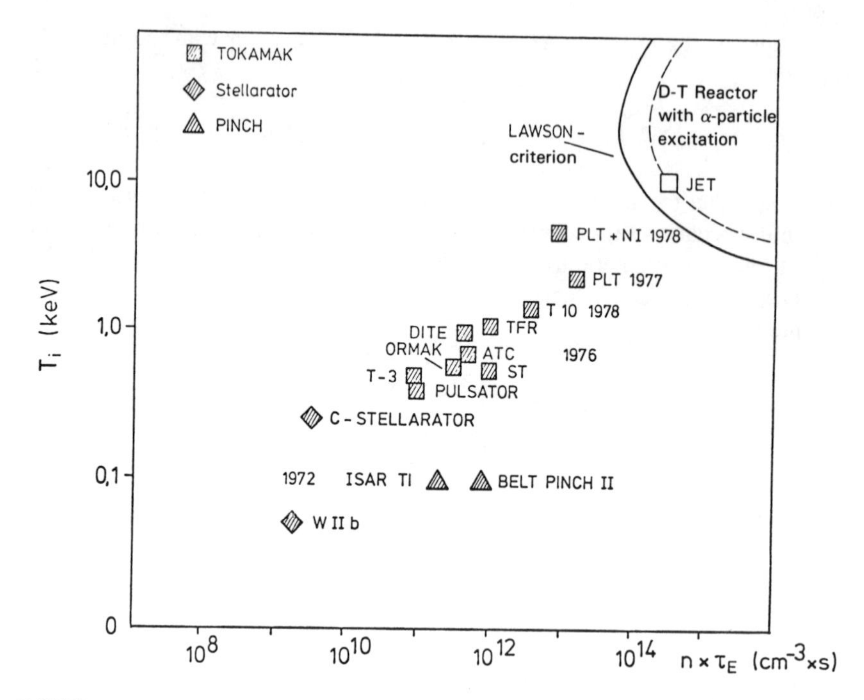

FIGURE 3.7. Lawson diagram of the D–T fusion reaction. Depending on the temperature of the reaction partner T_i (keV), the product of the particle density n (cm^{-3}) and energy inclusion time τ_E (sec) must reach a value such that the energy gained from the fusion reaction is greater than the radiation lost as a result of the continuous radiation and synchrotron radiation, as is possible within the limiting line in the figure. The symbols that denote the tokamak fusion machine, stellarator, and pinch are at top left; the respective laboratories are written next to the measured experimental results obtained (situation at the end of 1978).

ductivity increases so that another heating mechanism has to be introduced; e.g., some form of radiant heating such as irradiation by means of electromagnetic energy of a certain frequency for neutral-particle injection (NI), as exemplified by the injection of highly energized D atoms that are ionized and thermalized in the preheated plasma. This is the method that has given the most favorable values compared to the Lawson criterion in the magnetic-confinement method (see Figure 3.7). Apart from possible plasma instabilities, this concept, however, suffers from the problem of impurities. As the plasma burns in a vacuum vessel, and has a density of about 10^{14} cm^{-3} with temperatures in the kilo-electron volt region, the confinement times are in a range of a few seconds, and under these conditions, interaction with the vessel occurs, introducing impurities into the plasma, which come from ions of higher atomic number. These ions (e.g., Ni, Fe, Cr from the steel walls) are only partly ionized, so that increased radiation losses occur by continuous radiation. Apart from this, continuous radiation losses

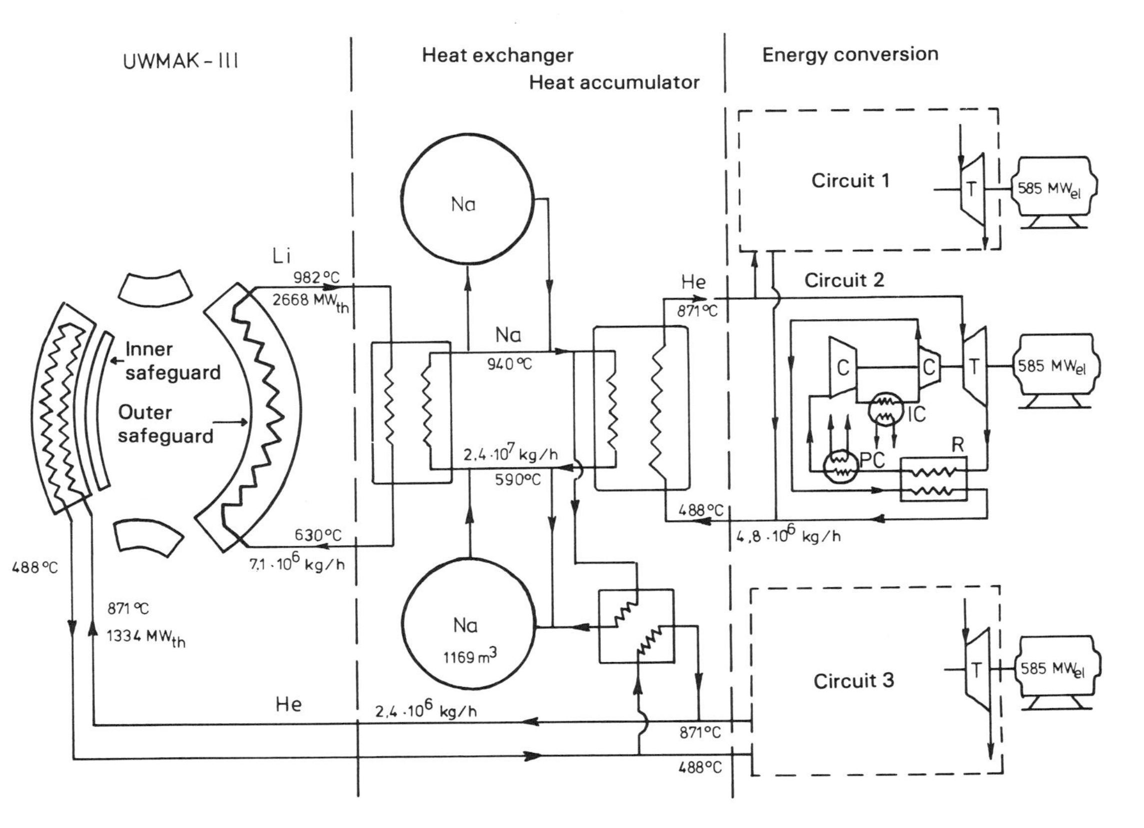

FIGURE 3.8. Schematic diagram of tokamak construction UWMAK III. From Badger *et al.*[25]

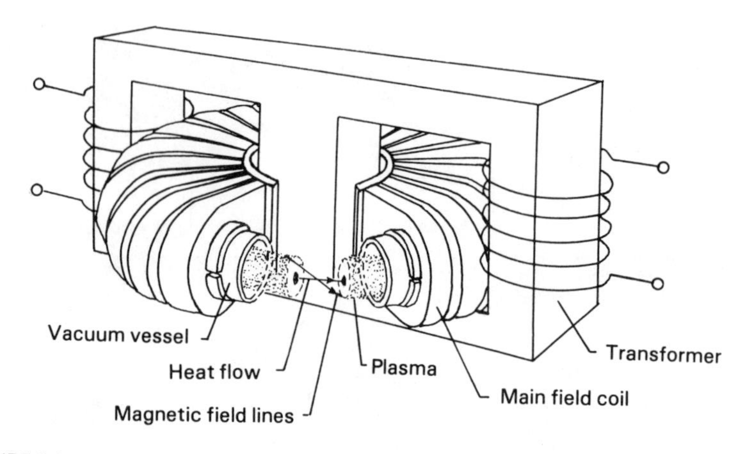

FIGURE 3.9. Construction of a tokamak. The plasma forms the secondary coils of the transformer; it is contained in the vacuum vessel within the main field coil. The annular plasma in the tokamak, with the help of the transformer, induces a toroidal plasma stream, through which is created a so-called torroidal magnetic field, running around the small diameter. The superimposition of the torroidal field and the main field leads to the twisting of the field lines.

are elements of higher atomic numbers. Reaching the Lawson criterion is thereby made more difficult (see, for example, Behrisch and Kadomtsev[27]).

Inertial confinement tries to attain the condition for fusion in the following way: The heating of the fuel (D–T mixture) is done so quickly, and with such compression, that despite the short confinement time, a nuclear reaction begins. The densities of the compounds that are attained in this compressive action are 10^{21} cm^{-3}), and the temperature is in the region of 10 keV.* The energy for both heating and ignition is to be attained by means of lasers or electron or ion irradiation. The experimental method of attempting to obtain this is to put D and T mixtures in glass spheres (with diameters in the micrometer range) and to bombard these spheres with a series of radiation pulses, insofar as possible to an equal degree from all directions. The plasma is formed and heated extremely rapidly in the outer layer and causes a shock wave to travel into the inner part of the particle, which immediately compresses and ionizes the D–T mixture. The parameters that have been reached in experiments of this kind are comparable with the best that the tokamak has so far shown.[28] Typical problems that arise

* Translator's note: According to W. Anthony Cooper of the Oak Ridge National Laboratory (1983), no experiments had ever been made at the time with 14 MeV neutrons, so that the fusion program (worldwide about $1 billion per year) is going forward without any direct examination of whether the materials available will be adequate. The difficulty is that until fusion has been made to occur for a significant time, it will not be possible to have a 14-MeV neutron supply by which experiments on materials problems of fusion reactors will be made.

in this approach to fusion are (once more) the instability of the plasma and anomalies in light and energy absorption and, furthermore, the technology associated with the target fuel.

The main long-term problems of all the actual concepts that are based on the D–T reaction are questions of cost. These may involve the cost of the fuel and the investment costs.[29,30] Severe material problems are inevitably connected with the 14 MeV neutron bombardment. While the radiation danger connected with time presence of tritium may be kept low, on one hand because of controllable chemistry and on the other because of the rather short half-life, neutron radiations arising from nuclear reactions cause the reactor material to become activated. It will therefore be necessary to make an extremely careful choice of construction materials. Contrary to the fusion reactor, only relatively short-lived and solid, activated substances are formed. Even more vital are the possible material modifications in the inner part of the reactor zones, in which the appropriate temperatures for energy decoupling must reign, and this is just where the neutrons have to be absorbed—neutrons that produce the temperature. There are also secondary neutrons and other particles (protons, alpha particles) and hard gamma radiation that are produced in nuclear reactions. These problems have led to the idea that in new reactor concepts, relatively low temperatures (<1000K) will be used. In doing this, the overall efficiency will fall and thermal pollution would rise. Thus, with one possible arrangement, one must tolerate, at 6 GW_{th} and 2 GW_{el}, thermal losses of about 4 GW_{th}, in the event that better ways are not found of using the heat in all temperature ranges, so that the Carnot processes are avoided.

In summary, it can be said that there is no single idea, or development, for a fusion reactor, and two very distinct fusion reactions, either the D–T or the D–D reaction. At the moment, only the D–T reaction is being worked on because the ignition temperature of 40×10^6K is about 10 times lower than that of the D–D reaction. Even though it was earlier regarded as somewhat doubtful whether the high values for the ignition temperature and the product or particle density and containment time could be reached, after 1978, a favorable approximation to the Lawson criterion (see Figure 3.7) was successfully obtained using the Tokamak principle (see Figure 3.9). It is therefore now regarded as quite possible that ignition may be achieved for the first time before the end of the century. Conversely, some difficulties have been underestimated, e.g., the effect on the materials in the inner reaction zone of the 14-MeV neutrons, secondary neutrons, protons, α particles, and hard gamma radiation from the nuclear reactions. It is on these grounds that the temperature of the cooling medium [e.g., the metal tubes with circulating liquid lithium (see Figure 3.8)] is now planned to be kept at less than 1000K. The result will be that the overall efficiency of the reactor will decrease considerably and the thermal pollution of the environment will be greatly increased—to values that correspond to those for conventional power plants. Even if all of these difficulties can be overcome, there will still be very considerable problems concerning the materials available for D and T fusion reactors, and the fact is that these reactors might not have a working life as long as that of the competing coal-burning power plants and the light-water reactors.

To attain the visions that have sometimes been projected for fusion reactors, in which mankind is supplied for a virtually infinite period on the basis of fusion using hydrogen isotopes from the oceans, the D–D reaction would have to be used. The advantage here is that no breeder reactions are necessary, and the reaction products are much easier to handle. However, the ignition temperature would then have to reach one billion degrees, and therefore any such development is not to be regarded as likely in the near future.

3.5. FAST BREEDER REACTOR

An optimal use of uranium supplies will not be reached by utilizing the light-water reactor that is now used in all parts of the world, because it utilizes only the relatively rare uranium isotope ^{235}U. The uranium isotope ^{238}U, which is in plentiful supply, cannot be fissioned by use of thermal neutrons. Natural uranium is a mixture of ^{238}U (99.27%) and the fissionable isotope ^{235}U (0.72%), and a small percentage of ^{234}U.[31] Light-water reactors need the easily fissionable ^{235}U for their operation. A small conversion of ^{238}U into fissionable plutonium also occurs in ordinary reactors, but this increases the use of nuclear fuel by only about 0.3%, and thus the use of present uranium supplies in light-water reactors (without taking into account any reprocessing of the spent fuel elements) is limited to about 1%.[32] According to present estimates, the economically and technically usable uranium reserves would last about 60 years,[18] and thus any energy scenario that relies on this type of reactor is going to run into difficulties in respect to the availability of fuel in a relatively short time.

To solve the problem of the future availability of nuclear fuels, a somewhat different concept—the "fast breeder reactor"—is now being considered. The essential characteristic of this concept is that the reactor produces more nuclear fuel than it uses. What is meant by a fast breeder reactor is a reactor in which not thermal but fast neutrons maintain the fission chain reactions. Fast breeders do not contain a moderator because the aim is to avoid any decrease in velocity of the fast neutrons that arise from fission. The incentive in of a fast breeder is the possibility of using the breeding process, i.e., the conversion of ^{238}U into ^{239}Pu. Thus, when the uranium isotope ^{238}U is bombarded with fast neutrons, it gives rise to ^{239}U, and this then undergoes conversion, utilizing two beta particles, to fissionable ^{239}Pu. This conversion process also takes place in a light-water reactor, but with a very small yield. The yield is much greater in the fast breeder because, owing to the absence of the moderator, there are more neutrons set free (per absorbed neutron) than are available for the breeding process, and these then yield fissionable ^{239}Pu from ^{238}U by means of neutron capture.

The fuel in a fast breeder is correspondingly a mixture of ^{239}Pu as fissionable material and ^{238}U as the breeding fuel. The energy-yielding chain reactions that occur during the functioning of a breeder reactor are concerned with the fission of ^{239}Pu brought about by fast neutrons. The breeding process that goes on at the same time is the conversion of ^{238}U into ^{239}Pu. The principal statistical balance from the chain reactions that set free the energy and the breeding process can be simplified as follows:

$$2^{239}Pu + 3^{238}U = 3^{239}Pu + \text{fission products} + \text{energy}$$

Thus, more fissionable material ^{239}U is obtained from the ^{238}U than is used up in the chain reactions that produce the energy.[33] Hence, the breeder reactor would theoretically be able to burn all the uranium originally available (i.e., ^{235}U and ^{238}U) completely and to convert all its energy to heat. In practice, however, only about 60–70% of the free energy in uranium can be utilized. The energy equivalent of one ton of uranium used in a light-water reactor is 6.8×10^5 GJ, but when used in the breeder reactor, it multiplies itself by 72 times to 4.9×10^7 GJ.[34] On this basis, then, the uranium reserves known at present to be economically and technically useful would last not merely for a few decades, but for many centuries (according to recent estimates, about 3600 years).[18]

Like other nuclear reactors, fast breeders are heat machines. The controlled nuclear fission that takes place in the nuclear fuel gives rise to heat, and this heat is then taken off by a heat-transfer medium and made to produce steam for use in turbogenerators.[32] The nuclear fuel consists of plutonium oxide ($Pu0_2$), which is mixed in a ratio of 1 : 5 with uranium oxide (UO_2). The fuel elements, consisting of fuel rods bound together, form the fission zone in the reactor core. Surrounding this zone is a so-called breeding mantle, which consists of fuel elements made of breeding rods bound together in a similar way to the fission zone. At the beginning of the reactor operation, this mantle contains no fissionable material. In the course of the reactor operation, the high flux of fast neutrons leads to the conversion of ^{238}U to ^{239}Pu; the breeding process takes place partly in the uranium of the fission zone, but mainly in the uranium of the mantle.

The high thermal output in the heart of the fast breeder has to be carried off in a suitable heat-transfer medium to produce steam. In all breeder reactors that have so far been built, or projected, liquid sodium is used as the heat-transfer medium. The sodium is raised from 370 to 540°C by being passed through the reactor core. This relatively high temperature allows the (Carnot) energy conversion efficiency of the plant to be as high as 40%, about as high as is reached by modern coal-burning power plants. For heat transfer, primary and secondary circuits are used in several parallel arrays for reasons of safety.

Research and development projects, as well as the operation of breeder reactors, are in progress in several countries. The first German sodium-cooled experimental reactor (output: 20 MW_e) went into operation in 1972. Since 1973, the sodium-cooled fast breeder SNR 300 (output: 300 MW) has been under construction at Kalkar on the Niederrhein. Completion is expected toward 1990. The costs will be more than several billion dollars.

The concept of using the breeding process to considerably broaden the supply of nuclear fuels on a uranium base seems at first sight to be impressive and effective. However, whether the introduction of this reactor will occur in time to give a solution to the problem of the exhaustion of liquid fossil fuels must be regarded as rather doubtful, on political, technical, and economic grounds.

Questions concerning the sociopolitical acceptance of breeder technology are directly connected with entry into the plutonium economy. In this respect, the Ford Foundation has made seven recommendations[35] to the United States Government:

1. A clear decision should be made to postpone the manufacture and recovery of plutonium.
2. A winding down of the much-recommended breeder program and abandonment of commercialization should take place within the near future.
3. A lower priority must be given to nuclear energy in energy research and planning.
4. Promotion of nuclear energy within the country and outside it should be discontinued.
5. An orderly expansion of enrichment plants in coordination with realistic predictions of energy needs within the country and outside should take place.
6. The export of technology for plutonium production and enrichment should be banned, and an attempt should be made to get other suppliers of this material to take similar actions.
7. Nuclear exports should be allowed only when they are in accord with American security regulations and undertakings and when they correspond to the politics of nonproliferation.

Before fast breeders reach a state of development that corresponds to the light-water reactors that are available on a worldwide scale today, there will have to be a development effort to attempt solutions of the technical problems that are still unsolved.[32] The present research and development work is concentrated in the following areas[33]:

1. At the heart of the development of fuel elements and cladding material are the swelling and cracking that occur under radiation. Further, chemical interaction between the nuclear fuel includes fission that of products with the fuel rods.
2. In respect to the reliability of the whole plant, the critical components have been shown to be the steam producer and the superheater. A far higher average quality is needed for the numerous welded seams.

3. The safety measures in respect to accidents involving sodium–air and sodium–water reactions must be investigated. Among the most urgent work is that concerning the loss of cooling material (e.g., that which would occur if a tube broke) or a situation in which the sodium boiled.

The principal problems of this type of reactor are concerned with the increase in the multiplication of neutrons that accompanies the loss of coolant.[36] As a result of the consequent failure in retardation of neutrons, the focal point of the neutron spectrum is shifted to higher energies, and this leads to a rapid increase of fission rates and a raising of the criticality, which may cause very serious interruptions of operation.

Even if the development of breeder technology proceeds—and the scientific and technological problems are solved—it is still questionable whether it will be possible to make breeder reactors economical and safe in operation. Further, great costs will be involved in the building of energy-supply systems based on breeders. In this connection, the need for closing the nuclear fuel cycle for fast breeders must be taken into account in future planning.

As a result of all these questions, no realistic time scale for the introduction of fast breeders—or any energy system based on them—can be given at present.

3.6. HYDROGEN TECHNOLOGY

A third possibility for a large-scale solution of long-term energy supply is that of introducing a hydrogen technology. The future use of hydrogen as a secondary energy source would have the great advantage that the technical basis of its production, transfer, storage, and use already has a sound foundation, and hydrogen can be coupled to other energy media with advantage. In considering a hydrogen technology, one must distinguish between two kinds of schemes, according to the primary energy source: The one would be a nuclear–hydrogen technology, the other a solar–hydrogen technology.[37,38]

In the conventional view of nuclear power, the product is thought of as being electricity distributed by means of wires throughout the national network. However, nuclear technology could also be combined with the production of hydrogen in a nuclear–hydrogen technology for the future.[13] All the types of reactors discussed above would then be applicable. The use of electrolytic hydrogen producers, both those already known and those using techniques now being developed with electrical energy from light-water or fast breeder reactors, would appear to have a good future from both the technical and economic points of view.[39] On the other hand, thermochemical cycles using the process heat from high-temperature reactions do not now seem to have much future due to technical difficulties.

In respect to a solar–hydrogen technology, the idea is to use solar energy collected on a massive scale and thus to utilize the potential of the principal inexhaustible source. The secondary energy source is hydrogen, and the energy of the global radiation is thereby "condensed" and made into a transportable and storable fuel.[40] The needed energy for the production of hydrogen would be available from solar thermal or solar electric power plants that work on the power tower or the solar farm principle and are already in operation.[41,42] The concept of a hydrogen technology based entirely on solar electric (photovoltaic) power stations has been described by Dahlberg.[43,44] Since the concept of a hydrogen technology will be described extensively in this book, no further details will be given here.

3.7. TIME FRAME FOR THE INTRODUCTION OF A HYDROGEN TECHNOLOGY

The time period for the introduction and the construction of the hydrogen technology would be determined by:

1. The supply and range of present energy media, e.g., natural gas, oil, coal, and uranium.
2. The amount of money devoted to research and development in the direction of a hydrogen economy.
3. The actual introduction and construction of the energy-supply system.

As has already been stated above, natural gas, coal, oil, and uranium will be exhausted by about 2030* unless the supplies are extended by greater exploration, new technologies for obtaining difficult available fuels, conservation, or other possibilities for more rational uses of energy. From this point in the future at which fossil fuels begin to decline in productivity, energy supply should be assured by means of hydrogen technology.

In fact, there has already been considerable research and development in respect to supply systems with gases in pipes, and the knowledge of production, transport, storage, and use of hydrogen, together with the extensive experience with an energy system using natural gas, are all available for application to a hydrogen technology, though they must be adapted to the specific needs and differences that the use of hydrogen would entail. Thus, the limiting factor for the introduction of a hydrogen technology can be regarded as the extent of economically available natural gas, oil, coal, and uranium without necessarily considering economic factors.† This statement, however, must be seen in light

* Translator's note: The question of climatic change may lead to the need for legislation to reduce the discharge of CO_2 into the atmosphere at an earlier time than given by consideration based on exhaustion alone.

† Translator's note: This estimate does not take into account the important aspect of the energy needed to obtain energy from the reserves discussed. It is clear that the ratio (energy available from a source/energy used to obtain this energy) must be greater than unity. Unfortunately, it has recently been shown that this ratio will sink below unity before the year 2000.

of the fact that the introduction and construction of a hydrogen technology would require very high capital investment. One must ask, then, whether the needed capital would be available, and if it is not, what is the alternative?

According to the material given here, it will be at least 40 years before a hydrogen technology would be useful on a large scale.* About half this time could be used for research and development work and the rest for construction. It is clear that there are difficulties of inertia that would prevent the present energy-supply infrastructure from being converted in practice to one using inexhaustible energy sources. Thus, the more expensive the existing system becomes, the more lengthy and difficult will be the introduction of a newer and even more expensive system. From this point of view, and from that of energy politics, it is very important indeed to utilize the 40 years left to prepare for the transition to inexhaustible energy systems and to begin the building of a new energy-supply system within this relatively short time.

REFERENCES

1. Die Kohlevorräte der Welt [The coal supplies of the world], *Brennstoff-Wärme-Kraft* **33**(6):249–250 (1981).
2. Global 2000—Der Bericht an den Präsidenten [Report to the President], Verlag Zweitausendeins, Frankfurt (1981).
3. H. F. Wagner: *Brennstoff-Wärme-Kraft* **33**(1):7–12 (1981).
4. XI. Weltenergiekonferenz [World Energy Conference], Munich, 8–12 September 1980, Tagungsbericht [Report on the meeting] (1980).
5. W. Häfele *et al.*: *Energy in a Finite World—A Global Systems Analysis*, Ballinger, Cambridge, Massachusetts (1981).
6. R. Gerwin: *Die Welt-Energieperspektive [The World Energy Perspection]*, Deutsche Verlagsanstalt, Stuttgart (1980).
7. Primärenergieverbrauch in der Bundesrepublik Deutschland, Pressemitteilung der Arbeitsgemeinschaft Energiebilanzen [Primary Energy Use in West Germany: A Report of the Working Group on Energy Balances], Düsseldorf (1981).
8. W. Müller-Michaelis: *Brennstoff-Wärme-Kraft* **33**(4):128–130 (1981).
9. H. D. Schilling, U. Krauss: *Brennstoff-Wärme-Kraft* **33** (4):130–134 (1981).
10. Ein-und Ausfuhren der BR Deutschland an Kernbrennstoffen 1980 [Import and Export of Nuclear Fuels in West Germany 1980], *Atomwirschaft—Atomtechnik* **26**(11):610–611 (1981).

* Translator's note: The year given refers to the time at which the production rates of total fossil fuels—largely coal—are expected at this time (1986) to begin to decrease. However, this estimate neglects the recent prediction that the ratio (energy obtained from a fuel/energy needed to obtain this fuel) will decrease below unity in the next decade for all fossil fuels; it also neglects the increase in temperature that would take place if the use of fossil fuels were to be continued after they became energy-losing entities. More recently, it has been calculated that this increase in temperature will lead to the melting of the polar ice and thereby create a significant increase in sea level the world over. Although the precise extent of the final magnitude of this increase has not yet been exactly determined, it would eventually (~2100) be on the order of several meters and would, at the very least, make the world's ports unusable. This could be avoided only by reducing discharge of CO_2 into the atmosphere as soon as it is politically possible to do so.

11. Dritte Fortschreibung des Energieprogramms der Bundesregierung, [Third Energy Program for West Germany], *Brennstoff-Wärme-Kraft* **34**(1):1–4 (1982).
12. A. Grütz: *Elektrotechnische Zeitschrift* **103**(2):72–73 (1982).
13. W. Häfele: *Atomwirtschaft—Atomtechnik* **25**(8/9):416–421 (1980).
14. M. K. Hubbert: Energy sources, in: National Academy of Sciences–National Research Council, *Resources and Man,* pp. 157–242, W. A. Freeman, San Francisco (1969).
15. M. K. Hubbert: Energy resources for power production, in: International Atomic Energy Agency (IAEA), *Environmental Aspects of Nuclear Power Stations,* pp. 13–43, IAEA, Vienna (1971).
16. M. A. Elliott, N. C. Turner: Estimating the future rate of production of the world's fossil fuels, American Chemical Society Division of Fuel Chemistry Meeting, Symposium on Non-Fossil Fuels, Boston (April 13, 1972).
17. W. L. Borst, J. Fricke: *Physik in unserer Zeit* **10**(2):35–42 (1979).
18. Energievorräte in der Welt and ihre Reichweite [Energy Supplies of the World and Their Extent] (nach: Bundesanstalt für Geowissenschaften und Rohstoffe), *Kernenergie und Umwelt,* No. 3, p. 1 (1982).
19. E. Scholand: *Brennstoff-Wärme-Kraft* **33**(4):134–138 (1981).
20. Primärenergieverbrauch 1981 stark rückläufig [Primary energy use strongly decreased], *Brennstoff-Wärme-Kraft* **34**(2):50 (1982).
21. W. Sassin: *Sci. Am.* **243**(3):106–117 (1980).
22. W. Häfele, W. Sassin: *Science* **200**(4338):164–167 (1978).
23. J. D. Lawson: *Proc. Phys. Soc.* **70B**(1):6–10 (1957).
24. R. F. Post: *Physics Today* **26**(4):30–39 (1973).
25. B. Badger *et al.*: UWMAK III, Report UWFDM-150, University of Wisconsin, Madison (1975).
26. J. W. Davis, G. L. Kulcinski: *Nuclear Fusion* **16**(2):355–373 (1976).
27. R. Behrisch, B. B. Kadomtsev: *Plasma Phys. Contr. Nucl. Fus.* **2**:229 (1974).
28. R. W. Conn *et al.*: SOLASE, Report UWFDM-220, University of Wisconsin, Madison (1977); Proc. 7th Int. Conf. Plasma Phys. Contr. Nucl. Fusion, Innsbruck, International Atomic Energy Agency, Vienna (1978).
29. R. Bünde, W. Dänner, W. Hofer, *et al.*: *Brennstoff-Wärme-Kraft* **26**(11):467–472 (1974).
30. W. Häfele, J. P. Holdren, G. Kessler, G. L. Kulcinski: IIASA Report R-77-8, International Institute for Applied Systems Analysis, Laxenburg (1978).
31. W. Koelzer: *Lexikon zur Kernenergie* [*Lexicon of Nuclear Energy*], Kernforschungszentrum Karlsruhe GmbH, Karlsruhe (1981).
32. G. Heusener, Schnelle Brutreaktoren, in: P. Borsch, W. Freier, E. Münch: *Perspektiven der Kernenergie* [*Perspectives in Nuclear Energy*], pp. 19–28, Jül-Conf-32, Kernforschungsanlage Jülich GmbH, Jülich (1979).
33. Bundesministerium für Forschung and Technologie, Zur friedlichen Nutzung der Kernenergie [The Peaceful Use of Nuclear Energy], Bonn (1978).
34. Energiereserven ohne und mit Schnellem Brüter [Energy reserves, with or without fast breeders], *Kernenergie und Umwelt,* No. 2, p. 1 (1982).
35. S. M. Keeny *et al.* (eds.): Nuclear Power Issues and Choices, Report of the Nuclear Energy Policy Study Group Sponsored by the Ford Foundation, Ballinger, Cambridge, Massachusetts (1978).
36. H. Küsters: Reaktorphysik [Reactor physics], in: H. Gobrecht (ed.), *Bergmann-Schaefer—Lehrbuch der Experimentalphysik* [Bergmann-Schaefer—Textbook of Experimental Physics] Vol. IV, Part 2, pp. 1273–1370, Walter de Gruyter, Berlin (1975).
37. P. Brennecke, H. Ewe, E. Justi: In H. Selzer *et al.* (eds.): Tagungsbericht des 3. Internationalen Sonnenforums [Proceedings of the Third International Solar Energy Forum], Hamburg, 24–27 June 1980, pp. 297–306, DGS-Sonnenenergie Verlags GmbH, Munich (1980).
38. P. Brennecke, H. H. Ewe, E. W. Justi: Status and near-term development goals of a solar hydrogen economy, *Revue Internationale d'Héliotechnique,* 2e Semestre, pp. 52–57 (1981).
39. E. Wicke: *Chemie-Ingenieur-Technik* **54**(1):41–52 (1982).

40. E. Justi, P. Brennecke, J. Kleinwächter: *Abhandlungen der Braunschweigischen Wissenschaftlichen Gesellschaft* **32:**153–185 (1981).
41. W. Palz, A. Strub, J. Gretz, *et al.*: in: H. Selzer *et al.* (eds.): Tagungsbericht des 3. Internationalen Sonnenforums [Proceedings of the Third International Solar Energy Forum], Hamburg, 24–27 June 1980, pp. 3–20, DGS-Sonnenenergie Verlags GmbH, Munich (1980).
42. W. Grasse: *DFVLR-Nachrichten,* No. 34, pp. 6–10 (1981).
43. R. Dahlberg: *Elektronik* **29**(24)**:**49–56 (1980).
44. R. Dahlberg: *International Journal of Hydrogen Energy* **7**(2)**:**121–142 (1982).

CHAPTER 4

Direct Energy Conversion

4.1. CONVERSION INSTEAD OF PRODUCTION

According to the First Law of Thermodynamics—the energy principle—one can neither destroy nor produce energy in a closed system, but only convert it from one form to another. The First Law is also sometimes called the Law of the Impossibility of Perpetual Motion. The First Law is exactly valid down to particles of the smallest known dimensions, and so universal that patent authorities always reject inventions if they claim to concern the *production* (rather than the conversion) of energy. Radioactive energy is not a contradiction of the First Law, because when it seems that a certain amount of radioactive energy is produced, there is in fact a small amount of mass destroyed, and the two are related by the well-known Hasenöhrl–Einstein equation: $E = mc^2$.

4.2. DIRECT AND INDIRECT ENERGY CONVERSION—THE DEC MATRIX

In the case of home space heating, the desirable type of heat is of the low-temperature variety, and this can be collected almost directly with photothermal panels that react appropriately to the sun's warmth. This direct conversion from available energy to useful energy of the same kind is unusual, it being far more common to find that the type of energy that is needed is quite different from the form of energy that is available. For example, solar energy is available primarily as optical energy, but it is desired in the form of electrical energy. Again, this electrical energy may be needed in another form, e.g., in the form of hydrogen, whereby it may be stored.

In general, it is necessary to be able to convert a form of primary energy into a form of secondary energy. It must be understood, however, that as a principle, this conversion can never be achieved with 100% efficiency. Because each step loses energy, it is always necessary to attempt to make the conversion in a single step, so that the energy loss is diminished by only a one-step, instead of a multistep, conversion process. If there is only one pair in the conversion process (i.e., if a certain form of primary energy is to be converted into a certain form of secondary energy), then these processes can be portrayed according to Justi in a two-dimensional block diagram that represents direct energy conversion (DEC) processes (i.e., the DEC matrix).[1]

Listed vertically in the left-hand column in Figure 4.1 are the five most usual forms in which energy appears: mechanical, thermal, optical, electrical, and chemical.

The same forms of energy are listed horizontally in the top row of the matrix, where they represent the forms needed. In the intersecting boxes are listed the devices that utilize the physical conversion effects. For example, one finds the direct conversion of mechanical (1) into electrical (4) energy in the dynamo and the microphone (box 1.4). Less well known are the means for converting thermal energy (2) into electrical energy (4), which are in box 2.4;

	MECHANICAL ENERGY (1)	THERMAL ENERGY (2)	LIGHT ENERGY (3)	ELECTRICAL ENERGY (4)	CHEMICAL ENERGY (5)
MECHANICAL ENERGY (1)	Simple machines (1.1)	Frictional heat Heat pump Refrigerator (1.2)	Friction Luminescence (1.3)	Dynamo machine Microphone (1.4) MHD generator (1.2.4)	(1.5)
THERMAL ENERGY (2)	Heat-power machines (2.1)	Absorption cooling machine (2.2)	Incandescent lamp (2.3)	Seebeck effect Thermionic diode (2.4)	Endothermal chemical reactions (2.5)
LIGHT ENERGY (3)	Radiometer (3.1)	Light absorption (3.2)	Fluorescence (3.3)	Barrier layer Photocell Solar cells (3.4)	Photosynthesis Photodissociation (3.5)
ELECTRICAL ENERGY (4)	Electric motor Electrical osmosis MHD pump (4.1)	Peltier effect Thomson effect (4.2)	Spectral lamps Fluorescent tubes (4.3)	Storage in condenser or pump storage plant (4.4)	Electrolysis Electrodialysis (4.5)
CHEMICAL ENERGY (5)	Muscular osmosis (5.1)	Exothermal chemical reactions special combustion (5.2)	Chemiluminescence "Lighting bugs" (5.3)	Galvanic special fuel elements (5.4)	Reformer reaction in fuel cells (5.5)

FIGURE 4.1. Direct energy conversion matrix. From Justi.[1]

thus, there is the thermoelectric Seebeck effect. The mirror-image effect in box 4.2, then, is the inverse thermoelectric effect of Peltier cooling and heating, and in box 4.1, the electric motor, the electro-osmosis as well as the MHD pump.

The boxes along the diagonals are also quite meaningful. Thus, in box 1.1, mechanical energy is used, as would be the case, for example, in the operation of simple machines or windmills. Thermal energy remains unconverted in box 1.2 in devices that store heat, heat pumps, and machines that use cold. Box 4.4 concerns the storage of electrical energy in condensers, a method that has considerable importance because, in the ideal case, the method could yield as much energy as it consumes; i.e., it could have an efficiency of 100%.

In this respect, it seems reasonable to ask why an accumulator, the universally known starter battery, has an energy-storage efficiency of only 60%. Here, the charging current is used at the cathode (the negative plate) to reduce lead sulfate ($PbSO_4$) (i.e., the sulfated plate) to lead, with the help of protons, which migrate under a potential gradient between the electrodes. Conversely, at the positive electrode, the so-called anode, an oxidation takes place by means of sulfate ions to produce lead oxide. In this way, an electrical potential difference is set up between the two electrodes. This brief description serves to demonstrate the general principle that dissymmetry in the chemical sense is a prerequisite for potential difference.[2] [Then, in the succeeding discharge of the battery, the electrical energies which have been stored in the two different chemicals produced, are reobtained, though after losses due to the nonideal behavior of the electrochemical reactions of the interface.] Research has shown the net efficiency to be about 60%.[3]

In respect to much of the energy wastage that goes on around us at present, the origin is the irreversibility of many of the energy changes that the First Law would allow. One can let a weight of $M = 1$ kg fall from a height of $h = 1$ m on a nonelastic plate. In this process, the potential energy (Mgh) is converted into heat by means of the equation ($M \times C_p \Delta T$). The measured value of $\Delta T = 1/427$ shows that 1 kcal is equivalent to 427 m $\times$ kg. The First Law would allow the reverse process to take place, i.e., 1 kg weight spontaneously rising to 427 m, bringing about a cooling of the environment by 1 kcal.

Let us consider a further experiment. We introduce 10 kg of water into a thermos flask containing an immersion electric heater. We pass 1 kW through the heater for 1 hr (e.g., 220 V at 4.54 amps). We measure an amount of heat equal to $\Delta T = 86°C$, and from this we see that the electrical equivalent of heat is 1 kWh = 860 kcal. The First Law would be consistent with the idea that the same immersion electric heater, dipped into 10 kg hot water, would give off 1 kWh of electrical energy, with the water cooling by 86°C. From experience, we know that the result of an experiment to manifest this process would be disappointing. The conclusion is that one can convert mechanical, optical, electrical, and chemical energies among each other, and also convert them to heat; on the other hand, heat can be reconverted only partly to the other forms of energy.

This conclusion is, in reality, a form of the Second Law of Thermodynamics, which limits the First Law by giving rise to a distinction between high-value (i.e., freely convertible) energy and low-grade (i.e., only partly convertible) energy. The situation is quite similar to that known to travelers who have freely convertible and controlled convertible currencies. The equivalent of $1 from West Germany is easily convertible into $1 from East Germany, but when one tries to change East German currency back into West German, one loses two thirds to three fourths of the original value. Correspondingly, a steam engine can, according to the direct–indirect matrix of Figure 4.1, convert heat (2) into mechanical (1) and finally electrical energy (4). If we then allow a stream of heat of 860 kcal/hr to flow back over the heat engine, neglecting the limitations imposed by the Second Law, we ought to get the equivalent of 1 kWh back again into electrical energy, just as the immersion heater gave heat from the current put through it. However, reality again disappoints us, for only 29% (for a nuclear power plant) or 38% (for a coal-based power plant) is obtained, and 71 and 62%, respectively, of the unconverted heat is given off to the surrounding atmosphere.

We have gone into this in some detail, particularly for readers who may not be very familiar with these principles of physics. One of the major reasons for the present energy crisis lies in the irreversibility that we are discussing. Thus, one should not produce low-temperature heat such as warm water, space heating, or process heat from electrical energy (compared with its production by direct combustion) because if one uses the indirect method, i.e., electrical heating, 62–71% of the energy is going to be lost. When one realizes that about 40% of our low-temperature heat is being produced electrically, one can see the seriousness of the situation. Through the appropriate arrangement of the direct energy conversion matrix in Figure 4.1, one can note the limited convertibility of heat (2.2) into mechanical (2.1), optical (2.3), electrical (2.4), and chemical (2.5) energies, this convertibility being indicated by the line that divides row 1. The height of the box above this line, as a percentage of the height of the whole box, is equivalent to the thermal conversion efficiency η; below the line, it is $1-\eta$, the nonconvertible part of the energy. The decreasing height of the top part of the boxes brings to mind the fact that coal power plants in the United States convert about 41% of the energy they use to useful power, but in the Federal Republic of Germany, there is a lesser efficiency of conversion of heat to electrical energy (about 38%) because the Federal Republic of Germany has a much higher population density than the United States, and therefore much less cooling water available.

What happens if one tries to make the difference $T_1 - T_2$ in the working of a heat engine as large as possible, not by diminishing T_2, but by raising T_1? It must be remembered that the usual upper limit of temperature in heat engines is $T_1 = 545°C$ when the vapor pressure is 245 atm. This limit is effectively dictated by the heat conductivity of carbon steels, the price of which is still commercially acceptable. In this case, the ideal Carnot value for efficiency would be $\eta_c = (545 - 30)/(545 + 273) = 63\%$. This is about twice as great as the actual value of 38%. In fact, it is aircraft technology—with its higher level of financing—that has produced gas turbines utilizing austenitic stainless steels, which cost about 100 times more than the cheaper carbon steels. These steels allow a maximum temperature of about 900°C, and this would allow an ideal Carnot efficiency for the working of a heat engine of $\eta = 870/1173 = 74.2\%$. Apart from this, there is an increase in efficiency of heat exchange (see Figure 4.1, box 1.1), and the overall efficiency is about 65%.[4] In fact, in the Jülich research establishment, a prototype heat engine with a corrosion-free helium atmosphere has been operating for years at temperatures up to 1100°C.

The conversion of heat to electricity cannot be followed so easily and clearly in the direct energy matrix as in the bar graphs of Figure 4.2. The top row of this figure represents a coal power plant, the first vertical bar of which, labeled Ch, represents chemical energy, i.e., the heating value of the combustion of $C + O_2$ with ΔH. Coal has a standard heating value of about $\Delta H = 8200$ kcal/kg $= 9.50$ $kW_{th}h/kg$. Thus, 100 kW_{th} would be produced by the oxidation of 10.5 kg standard units of coal. The combustion apparatus is therefore optimized so that when it is at 75% of its full load, it reaches a value of $\eta = 92\%$. This is shown in the second bar, W, as having a value of 92 $kW_{th}h$. In the third block, M, showing the working of the machine, the efficiency value is only 40% (conversion of heat to mechanical work), so in respect to the power available, it is $92 \times 0.40 = 36.8$ kW. This, then, appears in the fourth block, E, with a 95% conversion into electrical energy (i.e., the mechanical-to-electrical step), so that the final value is 35 kW_{el}. It is clear that the principal loss occurs in the conversion of heat to mechanical work.

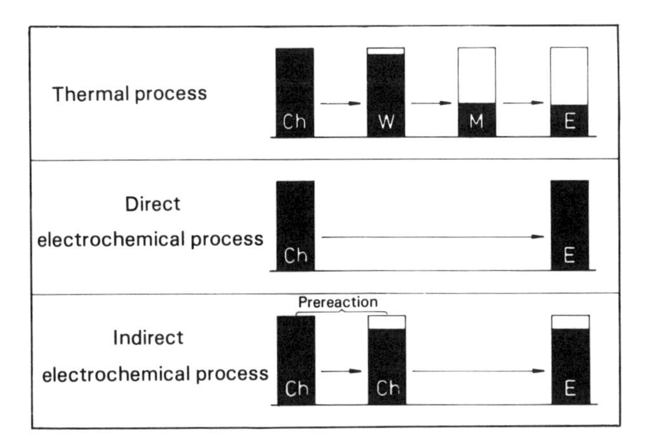

FIGURE 4.2. Bar graphs of direct and indirect energy conversion. Top: thermal power plant; middle: direct fuel cell; bottom: indirect fuel cell. From Justi.[1]

The middle graph shows the direct fuel cell according to the plans of William Ostwald[5] (the ninth winner of the Nobel Prize for chemistry). This represents the direct conversion of chemical energy (CH), according to box 5.4), into electrical energy in one step, avoiding the intermediate steps that are encountered in the thermal process. Thus, in this direct fuel cell method, a really high energy-conversion efficiency should be attainable. There must, of course, be some method to prevent the direct combination of C^{4+} and O^{2-}, for otherwise there will be considerable losses in "short-circuit heat" as with normal combustion of coal. Thus, the charge carriers have to be separated by certain geometric arrangements of the electrodes, keeping the positive and negative charges separate and conducting them off to the external circuits. The arrangement is similar to arrangements such as that in the well-known zinc–air cell introduced by Leclanché (Figure 4.3); in the arrangement that Ostwald suggested, however, it would have been necessary to heat the coal to 1000°C to ionize enough carbon per time unit, and this was the difficulty because of the considerable material problems and heat losses that were foreseen as arising in this arrangement.[6] For this reason, all effort in fuel-cell work goes into the indirect fuel cell, as shown in the bottom row of Figure 4.2. Here, there is a prereaction that is the main difference. Instead of coal, which is slow-reacting, one requires a fuel that has the property of reacting rapidly (and preferably in gaseous form). Thus, one can use the natural gas, CH_4, and re-form it outside the cell to hydrogen, which can in turn easily be converted to electricity. Figure 4.4. illustrates the demonstration model of a hydrogen–oxygen fuel cell that functions at room temperature at

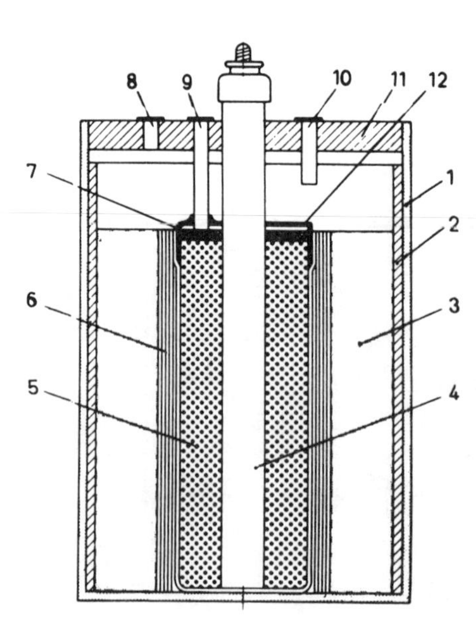

FIGURE 4.3. Cross section of an oxygen–zinc cell. (1) Insulated beaker; (2) zinc cylinder; (3) mixture of NH_4Cl and expansion medium; (4) carbon; (5) activated coal pump; (6) absorbent paper; (7) porous air distribution layer; (8) vent pipe; (9) air feed to positive electrode; (10) storage pipe; (11) top cover; (12) air chamber. From Euler.[8]

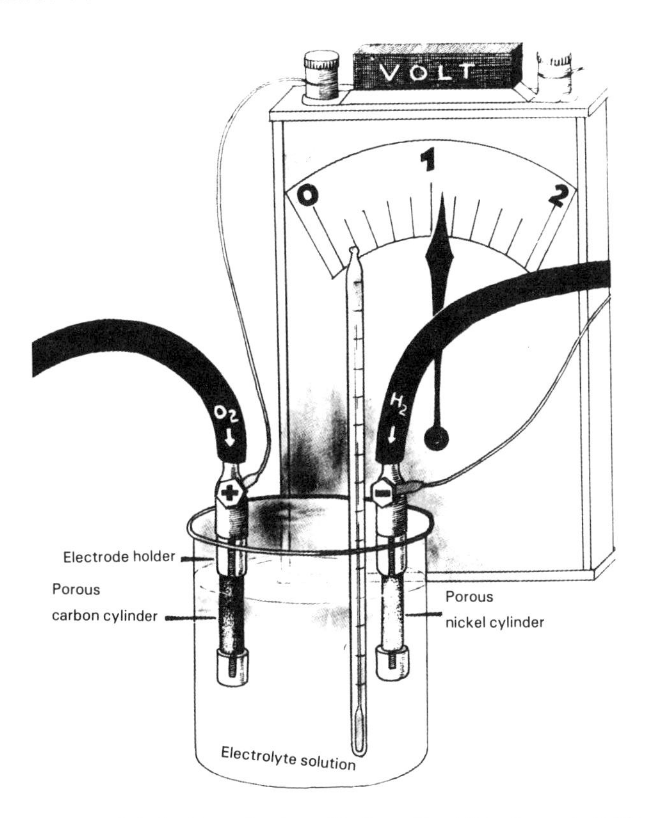

FIGURE 4.4. Diagram of the "cold combustion of hydrogen" publicly demonstrated in 1955. The voltmeter shows 1.0 V, with which 89% of the ideal original voltage of 1.23 V is achieved. This is equivalent to 91% efficiency. From Justi.[7]

atmospheric pressure and was first shown at a public meeting in 1955 by Justi.[7] If the free energy of the reaction of $H_2 + \frac{1}{2}O_2$ could be used reversibly, the potential associated with this cell would be 1.23 V, and the voltmeter is seen to be registering 1.1 V, or about 90%, for passage of the relatively small current values of 1 milliamp cm^{-2}. Thus, it is possible, even though for reduced charge densities—according to box 5.4—to convert chemical and electrical energy by means of fuel cells with a very high electrical conversion efficiency.

The direct energy matrix is not really scientifically complete, for we have not considered magnetic energy, or the conversion of mass into energy according to the Hasenöhrl–Einstein relation, $E = mc^2$. The matrix is instructive and useful because it enables us to establish whether a certain energy-conversion process can be carried out directly and gives a clear indication of the known physical effects that are associated with the applications that can be made.

4.3. SELECTED EXAMPLES OF DIRECT ENERGY CONVERSION EFFECTS

According to these fundamental discussions concerning energy conversion in general, and the direct–indirect effects in particular, let us select for more detailed discussion a few of the 25 physical effects in the direct energy conversion matrix of Figure 4.1 which might be applicable to solar energy uses in the future.

The following discussion is devoted to *physical* aspects. Only in cases in which the devices have actually been developed at a commercial level will there be discussions in later chapters giving a description of technical developments.

4.4. CONVERSION (AND PRODUCTION) OF WIND ENERGY

Wind energy is one of the oldest energy techniques used by man (windmills, sails). We shall discuss it here—though not elsewhere—as the first concrete example of energy conversion. It appears in the diagonals of the matrix as a direct-energy situation, the energy form not being changed throughout, thus justifying the term "ready-made" energy.

Wind energy has received renewed attention at several conferences in recent times. These are admirably documented in reference 9. The sudden increase in interest is all the more surprising because of the description given by H. Dörner[9] (p. 81) in his introduction:

> It is easy to see from the readily available literature that all the various ideas that are being usefully applied to wind energy today had already been suggested. There is among these systems nothing really new, and it seems unlikely that further new ideas will arise. Thus, vertical axis converters were being used in the Arab states (with bow-shaped sails) as early as 1271 B.C. At the turn of the century, all the various types of Dutch windmills were being used by the thousands. They were really efficient machines in operation at moderate rotation speeds. The euphoria that prevailed at the beginning of the Steel Age gave rise, remarkably enough, to a diminution in efficiency because of the use of the limited-size Rosetta type of windmill (Western type). These were multibladed machines that ran rather slowly and were effective only for direct mechanical drives (as for example in water pumps).

New arguments for a renaissance of wind energy as a supply for mankind are based on the following considerations:

1. In the last decade, it has become possible through refinement in calculation, collection of more data, and evaluations of meteorological observations, using computers, of winds at low and high altitudes (as well as the speed of moving weather fronts), to determine the mean content of kinetic energy in the atmosphere worldwide as $E_{th} = 2.2\text{–}3.7 \times 10^7$ TWh/a. This is equivalent to a continuous production of power equal to 3.5×10^{15} W. It is equivalent to a heat engine working on solar energy with a thermal efficiency of about 2%.

About 3% of this, namely, 6.9–11 × 10^5 TWh/a could be taken out of the atmosphere by wind turbines without causing any marked alteration in the winds, although there would be an increase of frictional interaction with the land of about 10%. This amount of energy is about 30 times greater than the present total energy needs of mankind.

2. Progress made by modern aerodynamics guarantees an improvement compared with the windmills of the Middle Ages, which were themselves very efficient machines. Nowadays, one could make wind-energy converters with as much optimization and skill in construction as are put into aircraft (Figure 4.5).

3. Correspondingly, the possibility of combining wind-generated electricity and a hydrogen economy is attractive on the one hand because the transport and storage of hydrogen in the pipe system solves the difficult storage problem which fluctuating wind energy presents[10]; on the other hand, in electrolysis, there is no Carnot factor for the conversion of heat to work (38% in the conventional plants and 29% in light-water reactor nuclear plants). Thus, electrolytic efficiencies such as one finds in the eloflux process (Chapter 8) amount to about 80% and avoid the debilitating efficiency with which heat energy can be converted to electricity, one of the greater difficulties in the use of hydrogen as a universal energy medium. Thus, wind and hydrogen make an ideal marriage in which each of the partners wins.

4. The former problem that all wind-energy converters ought to produce

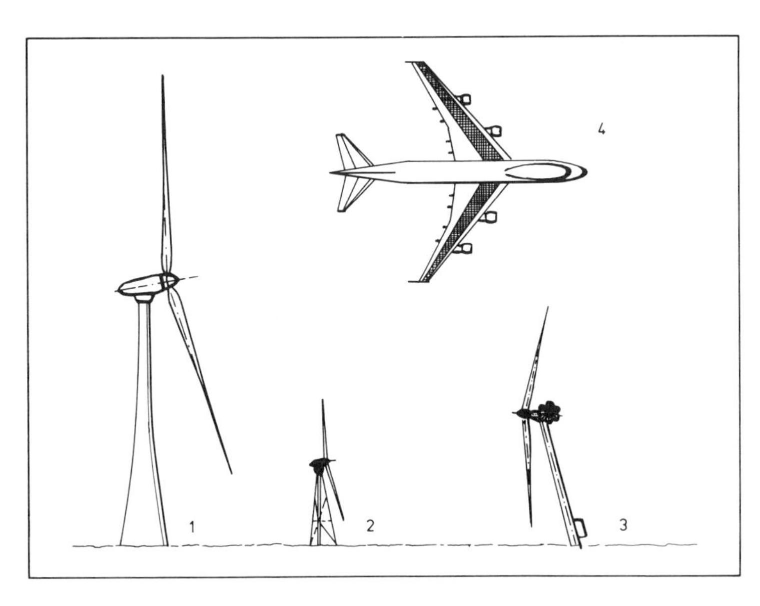

FIGURE 4.5. Size comparison between a modern Boeing 747 jumbo jet (4) and wind-energy converters of 0.6 (2), 1 (3), and 3 MW (1). From Hütter.[9]

alternating current of exactly 50 Hz to be acceptable for the grid has recently been solved by the use of thyristors, i.e., electronic converters with no moving parts.

From the standpoint of energy-conversion theory, it is interesting that despite the mechanical characteristics of wind energy, it can be only partly removed from the atmosphere. Betz was the first to show that the theoretical maximum efficiency for drag- and friction-less wind energy conversion is 59.26%. The losses that result in the "production" because of the finite number of plates, drag, and friction decrease this limiting value to 46–50%; if the windmill is run at a constant rate of rotation, the losses increase because of adjustment, e.g., as friction in the gears, in storage, and in the generator, so that the overall efficiency drops to 28–32%.

Due to the smaller frictional coefficients on the open sea ($c_{\text{friction}} < 0.005$), winds coming in from the sea reach high mean velocities and die away some 150–200 km inward from the coast to a constant velocity and are displaced to high altitudes. Because wind speeds of less than 4 m/sec are not economically useful, Germany, with its short stretches of coastline, has to use the mountains, in particular the Alps. Figure 4.6 is a map of Europe on which the places where the wind velocity reaches more than 4 m/sec are shaded. It is immediately apparent that England and Ireland together have the lion's share (38.6%) of the wind potential, which is 2400 TWh/a, and that the Federal Republic of Germany has only about 229 TWh/a (4.8%). Thus, England could—at least theoretically—supply all Europe with wind energy. Despite this, the Federal Republic of Germany could obtain about 70% of its energy need autonomously if the shaded parts of the country were covered with wind-energy converters in the megawatt range at intervals of 1 km, or about 30,000 in all. It may be thought that the "visual burden" on the environment of such windmill farms seems acceptable, as illustrated by Figure 4.7, in which there is an attempt to show the environmental aspect of a large concentration of windmills. It does not differ much from that of high-tension cables; additionally, it should be noted that both small and medium-size wind plants operate almost noiselessly.

The technical development of modern wind-energy converters is far advanced. The German Ministry of Research and Technology has had a 100-kW wind-energy converter erected and tested on the Alb Mountains in Schwabia, and a three-volume report[11] has appeared. There has been an international agreement that three wind-energy converters with capacities of 0.6, 1, and 3 megawatts (see Figure 4.5) should be developed.

In summary, one can say that the chances of a wind-energy renaissance are probably increasing. What wind-energy converters do is to take kinetic energy from the effects of the solar heating of the atmosphere. Functioning as a heat engine with about 2% thermal efficiency (see Table 5.1), the wind generator has the capability of supplying about 30 times the energy consumption of mankind. The handicap under which they have functioned hitherto is in respect to the

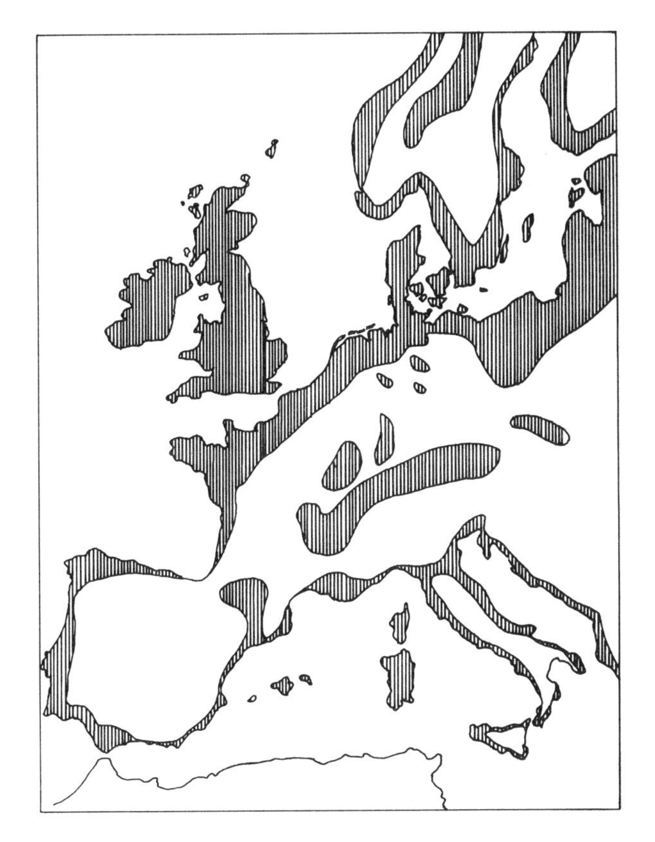

FIGURE 4.6. Map of Europe with areas shaded that have an average wind speed of more than 4 m/sec. From Hütter.[9]

pronounced fluctuations in amount and duration of output sporadic nature of wind, but hydrogen storage would deal with all that. The high efficiency of hydrogen electrolysis and the elimination of the usual Carnot cycle between the energy in such a source as oil and the final result in mechanical or electrical energy (which usually amounts to 29–38%) improve the situation. Alternatively, using modern thyristor technology, one could obtain phase-matching storage with a large number of wind-energy converters of constant grid frequency in an extensive distributing electricity-network, an arrangement that would have the appropriate storage characteristics. It must be understood, however, that it is economical to use wind energy only when the wind velocity is 4 m/sec or greater, as it is over islands (in particular England and Ireland), along coastlines (Belgium, Luxemburg, France, Spain, and Scandinavia), and on top of mountains (see Figure 4.6).

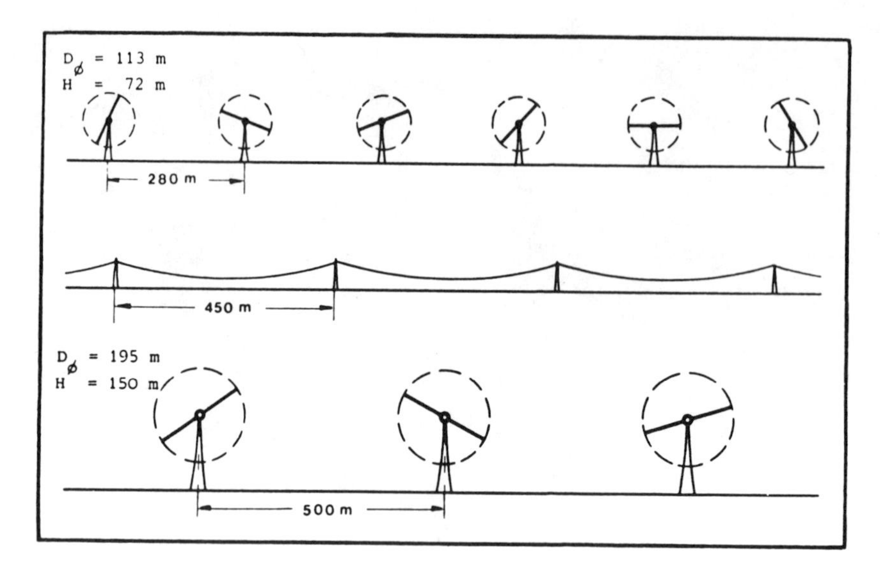

FIGURE 4.7. Comparison of the "visual environmental pollution" of connected wind-energy converters (top and bottom) with that of overhead power lines (middle). From Kramer.[10]

4.5. PHOTOVOLTAIC DIRECT ENERGY CONVERSION

One widely known device which directly converts light into electrical energy is the photocell in light meters. It was Adams and Day[12] who first constructed the selenium cell in 1877. Neither this nor the later copper oxide barrier layer cell invented by Schottky and Lange[13] could be used as a direct energy converter because the efficiency of conversion is less than 1%. The problem of turning these light–current devices into devices that provide acceptable efficiency of conversion, light weight, and reliability, and long life expectancy was first solved in 1953 by Chapin *et al.*[14] with the silicon barrier layer cell developed in the Bell Telephone Laboratories. The maximum 11% efficiency that was measured at that time was enough to raise expectations for a practical solar cell for solar energy. This device had the fortune to find a sponsor in NASA, and so far this agency has used this type of energy-conversion device in about 1000 space vehicles and makes about one million cells per year with a capacity of 50 kW. At present, there are three different kinds of cells (Figure 4.8):

1. Semiconductor homojunction cell of Fuller and co-workers, in which the barrier layer is between *n*- and *p*-doped layers of the same material, as exemplified by silicon.
2. Semiconductor heterojunction cell, in which the barrier layer is between two chemically different semiconductors, mostly $p\text{-}Cu_{2-x}S$ on $n\text{-}CdS$.
3. Schottky barrier, in which the barrier layer is between a thin and transparent semiconductor layer and a metal electrode. In the past, this was

p-$Cu_{2-x}O$–n-Cu, but more recently n-Si and p-Si[15] have been used, corresponding to the lower scheme in Figure 4.8.

Because of the requirements of space flight and the advantageous characteristics of silicon cells, these cells dominate the situation; for this reason, we shall continue to discuss the subject principally in terms of silicon.

Since this chapter is limited to an account of a few conceivable sources of readily available energy sources, descriptions of the actual physical arrangements of the best among them will be deferred until Chapter 5. We will therefore limit the discussion here purely to the scientific side. We shall see how—and why—the barrier layer solar cells are the most attractive and then, in Chapter 6, study their industrial development.

4.6. THERMOELECTRIC DIRECT ENERGY CONVERSION USING THE SEEBECK EFFECT

The spread of indirect current production in plants led to the view that direct conversion of heat to electricity was impractical, although box 2.4 of the DEC matrix shows the Seebeck effect discovered in 1822. However, the conversion

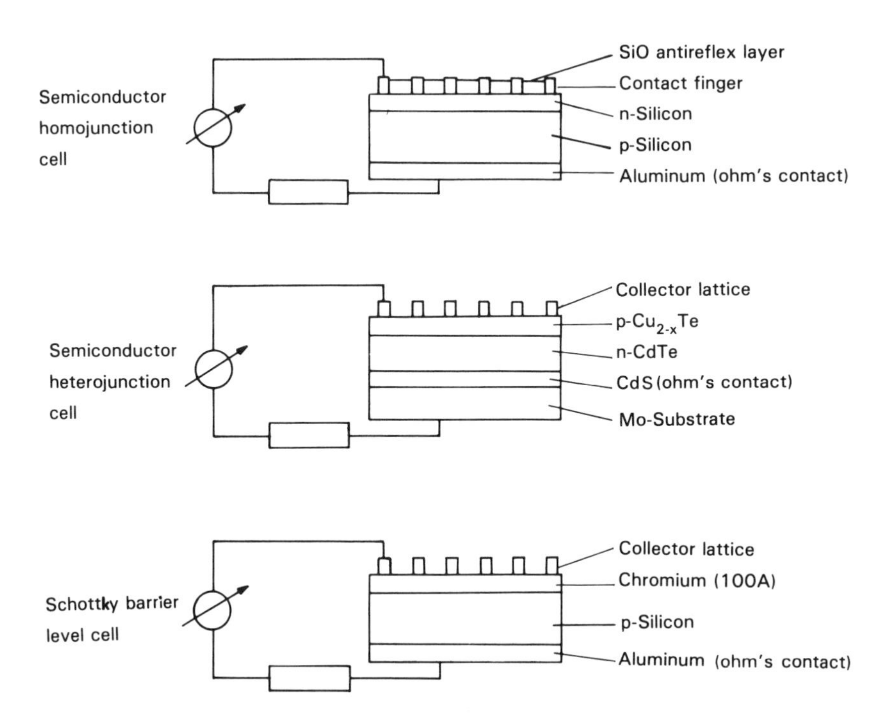

FIGURE 4.8. Schematic comparison of three types of solar cells. From Hecht *et al.*[15]

efficiency of this effect was regarded as so low that it could be used only for temperature measurements. The principle of thermoelectric energy conversion is shown schematically in Figure 4.9, which represents a circuit consisting of two different materials, conductors A and B, in contact, e.g., a copper wire and a constantan one. If both are soldered junctions at equal temperature, then the circuit to the right and to the left of the line *AB* is an example of mirror symmetry, and no potential difference can arise from it. If there were, for example, a potential on the right-hand contact, it would be compensated by one on the left, for otherwise a perpetual motion machine would be operative. However, if the soldered junctions are not at equal temperatures—e.g., if heat is added to the right-hand contact—then part of the heat flowing through each section of the system is converted into an electric current. Figure 4.10 illustrates a possible mechanism for the simplest case that one conductor (e.g., *A*, in this case copper) contains only negative- and the other [e.g., *B*,Cu–Ni alloy (constantan)] only positive-charge carriers, which again follows from symmetry considerations because otherwise no flow of current could occur. Figure 4.11 shows the well-known technique of thermoelectric temperature measurement in which one junction is kept at 0°C while the other is exposed to the temperature to be measured. The potential that now develops is calibrated, and therefore one can measure temperature from the calibration curve, knowing the potential difference developed. This principle, namely that a flow of heat can be turned into a flow of current without any moving parts or chemicals, is a remarkable phenomenon; that it could not in practice turn heat into electricity was attributable to the low efficiency. Thus, in the example just given, the thermoelectric copper–constantan couple gives rise to a typical potential difference of only 40 μV/°C, so that at a ΔT of 700°C, the energy-conversion efficiency (i.e., the electrical power divided by the heat flow) is only 0.2% instead of the Carnot efficiency factor of

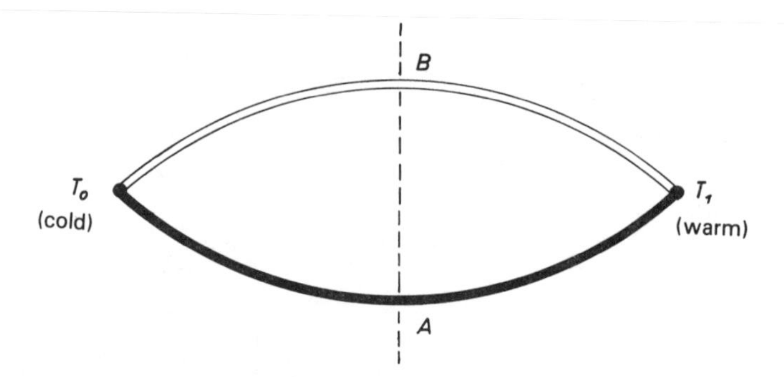

FIGURE 4.9. Diagram of an electrical circuit, consisting of two metals, a conductor *A* of *n*-type, e.g., Cu and conductor *B* of *p*-type [e.g., alloy (60 Cu/40 Ni)].

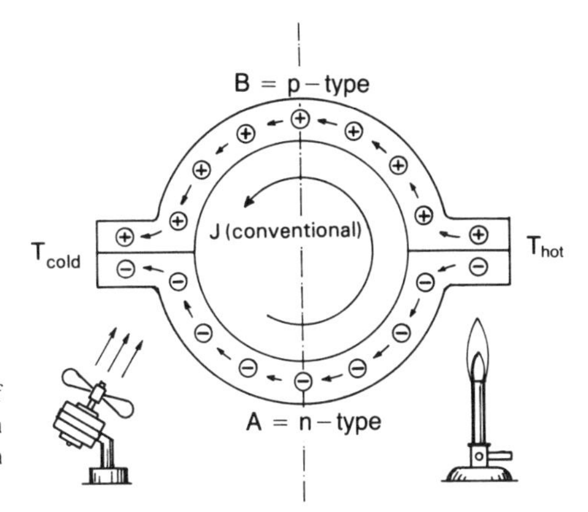

FIGURE 4.10. Graphic explanation of the occurrence of a flow of current in an circuit, consisting of two metals. From Justi.[1]

$\eta_c = 700/(700 + 273) = 72\%$. The thermoelectric performance is therefore $\varepsilon_{th} = 0.2\%/72\% = 0.28\%$.

In recent decades, due to the pioneering work of A. F. Joffe in Leningrad and of Justi and Lautz in Braunschweig, this loss factor has been reduced to about $\eta_{th} = 20\%$. It can be shown theoretically that one needs not only a maximal thermoelectric power (the Seebeck coefficient) α (μV/°C), but also a maximum electric conductivity, σ, so that the current produced will not dissipate itself in joule heat. Further, one needs a minimal thermal conductivity, κ, so that the ΔT between the two junctions will not be dissipated. These requirements are internally contradictory because according to the Wiedemann–Franz law, σ is

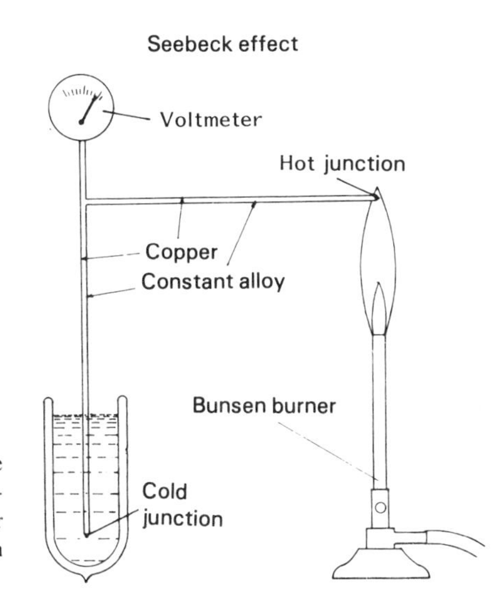

FIGURE 4.11. Principle of measuring the temperature of a flame relative to a fixed temperature, e.g., melting ice. The millivoltmeter is made of Cu; the effect arises in the Cu branch of the thermoelecric circuit.

proportional to κ and inversely proportional to an exponential function of α. Thus, one can see the cause of the low efficiency of most thermoelectric energy-conversion devices. Of course, this also gives rise to ideas for optimization. This theory shows that the three aforementioned parameters enter into the energy balance in the form of a product for the "effectivity" of a junction: $z = \alpha^2 \sigma/\kappa$. In fact, the authors cited above have shown in both theoretical and experimental investigations that the highest z values can be attained with substances having electron concentrations of about $10^{18}/cm^3$, in the transition region between metals and semiconductors. In particular, for medium-high temperatures—those produced by radiation with sun-light—thermoelectric materials, such as bismuth telluride (Bi_2Te_3), that can be optimized to give rise to z values of 1/300. Materials of this type owe their small κ values to a small symmetrical crystal lattice with open structure, so that they are fragile and can be neither rolled nor pulled, but must be used in the form of cylinders, cast from material in the molten state. Figure 4.12 shows the maximum efficiency for a short circuit $R_e = 0$ if the cold-soldered junction is at $T_c = 300$K and the warm soldered junction is kept at T_w, as a function of the so-called goodness factor. For the best value $z = 0.0033$, one can attain $\eta = 5.3\%$, or $\varepsilon_{th} = 21\%$ of the Carnot factor, $\eta_z = 100/400 =$

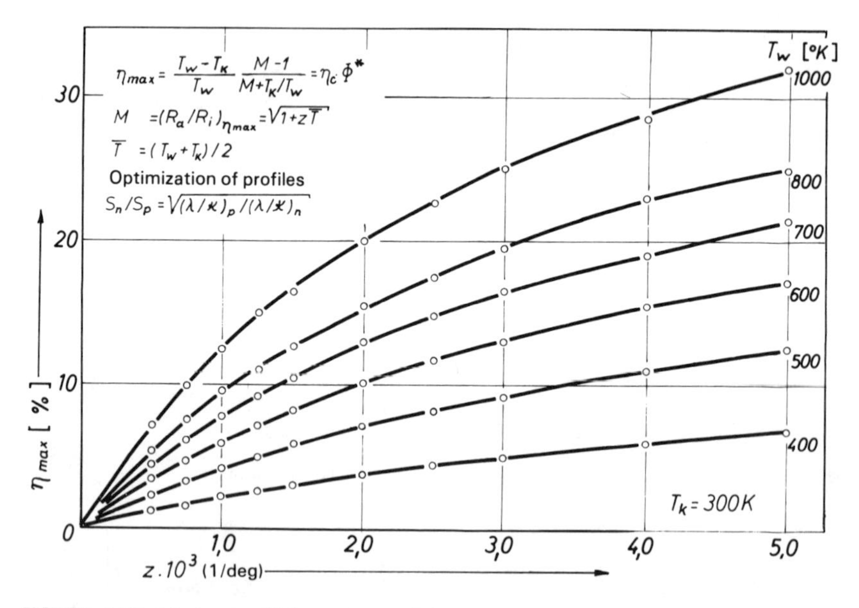

FIGURE 4.12. Maximum efficiency (η_{max}) of thermoelectric electricity production, using a one-unit Seebeck element, as a function of the effectivity $z = \alpha^2\kappa/\lambda(\text{deg}^{-1})$. Parameters: temperature T_w (K) of warm junction, cold junction is always at 300K external circuit load R_e not optimally adjusted. With $T = (T_w + T_c)/2$, $z \times T \simeq 1$, and $M = \sqrt{1} + T \simeq \sqrt{2}$, the optimum output $Q_{opt} = e_{AB}^2 \times \Delta T^2 M\ (R_e(M + 1)^2)$; with the given approximations $Q_{opt} \simeq 0.24 \times e_{AB}^2 \times \Delta T^2/R_e$. The maximum output is $Q_{max} = e_{AB}^2 \times \Delta T^2/(4 \times R_e)$ as a result of $R_i = R_e$. From Justi.[1].

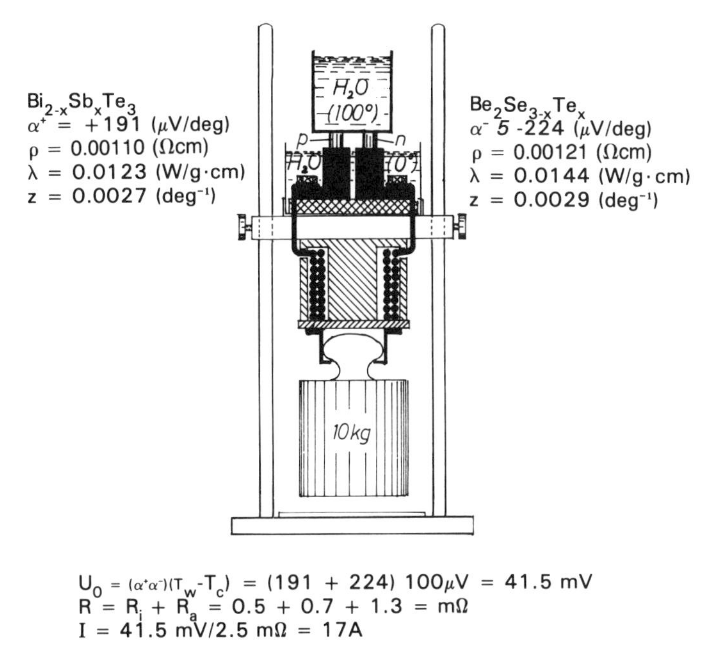

FIGURE 4.13. Semiconductor thermomagnet for demonstrating the great usefulness of a semiconducting thermoelectric couple for thermal electricity conversion, without moving parts, in the range of moderately high temperatures (approximately 40°–110°C). The conversion efficiency is about 5% (compare Figure 4.12). The thermoelectric current that excites the magnet is 14 amps. From Justi.[1]

0.25. This value drops further, as it does with all types of electrical generators, if one puts a load in the external circuit, so that the current must flow through the internal resistance R_i and the external resistance R_e. It can be shown that the corresponding decrease of the efficiency is by the factor $R_e/(R_i + R_e)$, and this is reduced to 0.5 when $R_e = R_i$. The current warms the hot junction and thereby recoups some of this heat as electricity, the rest being disposed of as in, for example, a dynamo. The Seebeck generator is at an inherent disadvantage compared with a heat engine because the metallic conductivity and the joule heating effect cannot be changed independently, as can the corresponding irreversible factor, conductivity and friction, in the mechanical analogue.

Although research in the field of thermoelectric generators is more or less stationary at present because the theoretical goals have been reached, there is hope that thermoelectric devices might be usable for photothermal solar collectors and will be competitive in this respect with photovoltaic solar cells. These reach efficiencies of around 14% as a practical maximum ($R_e = 0$!) at room temperatures, but at 60°C operating temperature in a strong sun, they are only double the value of the Seebeck cell. On the other hand, they are—at $250,000/kW—

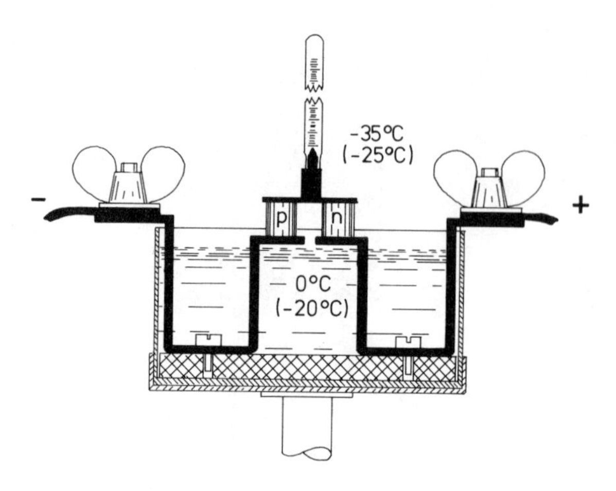

FIGURE 4.14. Demonstration experiment for the reversal of thermoelectric electricity production (Seebeck effect) to thermoelectric cooling or heating (Peltier effect). The same thermoelement is used as in the thermomagnet of Figure 4.13. It is soldered to two copper bands, which are brought to a temperature of 20°C in a water bath. The metal hot-water container in Figure 4.13 is replaced by a small Cu bridge with a soldered-on Cu pot, in which a thermostat is immersed and reads 20°C. If one passes through the thermoelement a D.C. current of appropriate strength using the polarity indicated at the wing nuts, the temperature gauge reaches the freezing point in 1 min and, if sufficiently insulated, decreases to approximately −28°C. If the polarity is reversed, the thermoelement draws heat from the water and passes it to the upper Cu bridge, so that the thermostat climbs quickly to 100°C. The first case is the electrical version of a refrigerator. The second case is analogous to a heat pump, and delivers 50% more heat than an immersion heater. This thermo-electric heating conserves energy and is noiseless and pollution-free, is desirable for solar-produced energy. From Justi.[1]

at least 100 times more expensive than the thermoelectric device. Thus, it is difficult to understand why more work is not being done on the development of thermoelectric converters. It would be easy to reverse the effect into a Peltier refrigerator or, alternatively, into a Peltier heat pump, and these devices yield per $kW_{el}h$, instead of the ideal equivalent of 860 kcal, 50% more, with nothing further needed.

Figure 4.13 presents some evidence advanced by Justi for the practicality of thermoelectric electricity production. Thus, one sees on one hand a warm-water reservoir at 100°C and a cold-water reservoir at 0–10°C. Between these there is a p and n thermoelectric element in contact with external and internal resistances R_e and R_i. This setup is appropriately joined by an electromagnet. The electromagnet, i.e., the coil of copper wire, is to be seen between the thermoelectric element and the 10-kg iron weight. In the magnet, there arises a current of—by the calculation shown in the figure—35 mV/2.5 mΩ = 14 amps, and this allows one to hold up an iron weight of 10 kg. The demonstration becomes quite convincing when the weight drops as a result of interrupting the current. A similar experiment is shown in Figure 4.14.

REFERENCES

1. E. Justi: *Leitungsmechanismus und Energieumwandlung in Festkörpern* [*Conduction Mechanisms and Energy Transformation in Solids*], Verlag Vandenhoeck and Ruprecht, Göttingen (1965).
2. F. v. Sturm: *Elektrochemische Stromerzeugung* [*Electrochemical Current Production*], Verlag Chemie GmbH, Weinheim (1969).
3. H. Bode: *Lead Acid Batteries*, Electrochemical Society Series, New York (1977).
4. K. Bammert: Sonnenenergie-Kraftwerke mit Gasturbinen [Solar-energy generating stations with gas-turbines], *Abh. Braunschweig. Wiss. Ges.* **28:**12 (1977).
5. W. Ostwald: *Z. Elektrochem* **1:**122 (1894).
6. E. Justi, A. Winsel: *Kalte Verbrennung-Fuel Cells[Cold Combustion—Fuel Cells*], Franz Steiner Verlag, Wiesbaden (1962).
7. E. Justi: *Jahrb. Akad. Wiss. Lit. (Mainz),* p. 200 Franz Steiner Verlag, Wiesbaden (1955).
8. K.-J. Euler: *Entwicklung der elektrochemischen Brennstoffzellen* [*The Development of Electrochemical Fuel Cells*], Munich (1974).
9. *Energie vom Wind* [*Energy From the Wind*], Proc. der DGS-Tagung in Bremen, DGS-Verlag, Munich, especially U. Hüter, pp. 1–16; M. Meliss, pp. 47–70; H. Dörner, pp. 81–91; H. Selzer, C. Cohrt, pp. 111–124; U. Bossel, pp. 275–304 (1977).
10. P. Kramer: *Der Wind als mögliche Energiequelle in der zukünftigen Wasserstofftechnologie* [*The Wind as a Possible Energy Source in a Future Hydrogen Technology*], Proc. 1. Deutsch. Sonnenforum Hamburg, Vol. 2, p. 581–590, Deutsche Gesellschaft für Sonnenenergie, Munich (1977); 178 (1978).
11. BMFT/AGF/ASA: *Energiequellen für morgen?* [*Energy Sources for the Future?*] Program studie des BMFT, Part III, *Windnutzung*, Umschau-Verlag, Frankfurt (1976).
12. W. G. Adams, R. E. Day: *Proc. R. Soc.* **A25:**113 (1877).
13. W. Schottky: *Z. Phys.* **31:**913 (1930); B. Lange: *Ibid.*, p. 139.
14. D. M. Chapin, C. S. Fuller, G. L. Pearson: *J. Appl. Phys.* **25:**676 (1954).
15. H. D. Hecht, E. Justi, G. Schneider: Schottky Type Si-Cr-Solar Cells, *Rev. Int. Héliotechnique*, 2nd semester, pp. 38–41 (1974).

CHAPTER 5

The Basis for the Use of Solar Energy

5.1. CHARACTERISTICS OF SOLAR RADIATION

Concern about the future of the world's energy supply has sharpened awareness of the difference between nonrenewable and renewable sources of energy. Sources of water power are inexhaustible and not harmful to the environment. They are renewed by the cycle of evaporation and rain, are driven by the sun, and will soon be in use on all continents. Wind energy also originates in solar energy and is therefore inexhaustible and—if it could be efficiently utilized—sufficient to meet the needs of all mankind some 30 times over. It is all the more unfortunate, therefore, that the geographic distribution and sporadic nature of wind energy make it a difficult source to use economically. Tides and geothermal heat, even though they do not arise from solar energy, are virtually inexhaustible, but are of restricted use because of their limited distribution.

Thus, there remain available for mankind, as inexhaustible energy sources, nuclear fission in breeder reactors, the fusion reactor, and solar energy. It was Justi[1] who published a quantitative analysis about 25 years ago in which he pointed out that if the net amount of solar energy reaching the earth could all be collected, it would meet the needs of mankind some 10^4 times over. The major point here is, of course, that our needs would be supplied without the pollution to which the present energy sources subject us. One can follow the calculation by which this result is reached in Table 5.1, which considers the balance between energy needs and solar input. Thus, the average German citizen needs about 4418 $kW_{el}h$ of electricity per year, and if one assumes that each of the four billion people in the world needs the same amount, the total world need for electricity amounts to 1.77×10^7 $GW_{el}h/yr$. To determine the total solar

TABLE 5.1
Yearly Maximum Electricity Consumption of Mankind and the Solar Energy Supply[a]

Consumption of electricity per capita and year in the Federal Republic of Germany 1974	4418 kWh/a = 4.418×10^3 kWh/a
World population 1974	4×10^9
Maximum electricity consumption	$(4.418 \times 10^3) \times (4 \times 10^9) = 1.77 \times 10^{13}$ kWh/a
Solar constant S	1.34 kWh/h $\times$ m^2 = 1.34 kW/m^2 = 1150 kcal/h $\times$ m^2
Earth circumference (by definition)	$U = 2\pi r$ = 40,000 km = 40×10^6 m
Earth radius	$r = (40 \times 10^6 \text{ m})/2\pi = 6.36 \times 10^6$ m
Apparent earth surface	$F = \pi r^2 = (6.36 \times 10)^{2\pi}$ m^2 = 1.27×10^4 m^2
Yearly solar energy supply	$S \times F = 1.34 \times 1.27 \times 10^{14} \times 2.4 \times 3.65 \times 10^3$ $= 1.49 \times 10^{18}$ kW$_{th}$h/a
Solar energy supply/maximum electricity consumption	$1.49 \times 10^{18}/1.77 \times 10^{13} = 0.842 \times 10^5 = 84400$
Conversion efficiency of solar heat to electricity	T_h = 160°C = 20%
Maximum electricity consumption/conversion efficiency	$1.77 \times 10^{13}/0.20 = 8.85 \times 10^{13}$ kWh/a
Low temperature heat and transportation	$2 \times 1.77 \times 10^{13} = 3.55 \times 10^{13}$ kWh/a
Yearly total consumption of solar energy by world population	16×10^{13} kW$_{th}$/a
Total solar energy supply : total energy consumption	$1.49 \times 10^{18} : 16 \times 10^{13} \simeq 10000$

[a] From Justi.[2]

input, it is necessary to state a value for the so-called solar constant, S, i.e., the radiation intensity per second received by one square meter of the upper atmosphere where there are no losses by clouds, dust, or mist. This numerical value is given by S = 1.34 kW/m^2 or, in heat units, 1150 kcal/hr $\times$ m^2. If one then multiplies by the circumference of the earth, 4×10^6 m^2 (corresponding to 1.27×10^{14} m^2 area), then by 24 hr $\times$ 365 days = 8760 hr/yr, one gets the yearly solar input of 1.49×10^{18} kWh/yr, i.e., 84,000 times more than the yearly electricity need of the present world population. Of course, it must be remembered that the efficiency of collection of solar energy by means of steam-driven turbogenerators is only 20%. Thus, it is necessary to add 400% to the input side and another 200% to the need side for heating and transportation, which leads to the conclusion that the actual need is approximately 9 times the calculated need for electricity. If we look at the entire energy need of $(1.77 \times 9 =)$ 16×10^{13} $kW_{th}h/yr$, this is 1/10,000th of the solar energy input. This amount still only assumes that the total amount of solar energy reaching the earth (apart from that lost in conversion to electricity) is available, and it should be made clear that only a very small fraction of it could ever be collected. However, there should still be a surplus. Indeed, a generous surplus is important to avoid changes in the environment through the extraction of too high a fraction of the total radiation or wind energy.

Since these calculations were first made (at which time they were not well accepted), solar energy as a sufficient supply for human needs has become a more popular concept. In fact, were only about 1% of the otherwise unused arid zones of our planet to be covered with converters for solar energy with a 10% efficiency of conversion into electric current, the entire energy needs of the present world population could be met, assuming a relatively high standard of living. Another striking formulation states that the earth receives as much energy from the sun in three days as would be obtained by combustion of the entire stock of stored fossil fuels such as coal, natural gas, and peat and of all the forests in existence.

To arrive at something more than a simple global balance of energy, one needs detailed information about the character of solar radiation, e.g., its spectral distribution, as given in Figure 5.1. In addition, one must consider what losses are sustained through the absorption of solar energy by components of the earth's atmosphere (if absorption by water alone is taken into account, it proves to be some 25%). Figure 5.1 shows the measured distribution of solar radiation outside the atmosphere, the ideal curve computed from Planck's radiation formula, and the amount of radiation absorbed by H_2O, CO_2, and O_3. One can clearly identify the "window" for visible sunlight that extends from the violet end at 0.4 μm to the red end at 0.8 μm. One can also see the strong absorption by the H_2O and CO_2 bands in the near infrared—the use of which is one of the objectives of solar heat collectors—and then, at the extreme left, the UV-absorption in the stratosphere by ozone (O_3), which saves living organisms on the earth from destruction by photochemically active UV light.

Thus, on a cloudless day, with the sun directly overhead and shining on a completely horizontal area of the earth's surface, about 1 kW_{th}/m^2 would be incident on the plate of a collector, and at 10% conversion efficiency, approximately 1 $kW_{el}/10\ m^2$ of collector surface would be available. Various effects reduce this availability because most of the real conditions deviate from the standard conditions we have chosen.

1. Due to the rotation of the earth, the position of the sun relative to the flat surface changes, but this influence can be minimized by continously reo-

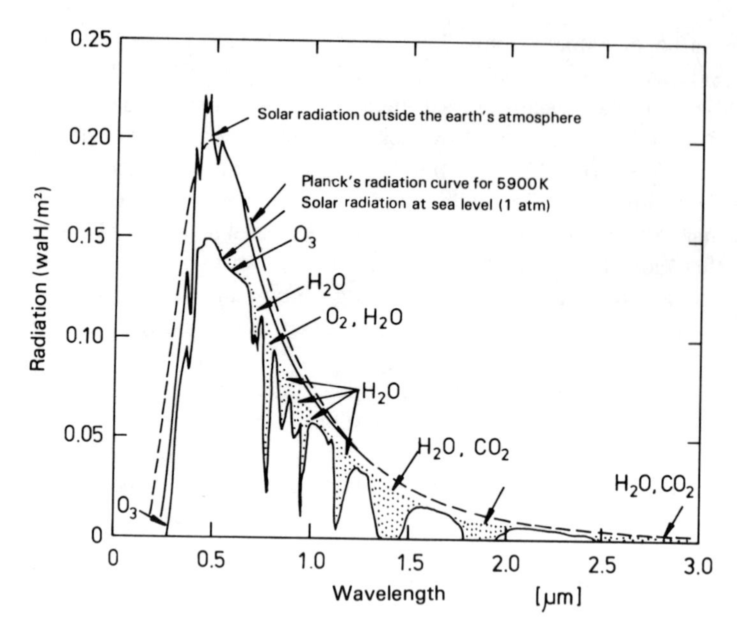

FIGURE 5.1. Spectral distribution curve of solar radiation. The stipled areas represent absorption by the indicated components of the atmosphere.

rienting the collector so that it is effectively perpendicular to the incoming radiation. It is, however, usually better—i.e., more economical—to replace this type of device with one that has constant orientation (flat-plate collector) such that a high percentage of available input is collected. This problem can be solved relatively easily in a general way by the use of elementary trigonometry. Thus, in respect to the incoming radiation, the situation acts as though only the components perpendicular to the flat surface are absorbed, so that for a given direction α, only the fraction $\cos \alpha$ of the radiation should be included. Thus, if one arranges all collectors so that the axes perpendicular to their flat surfaces point south at the position of the radiation maximum, the geometric efficiency factor is $2/\pi$, or 64%, as illustrated in Figure 5.2.

2. Correspondingly, one has to take into account the variation in the angle of the midday position of the sun with the changing of the seasons. One finds an optimized mean value for the slope of the collector to the horizontal, and the angle is between 10 and 15°, as shown in Figure 5.3, a figure verified by measurements. If one adjusts the collector twice a year, to the flatter position in summer and the higher position in winter, the collection will be 10–20% greater than if this adjustment is not made.

3. The reduction of solar radiation by clouds, dust, and water vapor plays an important role and can give rise to great differences in efficiency of collection even when the sites used are only a few kilometers apart. If one is going to set

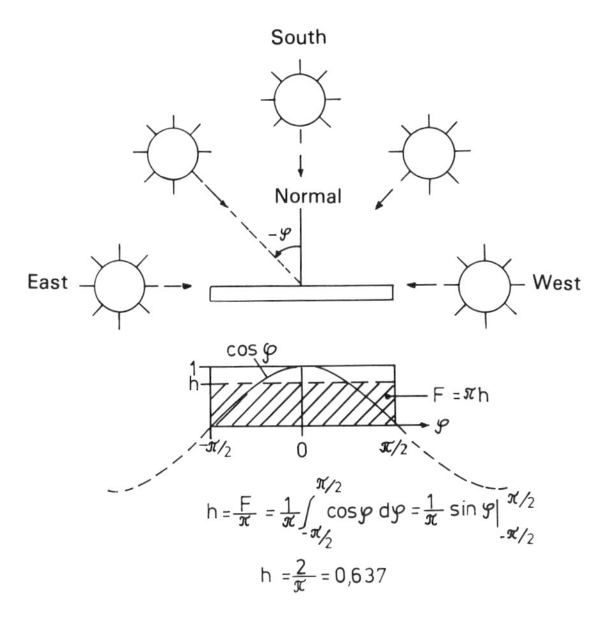

FIGURE 5.2. Trigonometric computation of the mean value h of the daily heat absorption of permanently fixed flat plate collectors, which are oriented toward the south.

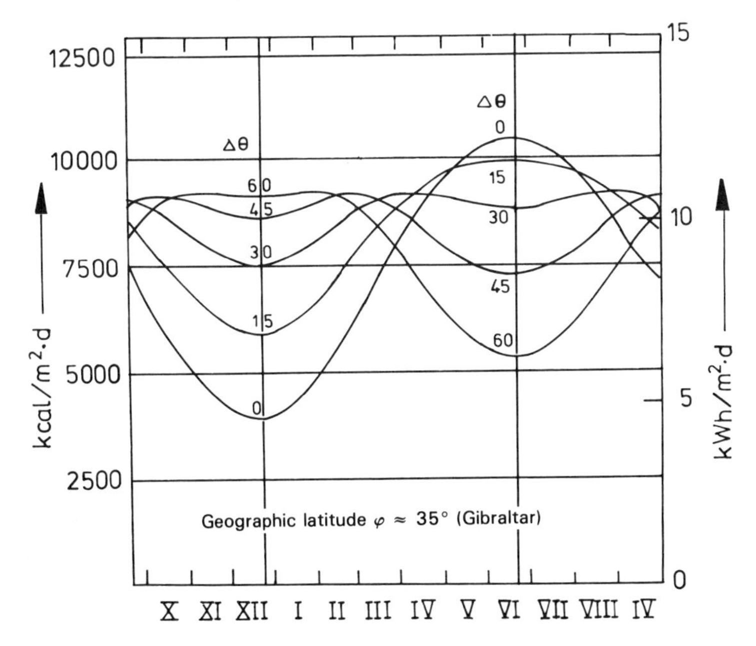

FIGURE 5.3. Daily solar heat collection Q of flat plate collectors oriented toward the south as a function of the season, geographic latitude θ, and degree of tilting of the collector toward the horizontal. Parameter: geometric inclination of the flat surface collector $\Delta\theta$ toward the horizontal direction according to measurements of NASA in Huntsville at 35°, the position of Gibraltar.[5]

up a practical plant for solar collection, one therefore needs radiation data over many years, and not simply the total incoming light; the data must be divided into diffuse and direct radiation, because it is only the latter that can be concentrated optically with mirrors. In the Federal Republic of Germany[6] and in Switzerland,[7] there are plenty of data, taken over many years, and the mathematical model of Figure 5.4[8] is an instructive map for Europe and Africa, giving the hours of daily sunshine. These data need extension, but they are often in themselves quite useful. Thus, it can be seen that Spain has about the same total

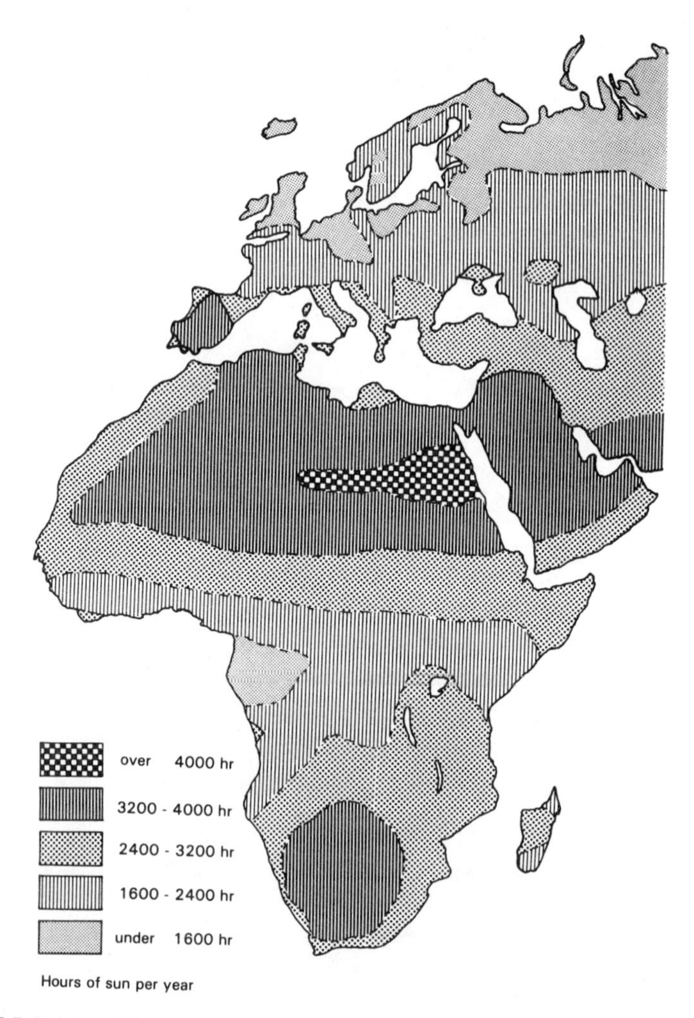

FIGURE 5.4. Map of Europe and Africa with different shading of areas showing varied hours of sunshine per year. From Justi.[8]

annual hours of sunshine (3200–4000 hr) as the southern part of North Africa and Arabia, while Portugal and the Atlantic Coast of Morocco have only 2400–3200 hr. It is easy for someone who flies frequently to understand the reason for this difference. The coasts in the latter locales are washed by cold sea currents coming from the south, and the resultant cooling of the air produces a very light mist that cuts down the incoming solar light. For similar reasons, equatorial latitudes are not very suitable for solar-energy-collecting areas because they are often relatively misty as a result of tropical rain forests.

There are tables available[9] which show the areas of highest solar radiation and the amounts of energy that could be expected, per year, from solar collecting stations set up in these areas. Of course, in the conversion of heat to ordered energy (mechanical or electrical), one must take into account the Carnot factor, $\eta_c = (T_1 - T_2)/T_1$. This means not only that a high condensation temperature T_1 in desert regions is important, but also that cold cooling water must be available (T_2). Such a heat sink is not easily attainable in these areas.

Photovoltaic cells are hardly suitable for use at high-percentage conversion of sunlight to electricity in desert regions,* although these direct conversion devices are not Carnot-limited. Nevertheless, the efficiency of the conversion of light to electricity in photovoltaic cells usually falls with increasing temperature because after thermal activation, the generation of photons is no longer so possible ("quasi-Carnot" processes). Hence it is highly desirable to have cold water or air available for keeping the photovoltaic cells at a reasonably low temperature. Thus, a reevaluation of the suitability of various areas for solar energy is necessary.

5.2. POPULATION AND LIVING STANDARDS

As has been shown above, the covering of about 10% of the earth's surface with solar energy generators of 1% efficiency can give us sufficient energy for our needs; in this way, it would be possible to support a world population of 10 billion people at 10 kW energy need per person. If we were to try to satisfy this need by some other means, e.g., fusion, the temperature of the earth's atmosphere would rise and the consequences would be similar to those of the greenhouse effect (see Chapter 1). Thus, production of more energy than can be obtained from solar energy collection must be avoided. This is a simple but fundamental recognition.

* On the other hand, Bockris was the first (in an internal report to the Westinghouse Corporation) to suggest setting up photogalvanic cells on floats in an ocean environment, with subsequent conversion to hydrogen, and sending this hydrogen back to cities (1962).

5.3. USE OF SOLAR ENERGY ON A SMALL SCALE

If solar collectors were put on the roofs of houses, they could make substantial contribution to the energy economy of the household, and the amount of energy that has to be transferred to houses over long distances from central plants would be reduced. In this way, the disruptions to which the energy supply of a country might be subjected would be reduced. It is possible that some of the excess energy from roof collectors could be used for other local purposes. The principal uses of small-scale solar collection are discussed below.

5.3.1. Solar Cookers

At the suggestion of Unesco, solar cookers were produced for developing countries such as India. The objective was to preserve valuable dung, which had continued to be used as a cooking fuel, for use as fertilizer in agriculture. However, it turned out that the cost of even the very primitive equipment needed (a few tens of dollars, Figures 5.5. and 5.6)[10] was too high for introduction into the poor sections of Indian society. The same fate befell later and cheaper Israeli-made equipment. At present, a more recent and more useful German device[11] that contains an inflatable parabolic mirror with a diameter of 1 m, made from metal-coated plastic, and useful for outdoor cooking, seems to have better prospects.

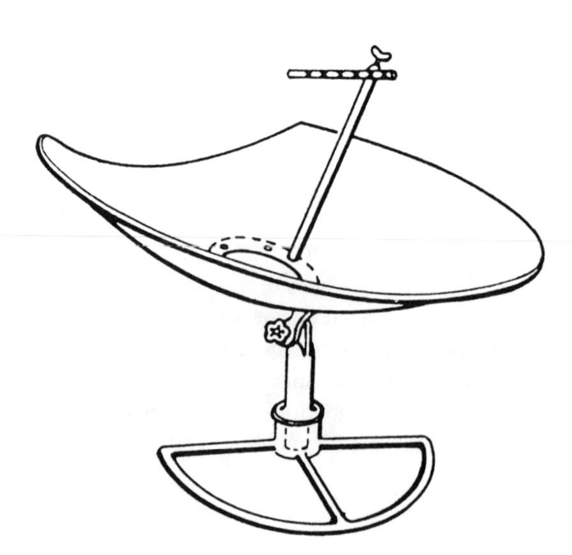

FIGURE 5.5. Schematic diagram of the construction of a cheap solar cooker of the National Physical Laboratory of India. From Robinson.[10]

FIGURE 5.6. Photograph of a solar collector used for cooking in Africa.

5.3.2. Water and Space Heating in the Household

Blackened sheets incorporating on their noninsulated sides welded copper tubes through which cold water is piped have been made by hand for many years. Such devices go under the name "Florida type" (Figure 5.7)[12]. They are expensive and not efficient enough.[13] Solar warm-air heating offers some advantages. The air that is to be warmed is conducted over the back of the rooftop collectors, and these can in addition be covered with solar cells which produce electrical power. Appropriate ducts take the warm air to be stored for use during sunless periods, in reservoirs such as heat-insulated water tanks, or in latent heat storers made from suitable salts, or in heaps of stones, which reheat the circulating cold air.[14]

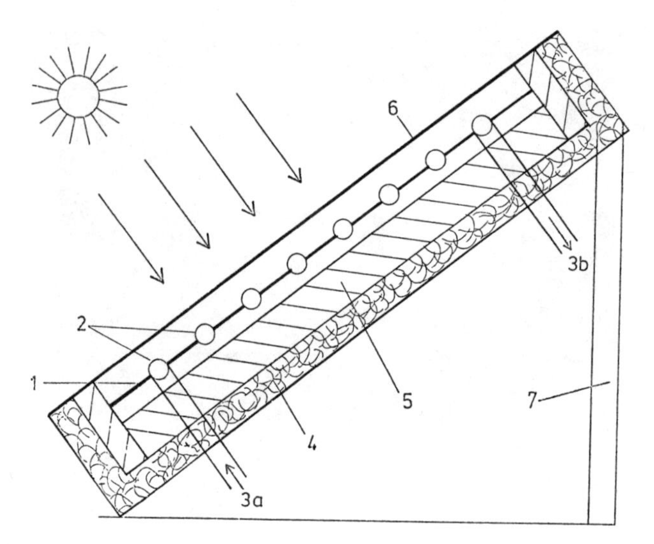

FIGURE 5.7. Cross section of a "Florida-type" flat plate collector for heating water. (1) Absorbent sheet; (2, 3a, 3b) water pipes; (4) box; (5) thermal insulation of sides and bottom; (6) spectral selective cover on the sunward side.

5.3.3. Air Conditioning and Complete Climatization of Living Quarters

In the hot regions of the earth, air conditioning is as important as heating in the colder regions. A combination of both is of interest for raising the quality of life, especially in Europe, which, despite its wealth, is underdeveloped in respect to air conditioning. On the one hand, according to Carnot's principle, heating involves a creation of disorder and is easier than cooling below the surrounding temperature, which involves a reduction of disorder. The same holds true for room heating and cooling. On the other hand, driving an air conditioner with solar energy facilitates the cooling process, since the maximum cooling demand is in the heat of the day, during maximum solar radiation, whereas for solar heating, the greatest heat demand is at night, when the sun is not shining. Thus, for solar cooling, the aforementioned solar heat reservoir is not necessary. Solar cooling systems are not new; in principle, they involve absorption of heat, as in small refrigerators, in which the gaseous cooling medium (e.g., an ammoniacal solution) is compressed noiselessly through electrical heat, instead of by means of a noisy mechanical compressor. The cooling effect is produced through subsequent expansion (Figure 5.8). The transition from electric heat to solar heat is quite practical, because only moderately high temperatures (about 100°C) are needed, and also profitable, because energy consumption in household cooling is relatively high. The technological task is to find a suitable cooling medium; a favored medium is an aqueous LiBr solution, which achieves, using the so-called ARKLA system, a cooling down to 7°C. with an efficiency of 67% using only 99°C heat.

Another possibility for residential air conditioning is that of heat loss (Figure 5.9A) through infrared radiation to the night air (especially on clear nights, when the air is noticeably cooler). For this, special IR-permeable windows are necessary. A material adjusted to the wavelength of a specific ambient temperature, e.g., polypropylene foil (proposed by Justi[15]), would be advantageous for IR windows.

5.4. METHODS FOR THE COLLECTION AND CONVERSION OF SOLAR ENERGY

Most people do not yet believe that liquid fossil fuels will be exhausted within decades, so that there has unfortunately been little general interest in the utilization of solar energy and research up to the present has been insufficient for us to be able to make a decision as to which of the following options would be preferred.

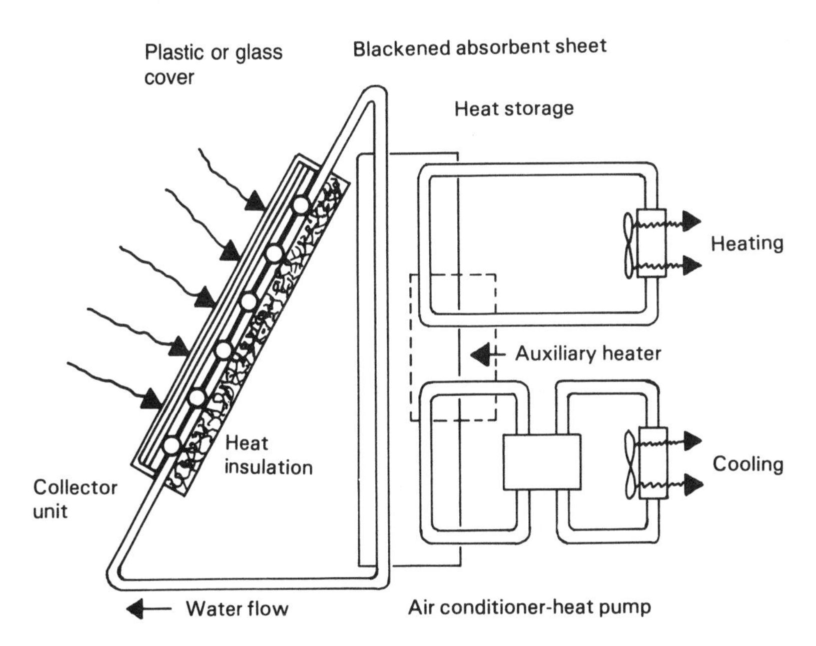

FIGURE 5.8. Schematic drawing of a completely climatic residental house with solar collectors, heat storage, auxiliary heater for the coldest months, and an air conditioner–heat pump unit, preferably constructed as a continuous absorber without moving parts and using an aqueous LiBr solution as cooling medium.

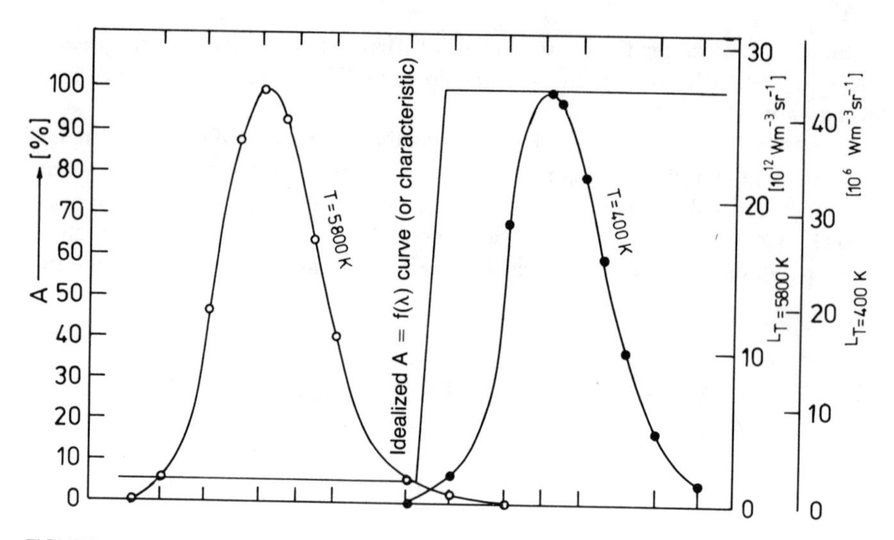

FIGURE 5.9A. Spectral distribution of heat reradiation of a black body at room temperature, valid for nocturnal cooling by reradiation.

5.4.1. Photovoltaic Conversion

This is the best-known method and can be understood by looking at Figure 5.15. Light quanta (called photons) enter a *p*-type Si semiconductor (i.e., one with an excess of positive electrons or holes) and excite electrons from the valency band to the conduction band. These diffuse to the *p–n* junction and give rise to a potential difference there. This potential difference is of the order of magnitude of 1 V and can be used in driving the current through an external load. A detailed consideration of this process is described in Section 5.6.

5.4.2. Photogalvanic (or Photoelectrochemical) Conversion

The method of converting light to electricity has been known for more than 30 years, but it has been studied very little until the 1980s. Hilson and Rideal gave a theory of this effect, assuming that the light was adsorbed by radicals of the intermediate products of reaction on the electrode surface (see Bockris[17]). In the direct photogalvanic effect, light is incident on an electrode and gives rise to the emission of electrons; this in turn energizes an electrochemical cell, from which power or fuels can be drawn. In an exactly simliar way, holes can be activated in semiconductor electrodes, and these are then available at the anode for the acceptance of electrons.

There are other electrochemical photoeffects. For example, photons incident on a solution excite molecules in the solution; the molecules degrade to lower energy states and transfer their electrons to electrodes in contact with the solution.

These photogalvanic effects have been researched much less than the photovoltaic ones; they do not need superpure single crystals as collectors (as photovoltaic devices do) and might therefore be the cheaper of the two types of converters. Kuhn[18] has carried out some admirable theoretical and experimental work in this field, concerning photoelectric conversion in molecular layer devices with the eventual goal of photolysis of water to form hydrogen.

5.4.3. Photothermal Effects

This effect was used long before it was identified scientifically by Duffie and Beckman in greenhouses and since the 1930s has been used to heat water and for space heating in Florida, as mentioned above. Figure 5.7 shows a cross section of a Florida-type flat collector, which consists of a wooden box with closed sides and bottom and heat-insulated. On the top—the sunward side—it is covered with one or more glass plates. Inside, there rests a copper sheet, painted black with a matte finish, and on the black side, there are attached copper pipes through which water circulates as the heat-transfer medium. The absorber is arranged, as was deduced above, at an angle of more than the geographic latitude, in a southerly direction and collects up to 63.7% of the thermal "harvest" if continuously rearranged in relation to the position of the sun. Let us make the simple assumption that the glass sheets are spectrally selective and absorb only the visible part of the sun, between $\lambda = 0.4$ and $\lambda = 0.8$ μm, whereas they do not radiate the infrared heat of the black sheets, which thus become warmer (cf. Figure 5.9A). This attitude involves the assumption that the six sides of the collector box do not dissipate the heat collected by the black sheet by means of convection and conduction. Assuming further that no heat is removed by water circulation, then the highest possible so-called "idling temperature" is achieved. Ideally, an absorber in radiation equilibrium with solar light could reach over 5000K, but in reality, equilibrium can be achieved only with the fraction of the solar radiation that reaches the collector, i.e., with 1.34 kW/m^2 = 1150 kcal/hr × m^2. The reradiation Q of the collector can be calculated in terms of the well-known Stefan–Boltzmann Fourth Law of Thermodynamics for black-body radiation. It comes to $Q = 5\,(T/100)^4$ kcal/hr × m^2 and for $Q = S = 1150$ kcal/hr × m^2, this means an equilibrium temperature between incoming and outgoing radiation of $T = 390$K or 117°C. This calculated value corresponds well with the highest values obtained by good examples of present technology in this field.[19]

Of course, the idling temperature is worthless from a practical point of view because one cannot take any work from it, but it is a kind of goodness factor that gives information concerning the extent to which it has been possible to suppress the loss of heat conduction, convection, and radiation and thereby increase the conversion efficiency. The loss mechanisms increase proportionally to the temperature difference, but the heat reradiation increases with the fourth power of the absolute temperature, and this implies special importance in respect to the suppression of infrared radiation. According to experience gathered in the Phillips Research Laboratories,[20] the back side of the glass sheets can be covered with a layer of indium oxide (In_2O_3) (Figure 5.9B). What this does is to decrease the reradiation, but allow the visible light to enter without diminution. Alternatively, one can choose some kind of window substance that itself has an [absorption (α)]/[emission (ε)] ratio higher than that of the usual glass, such as the ETFE Hostaflon foil developed by Justi (Figure 5.9C). Alternatively, one can cover the absorber itself with thin layers, and these, due to a special geometric structure, show an apparent exception in the IR range to Kirchoff's law for blackbody radiation according to which, at the same wavelength, the relationship of absorption to emission is 1. For example, photographers are familiar with coatings for camera lenses that have a thickness of about one quarter of the wavelength of light and at this thickness do not reflect light of that wavelength. If they are $\lambda/2$ thick, they polarize circularly. Thus, thickness changes in these coatings of only about 0.1 μm change their optical properties completely. However, it is of doubtful value to transfer the technology applicable to valuable lenses and mirrors of small diameter to large solar-energy collectors of many square meters in size, particularly since the latter must remain stable in the open air for many years without changing their thickness. A important factor is the cost—it must all be done for a cost of only a few dollars a square meter, or else the optical improvement of the collector will increase rather than decrease the net cost of energy conversion. Such a problem is a suitable challenge for solution in corporate laboratories in capitalist countries that have high standards of research and production and are subject to the rigors of competition. Among the more valuable solutions given are the aforementioned indium oxide layers produced by Phillips[20] and high-vacuum heat confinement, which uses the principle of the Thermos flask. Tabor,[21] in Israel, has utilized the "EBONOL" interference fringes invented there and applied them electrolytically (Figure 5.9B). Seraphin in the

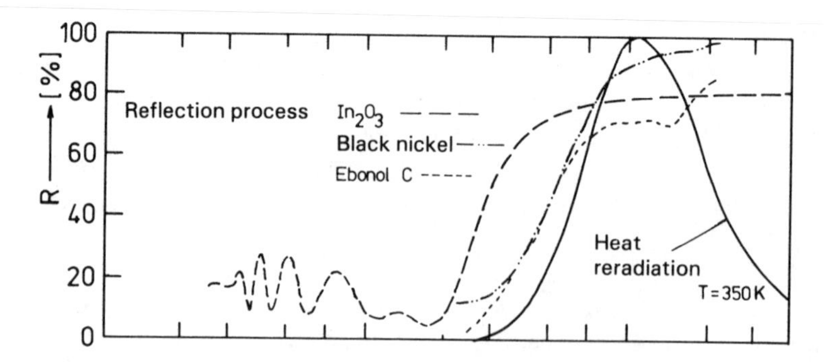

FIGURE 5.9B. Emission (ε)/absorption (α) of In_2O_3, black nickel (SOLAROX layer of Dornier-System) and the EBONOL layer of Tabor.

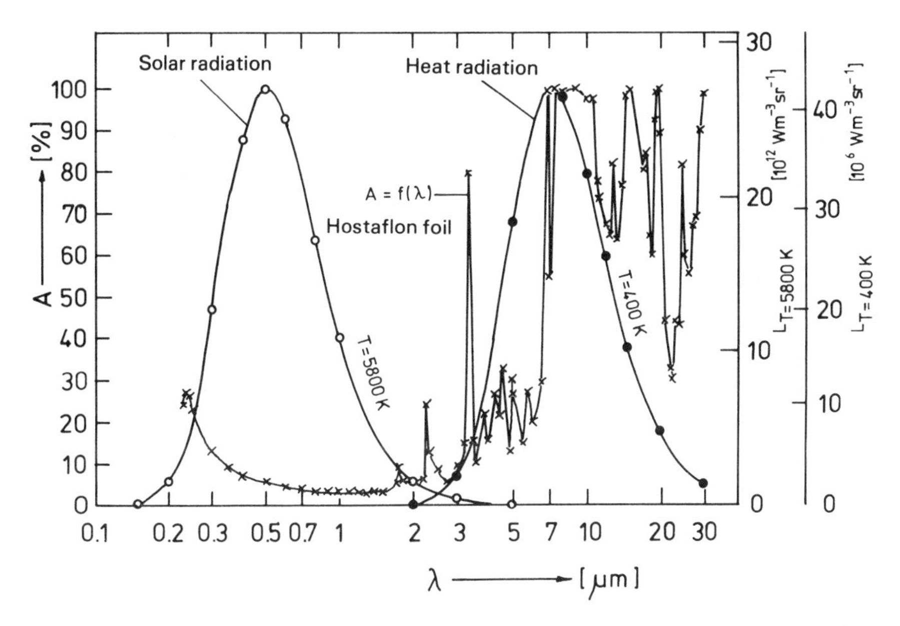

FIGURE 5.9C. Absorption A (%) of the Hostaflon-ETFE foil cover (left ordinate) and heat radiation of the sun and foil (right ordinate).

United States and Dornier systems in Friedrichshafen, developed and commercialized treelike or rodlike microstructures made from so-called black nickel, in which reradiation is reflected back and forth to the extent that it is finally absorbed (Figure 5.10). Thus, the Vereinigte Deutsche Metallwerke Werdohl[22] reached idling temperatures of 130°C in Germany using aluminum collectors blackened with "3M Black Velvet." In this case, water cannot be used as a heat-transfer medium because its pressure would reach over 4 bar and the aluminum sheets would not withstand this pressure. For this reason, Klöckner-Wärmetechnik GmbH converted to a synthetic organic fluid called "Gilotherm ADX10" manufactured by the French firm of Rhone and Poulenc. It has a low specific heat (half that of H_2O). It exhibits significantly increased temperatures even when exposed to the sun for short durations and only in quite moderate amounts. When it was used in the black nickel structures developed by Dornier,[23] idling temperatures of 180°C were reached, without the production of any excess pressure.

The synthetic organic heat-transfer medium "Gilotherm ADX10" represents a further development and modification of a known heat-transfer medium. Along with high-boiling hydrocarbon derivatives, glycol–water mixtures containing appropriate inhibitors have been marketed in recent years for use in solar-energy plants. In respect to the utility and useful lifetime of flat-plate collectors for making hot water or space heating, the corrosion behavior of the fluid is of decisive importance. Thus, in all the newer developments, the properties of

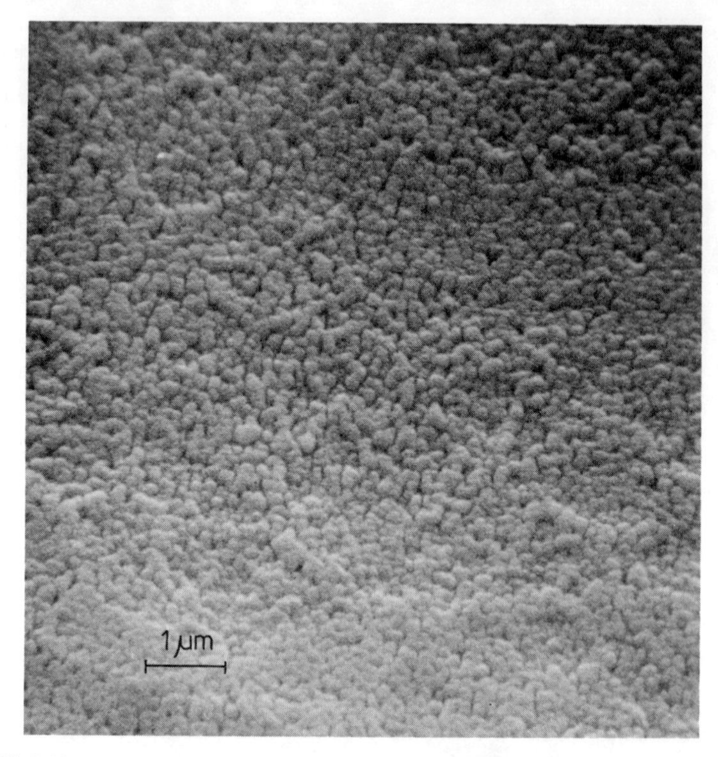

FIGURE 5.10. Electron-microscopic photograph of a SOLAROX layer on an aluminum rolling band absorber. Approximately × 10,000. From Scherber and Dietrich.[23]

systems in which more than one metal is in contact were taken into consideration to see that the metals did not form a galvanic couple. Among the more important new heat-transfer media which have their basis in glycol are PKL 100 and BP Thermo-Frost P. In Table 5.2, the most important thermal properties of water are compared with those of pure PKL 100 and pure BP Thermo-Frost.[24–26] From the values given there, it is apparent that the kinematic viscosity of both heat-transfer fluids decreases with increasing temperature. Thereby, with their use, there will be a self-regulation of the flow velocity and the collector circuit which is especially important for circulation pumps with strongly declining characteristic pump curves. Since the minimum temperature is −50°C, there is no danger of any destruction to the collector plant during the colder season. Corrosion processes are suppressed by addition of suitable inhibitors to both these fluids, and if corrosion is never completely prevented, it is at least considerably suppressed. A more detailed account of the complex corrosion mechanisms and the various possibilities of protecting solar plants from corrosion is given by Brennecke and co-workers.[27,28]

The alternative possibility of using selective black flat plate collectors and optimal heat-transfer fluids to operate low-pressure steam turbines at temperatures higher than 150°C without optical concentration has been underestimated. Actually, such a procedure is possible, for example, in the power-tower pressure turbines (cf. Chapter 6) at several hundred degrees centigrade. On the other hand, it is possible to convert only the directed part of the radiation, i.e., that which can be focused, and then the diffuse radiation is lost. In European climates, this is often more than half the total solar energy harvest; then in addition there are the costs associated with equipment for readjusting the mirrors.

5.4.4. Photosynthetic Effect

This can be used by growing suitable plants such as sugar cane and burning them, thus obtaining energy to use in heat engines. Alternatively, the material can be anaerobically heated and decomposed into methane, hydrogen, and higher hydrocarbons. Algae can be produced in suitably large amounts and then thermally decomposed, giving rise to hydrocarbons.[29] Photosynthetic concepts, on the other hand, are rather less attractive if one takes into account their poor efficiency in respect to the conversion of solar energy. This efficiency is only

TABLE 5.2
Comparison of Thermal Data of Water, PKL 100, and BP Thermo-Frost

Physical parameter	Heat carrier	Temperature								
		−20	0	20	40	60	80	100	120	140
Density ($kg\ m^{-3}$)	Water		999.8	998.2	992.2	983.2	971.8	958.3	943.1	926.0
	PKL 100	1114	1102	1089	1074	1060	1044	1027	1012	996
	BP Thermo-Frost	1082	1068	1055	1040	1024	1006			
Steam pressure (bar)	Water	0.001	0.006	0.023	0.073	0.199	0.473	1.013	1.985	3.614
	PKL 100						0.036	0.088	0.195	0.400
	BP Thermo-Frost			0.007	0.024	0.065	0.160	0.328	0.630	1.210
Specific heat ($Jg^{-1}\ K^{-1}$)	Water		4.217	4.181	4.178	4.184	4.196	4.216	4.245	4.287
	PKL 100	2.153	2.255	2.355	2.466	2.568	2.671	2.777		
	BP Thermo-Frost	2.488	2.602	2.708	2.825	2.934	3.048	3.156	3.268	3.379
Heat conductivity ($Wm^{-1}\ K^{-1}$)	Water		0.552	0.598	0.628	0.651	0.669	0.682	0.685	0.684
	PKL 100[a]									
	BP Thermo-Frost		0.251	0.243	0.234	0.226	0.218	0.210		
Kinetic viscosity (cSt)	Water		1.792	1.004	0.658	0.474	0.365	0.295	0.246	0.216
	PKL 100	700	145	42	17	7.7	4.5	3		
	BP Thermo-Frost	1500	220	53	19.5	7.8	4.2	2.5		

[a] No data on the heat conductivity of PKL 100 is available

about 1%, so that the total overall efficiency of conversion of solar to electrical energy comes to only 0.3% compared with the 10% available with photovoltaic conversion. Thus, the collecting areas will have to be about 30 times larger than those necessary with photovoltaic farms. The direction of research in this field is aimed at the selection of bacteria and algae containing enzymes that would give rise to greater efficiency of conversion of light to biomass as a result of catalysis. This is a fairly sensible goal, as shown in Table 5.3, which indicates that efficiencies of up to 4.7% can be reached.

5.4.5. Oceanic Thermal Gradients

These gradients result from solar radiation, and the energy they contain can be converted into electrical power. They represent—just as burning fossil fuel do—stored solar energy from that incident on the sea at an earlier time. The advantage is that they represent a *volumetric* concentration compared with *planar* dilution of solar radiation when it is collected on a flat surface. Additionally, they do not have the disadvantage of introducing more CO_2 into the atmosphere.

The temperature gradient in a tropical ocean can be obtained by recalling that at the sea surface, the temperature is more than 300K and at a depth of 1 km it is less than 285K. At the sea-surface water temperature, a low-boiling organic liquid can be heated and then work a heat engine, and cool water can then be pumped up from the ocean depths and condense the fluid back from the heat engine. There would then be a cyclical process for the extraction of heat from the surface water, conversion of a small fraction of it into mechanical power, and reinjection of the rest into the cold water of the deep sea (Figure 5.11).[30]

The efficiency of this hypothetical process is only about 2%. However, the great advantage is that it represents collection of solar energy from a *volume*,

TABLE 5.3
Maximum Growth Rate and Efficiency Estimates of the Conversion of Sunlight through Photosynthesis[a]

Plant	Country	Growth duration (days)	Growth rate ($g\,m^{-2}\ day^{-1}$)	Mean radiation $MJ\,m^{-2}\ day^{-1}$)	Radiation yield (%)
Swamp rush grass	Australia	14	54	1.4	4.2
Pindus radiata	Australia	4	41	25.4	2.7
Maize	California	14	32	28.8	2.0
Barley	England	—	23	20.3	1.9
Sugar	Hawaii	90	37	16.8	3.7

[a] From Report No. 17, Australian Academy of Sciences on Solar Energy Research in Australia.[3]

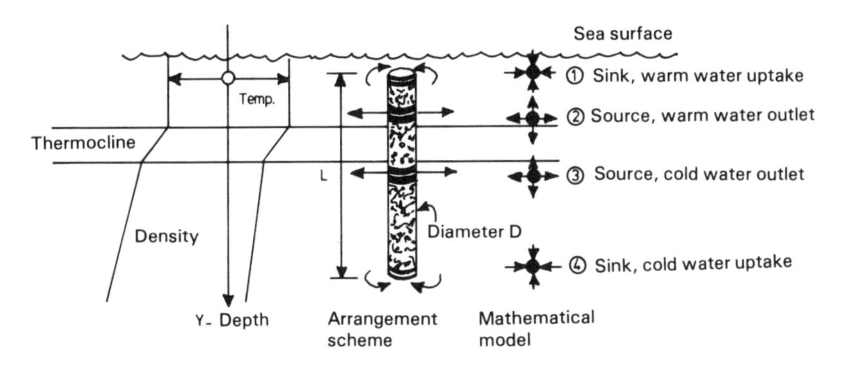

FIGURE 5.11. Topology of a solar–ocean energy configuration. *Left:* Development of temperature and density with water depth. *Middle:* Configuration scheme. *Right:* Mathematical model.

not just a *surface,* as occurs in photothermal processes, so that the low efficiency of conversion does not imply that there must be enormous areas devoted to collection. In the ocean-gradient concept, heat engines would be mounted on floating platforms and drive conventional turbines. The boiler would be above the sea surface, and the condenser would be directly under the boiler. Tubes would then bring cold water from the ocean depths. The electricity produced would be used for the electrolysis of hydrogen from sea water, and the hydrogen would then be conveyed to land in high-pressure pipes (cf. natural gas from the North Sea fields) on floating platforms and shipped.

The main handicap of all this is the relatively small ΔT, which would require heat exchangers of great size.

5.4.6. Wind Energy

As mentioned above, wind energy is another form of solar energy, and the total amount that can be considered is around 30 times the needed global energy. One of the advantages is that the energy can be converted to electricity mechanically without the Carnot factor. The negative side is the sporadic nature of the supply and the fact that the strong winds are localized on the coasts or seas, which suggests the use of floating platforms. The German Society for Solar Energy (DGS) had a meeting in 1977 that was devoted exclusively to wind energy. There was plenty of evidence at this meeting that modern aerodynamics is giving rise to new possibilities in a technology we were familiar with from earlier times. In addition, solid-state technology provides a satisfactory solution to the problem of inverting the energy into the national distribution network at appropriate frequencies and phase arrangements. Hydrogen enters the picture again because, as with all the sporadic sources, there is need for storage. An excellent review has been written by Kramer.[31]

5.4.7. Biophotolysis

This can be regarded as a separate form of photosynthesis in that one is not dealing with the chemical conversion of plants to biomass or the conversion of solar energy by direct combustion or through enzymatic partial gasification of green plants, a process that has very low efficiency (see Section 5.4.4. and Table 5.3). In the biophotolysis method, the intent is to siphon off the protons—which are produced photochemically in the chloroplasts by cleaving water molecules and are bound to an organic acceptor, ferredoxin—before they are used for the synthesis of molecular hydrogen. By this method, one could find a way to a higher degree of conversion and also discover an ideal energy carrier.[32,33] Benemann has used for this purpose squashed chloroplasts from spinach suspended in H_2O and has introduced enzymes that limited recombination of the hydrogen with oxygen and water. Solar radiation on this suspension gave rise to hydrogen bubbles. Unfortunately, the evolution of hydrogen bubbles ceases quite soon, probably because of the difficulty of avoiding the recombination of hydrogen with the oxygen. In lectures to the Hamburg solar energy forum of 1977–1978, Broda[34] and Kuhn[35] suggest that one should not use biological systems (cf. also Chapter 7), but instead develop and synthesize monomolecular layers, which are optimized for photolysis.

5.5. USE OF OCEANIC THERMAL GRADIENTS*

5.5.1. Principle

As a consequence of intensive solar radiation, there exists in the tropical oceans between the surface and the deep water very distinct temperature gradients. Figure 5.12 shows a typical temperature profile between surface water at about 25°C and that from the deep, which comes from the antarctic and is about 5°C.[36] With the help of the warm surface water, it is possible to get a low-boiling liquid (suitable liquids would be propane, butane, or mixtures of water and ammonia) to evaporate; to drive a heat engine with the vapor, thus producing electric current; and then to take the vapor and condense it with the cold from the deep-sea water. A schematic diagram of a power plant on a floating platform is shown in Figure 5.13.[37] Figure 5.14 shows a suggestion made by Lavi and Zener[36] for a solar power plant suitable for an ocean thermal arrangement. The warm surface water is brought into contact with an ammonia boiler and ammonia is evaporated from the ammonia–water mixture (8.7 bar vapor pressure at 20°C). This then drives a low-pressure turbogenerator. The expanded ammonia gas is cooled in the condenser by the cold water pumped up from the ocean depth, and condensed back into the water-circulation system.

* This section authored by Prof. H. H. Ewe, Hamburg.

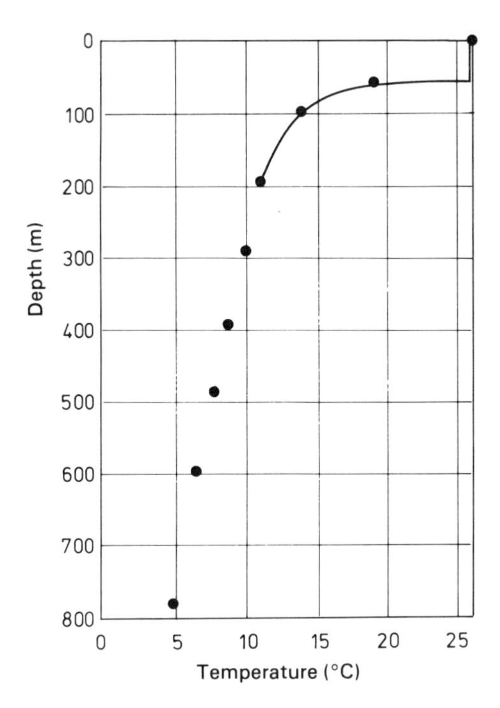

FIGURE 5.12. Typical temperature profile of tropical oceans. From Lavi and Zener.[36]

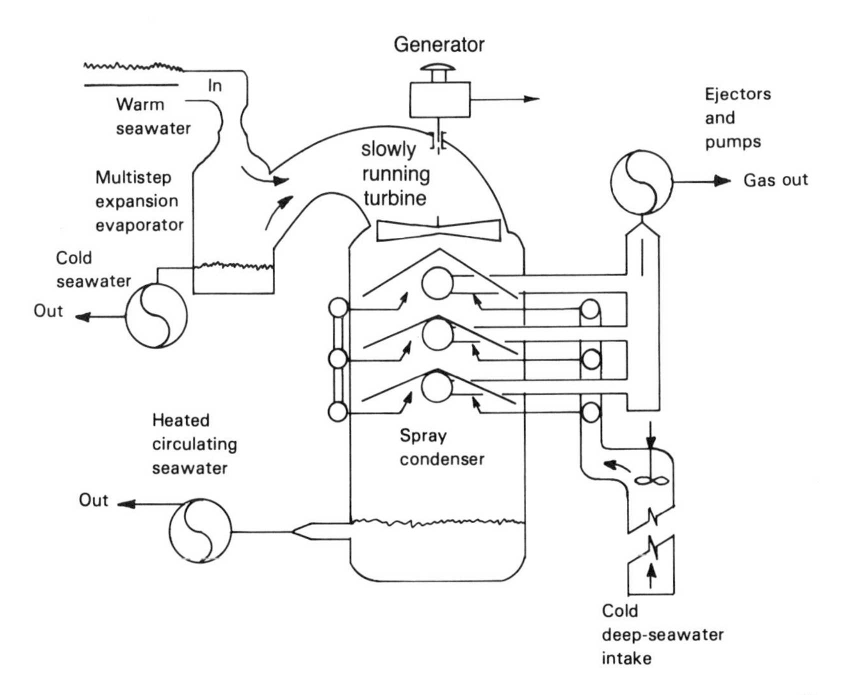

FIGURE 5.13. Diagram of an ocean heat power plant floating on the water. From Claude[37].

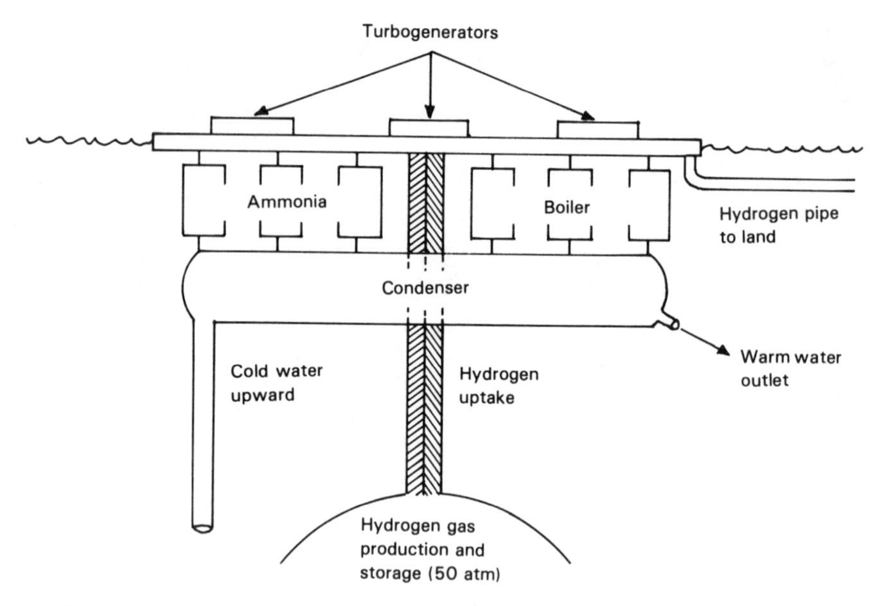

FIGURE 5.14. Block diagram of an ocean heat power plant with a dual medium system, i.e., water–ammoniate. From Bockris.[45]

5.5.2. Efficiency and Costs of Conversion

A pioneer in air liquefaction, Claude, tried in 1930 to install a 40-kW plant in Cuba, but the plant was destroyed by a storm.[37] Despite the absence of actual practical examples of these plants, some things can be said about them on the basis of the large number (around 30) of systems analyses that have been made for this type of converter.[38–40]

The most important problem is that large heat exchangers are needed because of the small temperature difference. When this difference is about 20°C between the warm and the cold water, only about 50% of it is useful because a temperature difference of about 10°C is needed for the heat flow in both heat exchangers. A usable temperature gradient of 10°C corresponds to a theoretical Carnot efficiency of 3.3% in respect to energy conversion to electricity. The practical losses during the operation of large water pumps reduce this to about 2%.

One of the advantages of these power plants compared with solar cells is that energy is collected from a *volume* of water, and this compensates for one of the greatest disadvantages of the collection of solar energy, namely, that it is always limited to the collection on *surfaces* of the solar energy that is incident on the earth at the time of collection. Lavi and Zener[36] have calculated that the performance of these power plants could reach 12 kW_{el}/m^2 of the area of the sea surface taken up by the power plant.[36] The investment cost would be about \$220–445/kW_{el} (1986 prices), and the cost of electricity about 0.8–1.8 ¢/kW_{el}.

The floating power plants would certainly have to be in highly insolated waters, e.g., in the Gulf of Mexico. The necessary transport of energy to the consumer would take place through hydrogen pipelines, or alternatively, hydrogen would be liquefied and taken away in barges, and in a few cases there might be electrical power lines from the plant to the shore.[41]

5.5.3. Technical Problems

The most important technical problems are connected with the large area of the heat exchangers. Because of the small difference in temperature between the surface and the depths of the ocean, the temperature difference of about 2°C at each boundary between water and vapor and the temperature gradients in the metal walls must be considerably reduced. This requires the manufacture of very long, thin-walled tubes from nickel alloys.[42] This in turn results in high investment costs, and it increases vulnerability to corrosion by seawater and also the danger of deposits of botanical and bacteriological origin. In particular, on the water side of the heat exchanger working at the higher temperature, a slimelike deposit develops. The metal is attacked by various fouling processes, which have a diminishing effect on the heat-transfer characteristics of the interface. The slimelike layer and also mineral deposits can be removed or reduced by high-pressure washing, by brushes that are moved by the flowing water, by chlorination, by coating with toxic paint or by other means.[43,44]

With the low pressures of ammonia vapor, the pipes can be fairly thin-walled, and this, of course, helps the heat transfer. But mechanical problems arise as a result of this thinness, and much attention has been devoted to corrosion protection. Thus, for example, titanium is a metal that should be used from the point of view of stability against corrosion, but in turn would be higher in cost.

Virtually all the systems analyses show that these problems can be overcome and that these ocean solar power plants may be (for so much depends on their lifetime) the cheapest of all methods for converting solar energy into electricity. To date, these optimistic estimates have been tested only in small trial plants, but they do show that there is considerable potential in this type of plant.

5.6. SILICON PROTECTIVE-LAYER CELLS

To understand the nature of the physical problems of these devices, it is first of all helpful to review the nature of light once more. Corresponding to the modern concepts of the dualism of wave and particle, which originated with de Broglie, there may be two parallel descriptions of one and the same thing. Thus, it is sometimes helpful to look at light as an electromagnetic wave consisting of a series of photons, each of which consists of an indivisible (quantumlike) piece of energy. Correspondingly, a light ray can be looked at as a system of electromagnetic waves of different wavelengths, just as it can be seen as a series of photons of different energy and moment. According to the Planck–Einstein relationship, the energy of the photons is given by $E = h\nu$, this h being a universal constant, i.e., the Planck constant of action (6.610×10^{-34} joule $\times$ sec), and ν being the frequency in sec. Usually, the energy is expressed in electron volts, and is given by $E = h\nu = h \times c/\lambda$, wherefrom the potential associated with the wavelength λ is given by $V = h \times c/(e \times \lambda) = 12{,}398/\lambda$ volts, when λ is the measured wavelength in angstroms (1 Å $= 10^{-8}$ cm). The term "stream of photons" is understood to mean the number of photons that pass through a vertical surface of 1 cm^2 sec^{-1} in a ray of light. If ϕ is the intensity of this light ray in watts cm^{-2}, then for a monochromatic stream of light $N = \phi$ $(h \times \nu) = \phi \times \lambda/(h \times c) = 5.04 \times 10^{14} \times \lambda \times \phi\,cm^{-2} \times sec^{-1}$. If a photon of sufficient energy is incident on a suitable semiconductor such as silicon, then it can give rise to the breaking of a bond there and create a hole–electron pair, as shown in Figure 5.15. Here, one sees in the band model how the energetics of this ionization occurs. On the left, one sees doping with a five-valent donor (Sb), thus making silicon into an n-type semiconductor; on the right, one sees the three-valent acceptor (N) doping of silicon into p-Si. In each case, the photon has to have at least 1.1 eV of energy, corresponding to an infrared wavelength of 1.13 μm, to give rise to a hole–electron pair through photogeneration. In semiconductor theory, it is easy to show that the product of the positive and negative charge-carrier density is constant at constant temperature and is equal to the square of the charge density in the intrinsic semiconductor, i.e., $n_i^2 = n \times p$ (n_i at room temperature is about 10^{21} cm^{-3}). For well-conducting n-Si, one has $n = 10^{17}$, and hence $p = 10^4$ cm^{-3}. If, therefore, one has a photon and generates

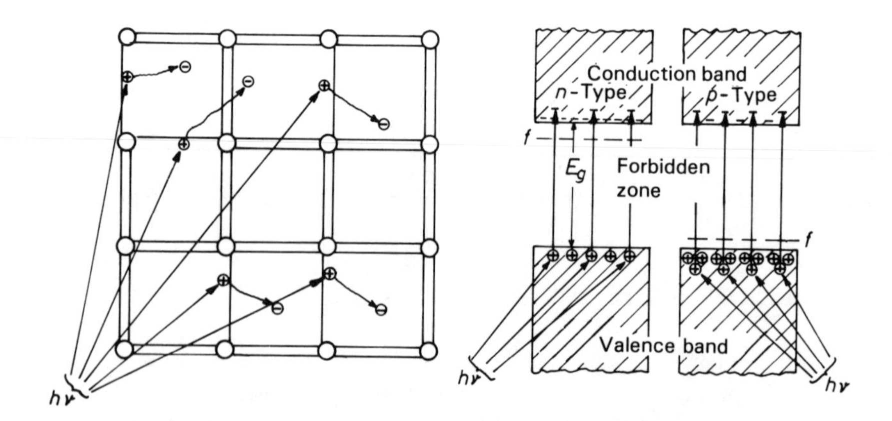

FIGURE 5.15. Reactions in an Si barrier-layer cell. *Left:* Photons with energy $h\nu$ within the Si matrix break bonds and set free hole-electron pairs (photogeneration). *Right:* Same process in the band model, converted in the left half to the n-type with support from a five-valent donor (e.g., Sb) and in the right half for the p-type with dopant, e.g., In.

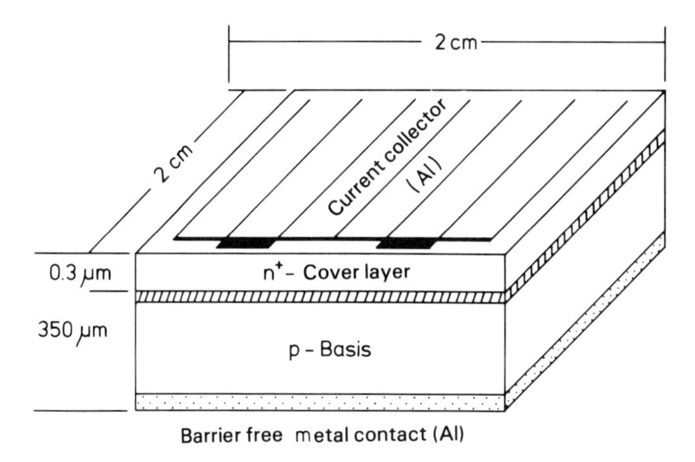

FIGURE 5.16. Cross section cut through a Si barrier-layer photocell on top of a 355-μm-thick *p*-Si layer. There is a 0.3-μm-thin n^+-Si cover layer. Electrodes: Al.

within the Si the same number of electrons and holes, the density of the majority carrier (*n*) is hardly changed, while the density of the minority carrier (*p*) is changed by 10 orders of magnitude. These photogenerated charge carriers will diffuse into the surroundings in about 10^{-4} sec until their concentration exceeds that of a thermal equilibrium.

The idea of a barrier-layer photocell is that opposite charges go in the opposite direction to protect them from recombination before they get into the external circuit and do work. This occurs because there is across the *n–p* junction of a semiconductor device a built-in permanent electrical field, as is well known through earlier work on solid-state rectifiers. In Figure 5.16, there is a schematic construction and numerical indication of the dimensions of a modern *n*-on-*p* solar cell that consists of a 0.355-mm-thick single crystal of silicon sheet, doped with about 10^{-4} % indium to make it *p*-conducting. On the surface of this sheet, created by the diffusion of phosphorus vapor, a flat *p–n* junction is produced. Ohmic contacts are produced by evaporation of aluminum, flat on the back side but comblike on the front side. Interference color gives the cell its characteristic blue appearance (Fischer).[46] To get the highest degree of efficiency, each photon should produce an electron–hole pair within the diffusion length of the *p–n* junction. Figure 5.17 shows the measurements of von Prince and Wolf,[47] which shows how the absorption of photons depends on their wavelength. The short-wavelength light quanta are absorbed in the surface layer, but the long-wave-length radiation produces hole–electron pairs in the core layer. What one tries to do, therefore, is optimize the diffusion processes by producing a surface layer about 0.3 μm thick, and this is a good compromise. Light of too long a wave-length is of no use for the production of electric power, and light of too short

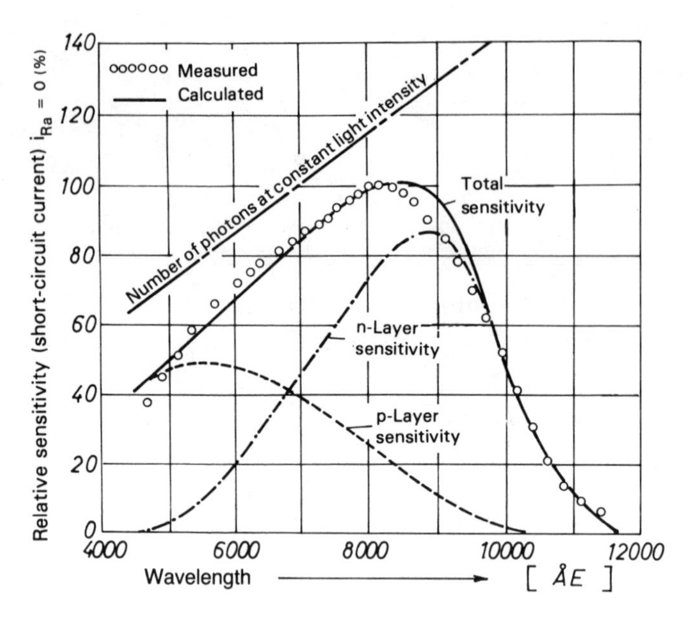

FIGURE 5.17. Charge-carrier photo generation inside a *p*-on-*n* silicon solar cell. The parameters of the curves are explained in the diagram. From Prince and Wolf.[47]

production of electric power, and light of too short a wavelength simply transformation of hole–electron pairs into heat, which is not only useless, but actually decreases the efficiency of the conversion process (Figure 5.18). It is therefore only a rather small region of wavelengths that gives rise to photocurrent production, and this of course therefore has to be adjusted to the energy distribution in the solar spectrum (see Figure 5.1) and requires a correct choice of the layer gap E_g (eV) in the semiconductor. Figure 5.18 shows in a somewhat simplified way (see Chapter 6) the maximum efficiency for photovoltaic conversion, as calculated by Rappaport,[48] as a function of the layer gap, the temperature being the parameter. Two important facts can be deduced from this diagram: First, the optimal band gap increases with increasing temperature, so that at room temperature, CdTe with an energy gap of 1.5 eV would be the most favorable semiconductor, while at 300°C, some material having an energy gap of 1.8 eV would be the best (one might obtain this from a ternary compound consisting of, say, GaAs and GaP). The second fact that can be deduced is that the reduction of the efficiency of silicon resulting from an increase in temperature (according to Figure 5.18, this is characteristic of semiconductors with large energy gaps, such as GaAs) can be compensated, but GaAs, cadmium sulfide, and cadmium telluride are unsuitable for other reasons. As Justi has pointed out, the decrease in the degree of efficiency of conversion with increasing temperature is by no means accidental, and is connected with the exponentially increasing thermal ionization that leaves nothing for photogeneration. The popular viewpoint that one could cover the deserts of the earth with solar cells that would then give rise to enormous amounts of electrical energy is difficult to accept because of the lack of cooling water, which is obviously not available in deserts. The author has calculated that photovoltaic electricity would not be sufficient to work refrigeration machinery for its own cooling needs, although solar cells are in principle direct converters of energy. In practice, however, solar cells are also heat engines that need a hot and a cold heat reservoir. For this reason, Justi has called them "quasi-Carnot processors," an expression that has the effect of guarding the practitioner from the illusion that they can be decoupled from Carnot-like considerations. In fact, a solar cell works at an efficiency of about 10%, which means that it rejects about 90% of the energy incident upon it as heat, just like a poorly operating heat engine (so that the availability of large amounts of cooling water to optimize conversion of the heat energy is necessary).

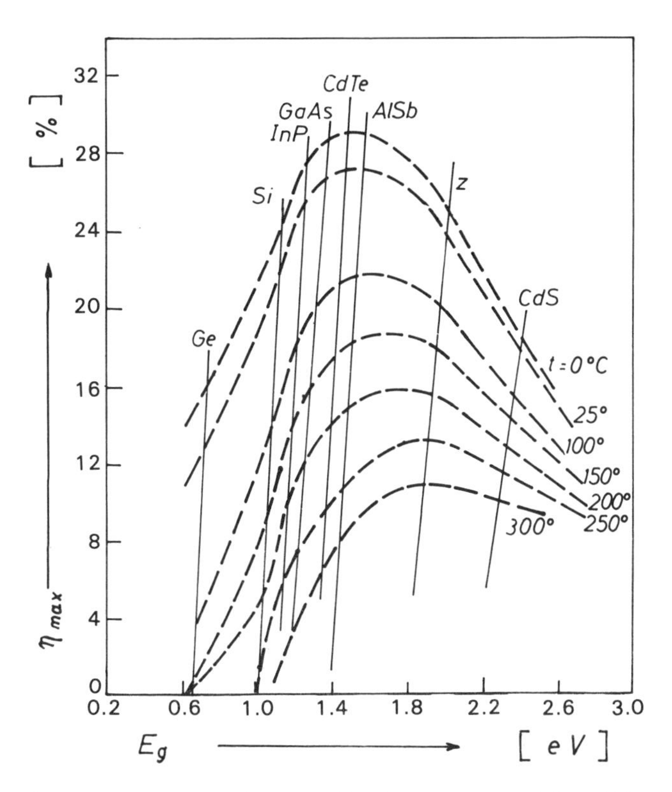

FIGURE 5.18. Maximum degree of efficiency of barrier-layer cells made from different semiconductors as a function of the band distance E_g (eV). Materials are indicated in the diagram. Parameter: working temperature. From Rappaport.[48]

Figure 5.19 shows in schematic form a voltage–current characteristic of an insulated barrier-layer cell. With optimized matching of the external load, the maximum efficiency is given by

$$\eta_{max} = \frac{\Lambda \times U_{mp}}{(1 + \Lambda \times U_{mp})} \frac{U_{mp} \times i_{mp}}{E_{av} \times N_{ph}}$$

where Λ is an abbreviation for $\Lambda = e/A \times k \times T$, A is a cell constant varying between 1 and 2, U_{mp} is the potential applied at the highest load, i_{mp} is the mean current density at the highest load, E_{av} is the mean energy of an incoming photon, and N_{ph} is the photon flow per unit area and time. The open-circuit voltage U_{max} depends on U_{mp} according to the equation

$$e^{U_{max}} = e^{U_{mp}} \times (1 + \Lambda \times U_{mp})$$

where e = base of the natural logarithm = 2.7183. In real photocells, $U_{mp} >> 10$, so that in practice, $U_{max} \approx U_{mp}$. The short-circuit current i_s depends on i_{mp} according to

$$i_{mp} = \Lambda \times U_{mp}(i_s + i_0)/(1 + \Lambda \times U_{mp})$$

where i_0 is the dark current density flowing through the junction in the direction of the barrier layer. The relationship among U_{max}, i_s, and i_0 is

$$e^{\Lambda} \times U_{max} = 1 + i_s/i_0$$

In real cells, i_s/i_0 is greater than 1. The short-circuit current can be calculated by $i_s = Q \times E \times n_{ph}$, where Q is the collection efficiency, i.e., the relationship of the charges flowing through the junction to the number n_{ph} of those photons that have enough energy to form hole–electron pairs. The current in the barrier-layer cells can be calculated from $i_0 = C \times \exp(-E_g/B \times R \times T)$, where C represents the influence of the specific resistance, effective mass, effective density states, and the minority carrier diffusion lengths for the semiconductors concerned and AB is an empirical factor. In modern high-performance cells, B = 1 is a fair approximation.[48]

Figure 5.20 shows the characteristics of the historic cell (a *p*-on-*n* silicon cell) that was made by Chapin *et al.*[49] In this diagram, one sees the open-circuit voltage U_0 (volts) on the left, as a function of the solar radiation intensity, on the abscissa; on the right, one sees the short-circuit current i_{Ra} in mA/cm^2. This diagram shows three things: First, it shows that increasing light intensity causes a rapid increase of the no-load voltage to a constant saturation value ($U_{\phi} = \infty = 0.6$V), and this in practice is an advantageous relationship because these cells are dependent on variable solar radiation and have to be connected

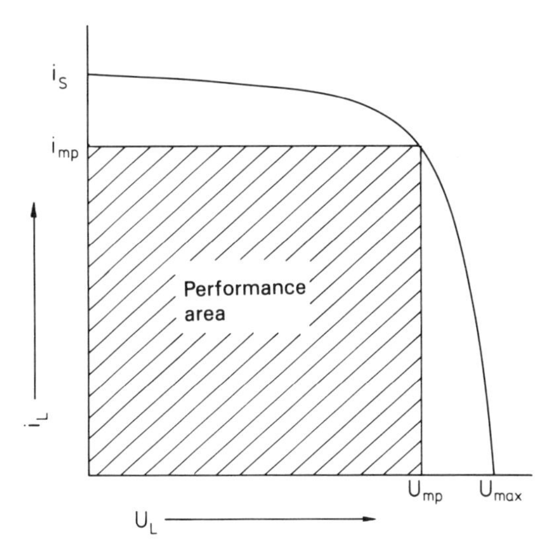

FIGURE 5.19. Schematic voltage–current characteristics of a barrier layer of any composition for the electrochemical–phenomenological computation of the conversion efficiency.

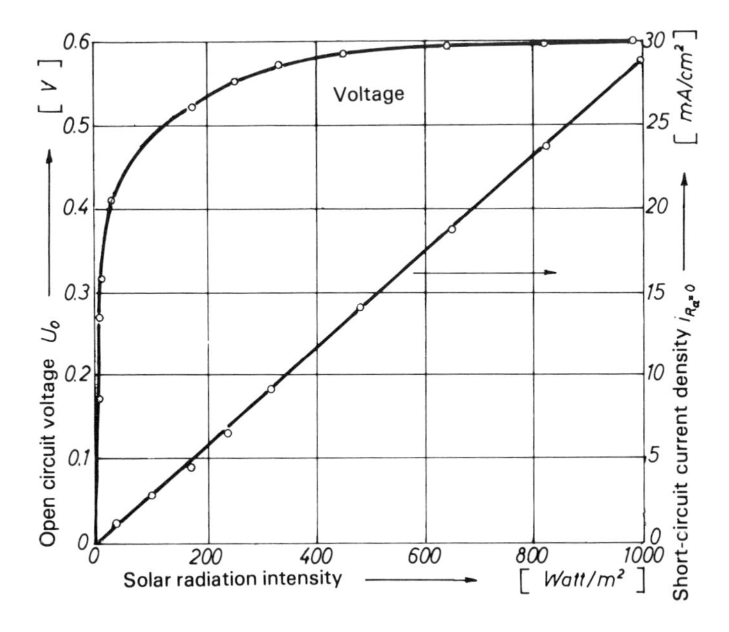

FIGURE 5.20. Measured characteristics of a Si solar cell as a function of the intensity of solar radiation. From Chapin *et al.*[49]

to accumulators. Second, it shows that the increase in the current density available is strictly proportional to the light intensity. Finally, the diagram shows the large contribution of the short-circuit current density, which in an equatorial latitude with 1000 W/cm^2 of solar radiation amounts to 30 mA/cm^2, which equals 300 A/m^2. In fact, with modern commercial types, this value is about 40% greater. Figure 5.21 shows the power output as a function of the closed circuit potential U varied by change of the external load. It can be seen that the performance maximum is 11 mW/cm^2 = 0.11 kW/m^2 at 0.47 V operating potential. If one divides this output by the effective solar radiation of 1000 W/m^2, one sees that the conversion of efficiency, which is plotted on the right-hand ordinate, gets up to about 11%, and indeed with modern industrial converters at present available, the efficiency rises to 14% (with reports from laboratories of 26% in 1987, from very small samples).

Since the primary solar energy costs nothing and the raw material silica is available to all people in limitless amounts, does not produce any poisonous material or radioactive products, and has a very long lifetime, photovoltaic electricity production is the ideal energy source for mankind in general. This would be immediately acceptable were it not that the present investment costs are too high. For space travel and satellites, it is possible to accept initial investment costs of a few hundred thousand dollars per kW$_{el}$ because the weight

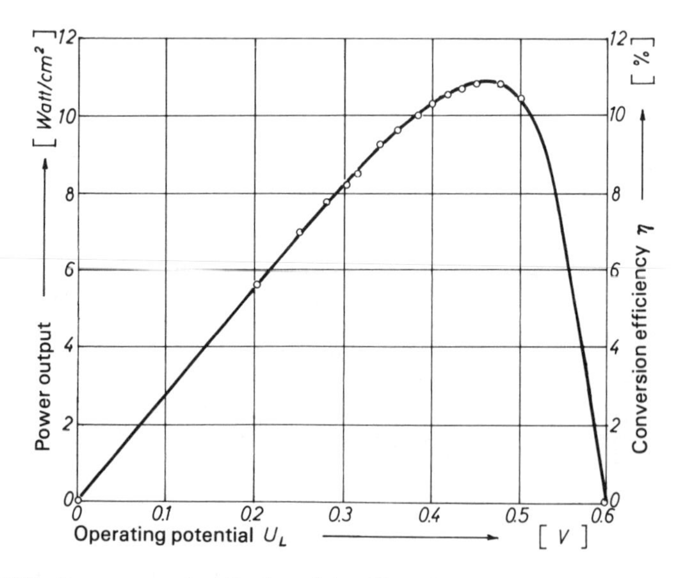

FIGURE 5.21. Power output of an Si solar cell (see Figure 5.17) as a function of the external load resistance R_a, which is dependent on the operating potential U (V). From Chapin *et al.*[49]

(a few kg/kW) is much less than that of any competitive producer of electric power. For commercial application, however, investment costs of more than about \$500/kW_{el} cannot be amortized. Hence, recent development in industrial countries has been concentrated on the idea of lowering these costs, e.g., avoiding the single-crystal manufacture of Si. Technical details are given in Chapter 6.

An alternative type of development is that of the manufacture of heterojunctions that use cadmium sulfide or cadmium telluride as the n-semiconductor. In the Federal Republic of Germany, such research work is being carried out principally at the solar energy facility at the Technical University in Stuttgart under the direction of Bloss and Hewig[50] with the support of the Bundesministeriam für Forschang und Technologie, and a report of the work of Bloss and Pfisterer[51] is now available. Here, the pioneer work of the Braunschweig school[52] on Cu_2S–CdS and Cu_2S–CdTe cells and also that on Cr-*p*-Si Schottky cells[53] must be mentioned.

REFERENCES

1. E. Justi: *Jahrb. Akad. Wiss Lit., Mainz,* p. 200, F. Steiner Verlag, Wiesbaden (1955).
2. E. Justi: *Sitz. Ber. Braunschweig. Wiss. Ges, Sonderheft 1, Beitr. Energiewirtschaft,* pp. 63–93, Verlag G. Goltze, Göttingen (1976).
3. Rep. No. 17, Austral. Acad. of Sci. on Solar Energy Res. in Australia (Sept. 1973), p. 22.
4. E. Justi, in: *Heizen mit Sonne [Space-Heating by Means of Solar Energy]* 1, Proc. Wiss. Tagung DGS Göttingen (Febr 1976), pp. 3–34, Verlag DGS, Munich (1976).
5. NASA Rep. No. M-TU-79-3 (1974).
6. J. P. Ratschow, in: *Heizen mit Sonne [Space-Heating by Means of Solar Energy]* 1, Proc. Wiss. Tagung DGS Göttingen, (Febr. 1976), p. 73, Verlag DGS, Munich (1976).
7. P. Valko, in: *Grundlagender Solartechnik [Fundamentals of Solar Techniques] I, Proc. Wiss. Tagung DGS Stuttgart* (Oct. 1976), p. 1; P. Kesselring, *ibid.*, p. 149, Verlag DGS, Munich (1976).
8. E. Justi, in: *Akademie Wiss. Lit. Mainz 1949–1974,* pp. 41–53, F. Steiner Verlag, Wiesbaden (1974).
9. K. Dehne: *Solare Strahlungsmessungen im Rahmen der Weltorganisation für Meteorologie (WMO) [Solar Radiation Measurements in the World Organization for Meteorology]*, 1. Deutsch. Sonnenforum Hamburg (1977), p. 15, Verlag DGS, Munich (1977).
10. N. Robinson: *Proc. World Symp. Appl. Sol. Energ.*, Phoenix, Arizona (1955). p. 58.
11. *Forsch. Entwickl, Lab. Prof. Kleinwächter, Lörrach,* Firmen-Druckschrift (1979).
12. H. C. Hottel: *Proc. World Symp. Appl. Sol. Energ.*, Phoenix Arizona (1955). p. 103.
13. J. A. Duffie and W. A. Beckman: *Sonnenenergie: Thermische Prozesse* [Solar Energy Thermal Processes], Munich (1976).
14. Heizen mit Sonne III—Speicherung [Space Heating with Solar Energy—Storage], DGS + SSES-Tag. Freiburg i. B. (Nov. 1977) Proceedings, Verlag DGS (1977).
15. P. Brennecke, H. H. Ewe, and E. W. Justi, in: *Proceedings of the Second International Solar Energy Forum*, Hamburg, July 12–14, 1978, Vol. 1, pp. 397–408, Verlag DGS, Munich (1978).
16. R. Hilson and E. K. Rideal: *Proc. R. Soc.* **199A:**295 (1949).
17. J. O'M. Bockris: *Modern Aspects of Electrochemistry 1,* Chapt. 4, Plenum Press, New York (1954).

18. H. Kuhn: *McGraw-Hill Yearbook of Science and Technology,* New York (1977), p. 69.
19. E. Justi: *Nachr. Chem. Technol.* **23**(16)**:**351–355 (1975).
20. R. Bruno, W. Hermann, H. Hörster, R. Kersten, and F. Mahdjuri: Hocheffiziente Kollektoren, Die Nutzung der Sonnenenergie [High efficiency collectors: the use of solar energy], Status-seminar des BMFT, Stuttgart (Sept. 1975).
21. H. Tabor: *Solar Energy* **6**(3)**:**112 (1962).
22. F. Steinberg: *Aluminium* **51**(7)**:**456 (1975).
23. W. Scherber and G. Dietrich: Selective Coating for Aluminium and Steel Solar Absorbers, Druckschrift d. Dornier System G.m.b.H., Friedrichshafen a. B. (1978); also *1. Deutsch. Sonnenforum* [The First German Solar-Energy Forum] Hamburg (1977), Chapt. 2, 18, pp. 363–412.
24. F. Kohlrausch: *Praktische Physik* [Practical Physics], Vol. 3, *Tafeln,* Verlag B. G. Teubner, Stuttgart (1968).
25. *PKL-Produkte: Wärmeträger für Solar- und Wärmepumpenanlagen* [P.K.L. Products: Heat Exchangers for Solar and Heat Pump Plants], Firmenschrift der BASF AG, Ludwigshafen.
26. *BP Thermo-Frost P,* Firmenschrift der Deutschen BP AG, Hamburg (April 1982).
27. P. Brennecke and E. Justi: Korrosionsprobleme bei der Verwendung von Aluminium im Kollektorbau [Corrosion problems in the application of aluminum in the building of collectors], Tagungsbericht *Grundlagen der Solartechnik I,* pp. 75–95, Verlag DGS, Munich (1976).
28. P. Brennecke, H. Ewe, E. Justi, and H. J. Selbach: Werkstoff- und Korrosionsprobleme in Solarenergie systemen [Materials and corrosion problems in solar-energy systems], Tagungsbericht *3. Internationales Sonnenforum,* pp. 243–252, DGS-Sonnenenergie Verlags GmbH, Munich (1980).
29. G. C. Szego, M. D. Fraser, and J. F. Henry: *2. Int. Solar Energy Forum,* Hamburg (1978), p. 1, Verlag DGS, Munich (1978).
30. Solar Sea Power, Semi-Ann. Rep. Covering Period Nov. 1973–Jan. 1974; Prep. by the Carnegie-Mellon-Univ. NSF Grant No. 39114.
31. P. Kramer: *1. Deutsch. Sonnenforum* [*The First German Solar-Energy Forum*], Hamburg (1977), **2,** pp. 581–590, Verlag DGS, Munich (1977).
32. A. Mitsui, S. Miyachi, A. San Pietro, and S. Tamura: *Biological Solar Energy Conversion,* New York (1977).
33. J. R. Benemann: *Hydrogen Production from Water and Sunlight by Photosynthetic Process,* University of California Press, San Diego and La Jolla (1973).
34. E. Broda: Grosstechnische Nutzbarmachung der Sonnenenergie durch Photochemie [Large-scale application of solar energy via photochemistry], *Naturwiss. Rundschau* **28:**365 (1975).
35. H. Kuhn: *Proc. 2nd Int. Solar Energy Forum,* Hamburg 2 (1978), p. 397, Verlag DGS, Munich (1978).
36. A. Lavi and C. Zener: *IEEE Spectrum* **10** (October 1973).
37. G. Claude: *Mech. Eng.* **52:**1039 (1930).
38. F. Haber: *New Scientist* **13** (December 1977).
39. Ocean-thermal energy conversion program survey (October 1976), Division of Solar Energy, ERDA, Washington, D.C., 20545.
40. G. H. Larvi: Alternate Energy Sources Conference, Miami, Florida, December 5–7 (1977).
41. A. Talib, A. Knopka, N. P. Biederman, C. Vlazek, and B. Yudow: Alternate forms of energy transmission from OTEC plants, Int. Solar Energy Conf., Vigyan Bhawan, New Delhi, India, Jan. 16–21 (1978).
42. H. Zettler: Beilage "Natur und Wissenschaft," *Frankfurter Zeitung,* Dec. 21 (1978), p. 25.
43. D. Perrigo and G. E. Jensen: *Proc. 14. Int. Conf. Ocean Thermal Energy Conv.,* New Orleans, Mar. 22–24 (1977), p. VII–3.
44. J. H. Nath, J. W. Ambler, and R. M. Hansen: *Ibid.,* p. V–56.
45. J. O'M. Bockris: *Energy: The Solar–Hydrogen Alternative*, The Architectural Press, London (1976).

46. H. Fischer, *Laser Optoelectronics Conf. Proc.* (1975), p. 1, *Festkorperprobleme* **18:**19 (1978).
47. M. Prince and M. Wolf: *J. Br. IRE* **18:**583 (1958).
48. P. Rappaport: *RCA Rev.* **18:**373 (1959).
49. D. M. Chapin, C. S. Fuller, and G. L. Paerson: *J. Appl. Phys.* **25:**676 (1954).
50. W. Bloss and G. Hewig: *Bild Wiss.,* No. 9, p. 110 (1976).
51. F. Pfisterer and W. Bloss: *Proc. 2. Int. Sonnenforum,* [International Solar Energy Meeting], Hamburg (1978), Vol. 1, p. 369, Verlag DGS, Munich (1978).
52. G. Schneider, Physik. Grundlagen der Solarzellen, 7. Lehrgang fur Raumfahrttechnik (Oktober 1968).
53. H. D. Hecht, E. Justi, and G. Schneider: Schottky-type Si-Cr solar cells, *Rev. Int. Héliotechnique.* (1974), 2nd semester p. 38.

CHAPTER 6

Solar Cells and Solar Power Stations

6.1. TECHNOLOGICAL PROBLEMS ASSOCIATED WITH LOWERING THE COSTS OF TERRESTRIAL SILICON SOLAR CELLS

From the earlier material concerning photovoltaic cells as direct energy converters of light to electricity (Chapter 4), the physical processes that take place in them, and the electrical engineering associated with their use (Chapter 5), it can be concluded that we are dealing with simple light-to-electricity converters that have no mechanically movable parts and do not require the use of chemicals. These cells work with an efficiency as high as 20%—a relatively high degree of efficiency—and it can be seen at once that the high concentration of silicon in the earth's crust could permit the mass production of silicon, the preferred semiconductor. The single difficulty in this picture is the (present) investment cost which is around $5000/kW_{el}$, which is still somewhat too high as far as the terrestrial production of electricity from solar energy is concerned. The price of silicon solar cells currently in use would have to be reduced by a factor of 3–5 to allow a rapid and complete breakthrough in solar electricity. It is easy to see the direction in which development must proceed. Much of the present cost goes into the preparation of superpure perfect single crystals. In this direction, it seems that efforts in the direction of lowering cost have become asymptotic, so that future efforts to achieve a reduction in cost should be concentrated on the production of less perfect silicon crystals. It is probable that such material would have a somewhat lower efficiency of conversion, but this is acceptable so long as the reduction in cost of the starting material is greater than the reduction in efficiency.

This was the central theme of the meeting of photovoltaic specialists in New York in 1978 and that of the European community in March 1979, in Berlin.[1] At these meetings, there were more than 120 lecturers. The great number of participants, however, reduces to some extent reduces the clarity of the picture of the trends that are developing in this work. For this reason, we shall concentrate on the work done in the Federal Republic of Germany by the firms of Wacker-Chemitronic and AEG Telefunken. A very attractive possibility is the use of polycrystalline instead of monocrystalline silicon. In one of the research projects of this work, the object was to prepare silicon with a really high degree of purification, so-called solar-grade silicon ("SoG-Si"), in place of the so-called semiconductor-grade material (ScG-Si) which has been used up to this time. The question whether it is better to produce a low-grade polycrystalline silicon for solar cells in place of the monocrystalline silicon depends on the possibility of making cells that have a conversion efficiency of greater than 10%. Apart from the efficiency factor, the cell made of the low-grade material must be suitable for large-scale production with a small degree of waste in the form of rejects, and the production method must be compatible with mass production. It was concluded at a 1974 conference in the United States devoted to the question of the conversion of solar energy for terrestrial purposes[2] that a 10% efficient polycrystalline solution would not be available before the middle 1980s.* Some contributions to this question have been made by a program funded by the Ministry for Research and Technology[3] which merges experimental work carried out in the Telefunken Research Center in Heilbronn under the direction of Fischer[4] and in the laboratories of Wacker-Chemitronic in Burghausen under Authier.[5]

6.2. FUNDAMENTAL CONSIDERATIONS OF POLYCRYSTALLINE SILICON SOLAR CELLS

The terms "polycrystalline" and "polysilicon" are applied to a range of crystallographic structures that arise as intermediate products in the processes used in the production of SoG silicon. In contrast to the structure of monocrystalline silicon, polycrystalline silicon is a conglomerate of individual crystal grains that are separated by varying grain boundaries. These crystal grains form

* Translator's note: The prediction has been borne out: Amorphous Si is now (1987) available for use in solar energy conversion with efficiency in practical large arrays having an efficiency of 5% and efficiencies in small samples of up to 12%. Instability due to loss of H has been overcome. Cost is headed downward, with DOE projections to 2000 of $0.50 per peak watt.

an inner surface of the working substance, and the large density of surface states on the boundary leads to effective recombination centers for photogenerated (i.e., not thermally dissociated) charge carriers. Apart from this, the layer near the border surface causes a nonlinear resistance gradient for the flow of majority carriers. Together, this recombination decreases the photocurrent, increases the dark current, reduces the shunt resistance, and increases the series resistance of the cells.

Polycrystalline silicon solar cells can be used either as *p–n* junctions or as Schottky cells.

Problems arise where the grain boundaries intersect with the space charge region. This gives rise to a diminution in the i–U characteristic of the cell.[6] Figure 6.1 shows various structures of polycrystalline cells. When the grains are statistically oriented, it is only the upper grains that contribute to the performance of the cell (cf. the cross-hatched grains in Figure 6.1), the grains below them being isolated with respect to any transition. If the material is in the form of columns or fibers—i.e., if the grain boundaries are preferentially perpendicular to the surface—the majority of carriers in the columnar grains intersect with the transition region, and the thickness of the layer can then contribute to the performance. In this case, the entire solar cell can be regarded as a parallel combination of small individual columnar cells. These small cells act in the normal way, except that the minority carriers can recombine on the side of the fiber as well as on both the front and the rear wall contact. In contrast to the usual crystalline silicon, which is oriented in all directions, silicon that is fibrous or columnar can give rise (sometimes in the form of a film) to quite effective arrangements. This type of material is called "non-monocrystalline silicon"[7] and it shows a columnar structure of the individual grains.

A complete three-dimensional analysis of this type of material is not yet

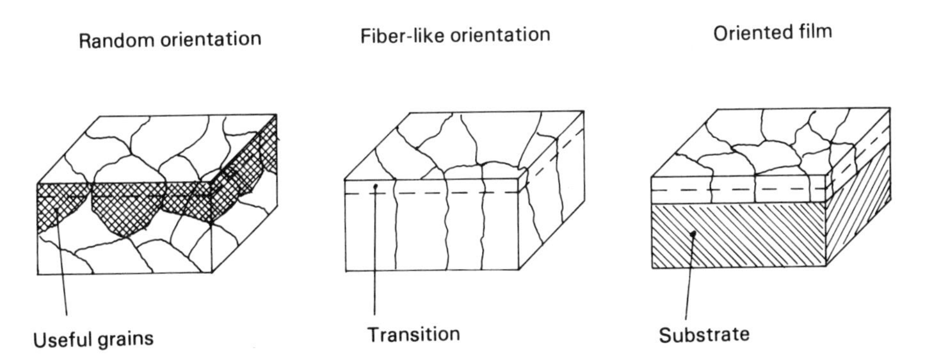

FIGURE 6.1. Different structures of "polycrystalline" solar cells. In randomly oriented silicon, only the upper grains are useful.

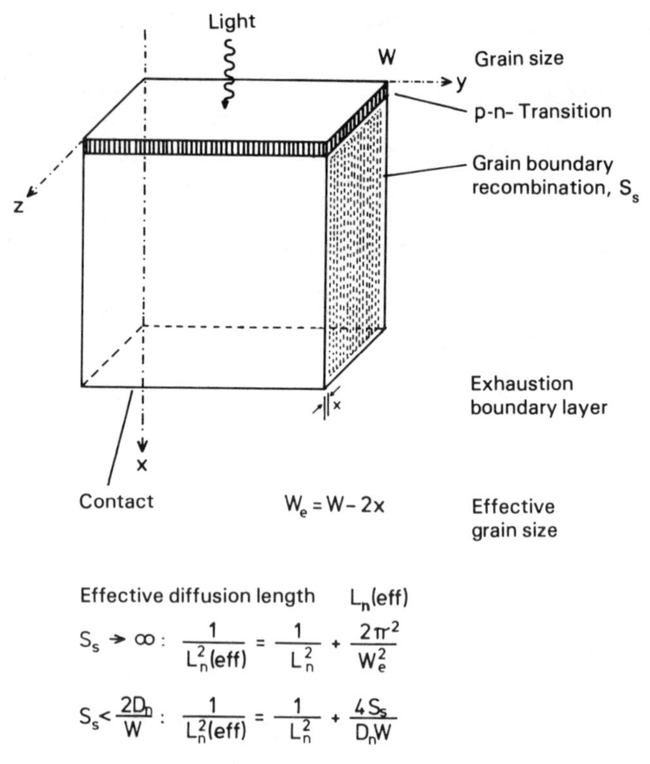

FIGURE 6.2. Simple model of an isolated Si grain. The "effective" diffusion length $L_n(\mathrm{eff})$ is determined through grain size and grain boundary recombinations.From Fischer.[4]

available, but various authors[8] have arrived at a partial solution to the production of minority carriers through the whole volume of a columnar grain. They use a lifetime that consists partly of the internal lifetimc and partly of the grain boundary recombination durations.[9] Figure 6.2 shows a simple model of an isolated grain. An effective diffusion length can be defined in a quasi-one-dimensional approximation, and this parameter depends primarily on grain size. Figure 6.3 shows the "effective diffusion length" as a function of grain size for various diffusion lengths and doping. There are two different cases of high and medium grain boundary recombination velocities. The thickness of the silicon must be taken into account, and the dominating influence of this parameter can then be easily realized. An effective collection of photogenerated charge carriers arises when the volume diffusion length reaches the thickness of the cell.[10] When the grain size is greater than the volume diffusion length and the thickness of the silicon, the photogenerated carriers have a high collection probability and the

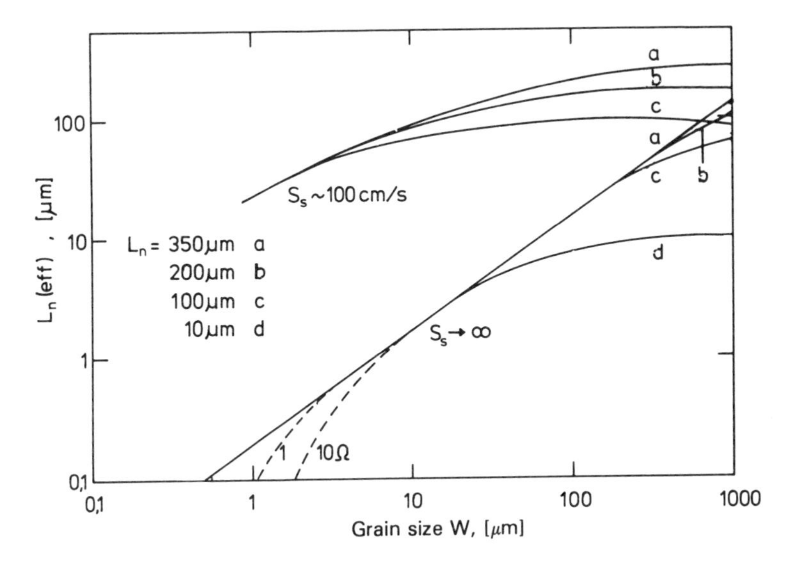

FIGURE 6.3. Effective diffusion length as a function of grain size for strong and medium grain boundary recombination. Parameter: Internal diffusion length and doping. From Shockley.[9]

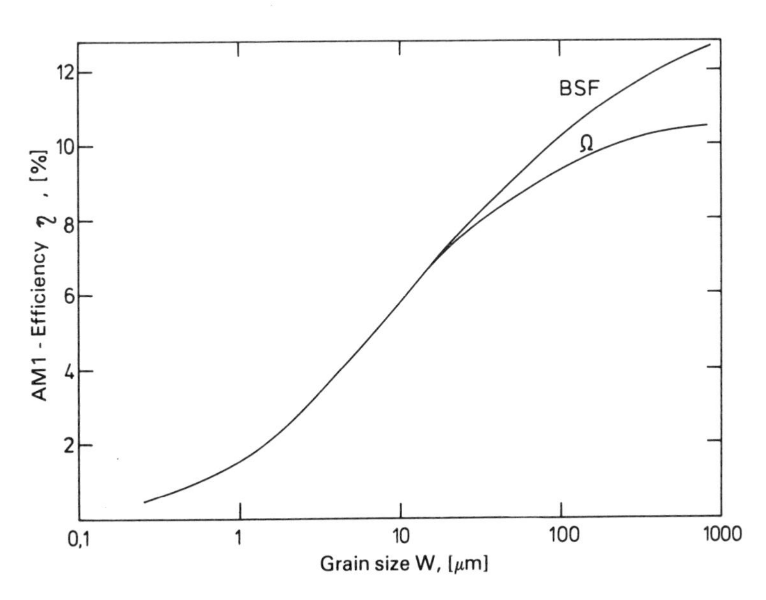

FIGURE 6.4. Electricity conversion efficiency η at AM1 in % as a function of grain size of an "nonmonocrystalline" cell. Parameter: n^+-p-type, thickness 350 μm, p_P = ohm × cm; BSF = back surface field; "Ω" = ohm contact. From Fischer and Pschunder.[10]

arrangement behaves as a single crystal. It is easy to see that the strong dependence on the size of the grains can be eliminated if grain boundary recombination can be reduced.

The dependence of the degree of conversion efficiency of a non-monocrystalline silicon cell on grain size is shown in Figure 6.4. The case of an ohmic contact can be described in this way: One can expect conversion efficiencies of more than 10% for the columnar oriented silicon cells having crystallite diameters of more than 100 μm. For this analysis, it is assumed that the influence of the grain boundaries does not extend to the exhaustion layer of the interface, which could cause a local scarcity of charge carriers or increased recombination current. These problems of the nonmonocrystalline material can be overcome by dimunition of recombination along the grain boundaries, using preferred doping of regions around them.[11] In Figure 6.5A and B, this effect is shown for normal and preferentially doped grain boundaries. In the latter case, photogenerated carriers that diffuse to the grain boundary no longer get lost due to recombination, but are collected as a photocurrent. This scheme allows a good increase of the efficiency, independent of the grain boundaries, and results in quasi-monocrystalline structures.

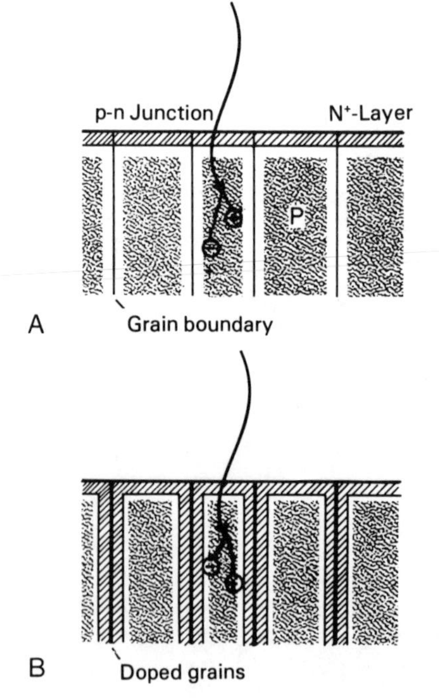

FIGURE 6.5. Schematic cross section through a rectangular crystalline cell. (A) Recombining grain boundary $S_s \rightarrow \infty$. (B) Collecting grain boundary $S_s \rightarrow 0$. From Soclov and Iles[8] and di Stefano and Cuomo.[11]

6.3. DESCRIPTION OF SILICON MANUFACTURE

Authier[5] has described the conventional and newer methods of preparing silicon samples for disks or solar cells in a block diagram (Figure 6.6). In the three-step process, the first step is common to the two processes, namely, production of highly purified silicon starting material, obtained by chlorination. The two processes take separate paths after this first common step. In the earlier methods, crystallization was carried out (see left side of Figure 6.6) by the well-known process of extracting the crystal from a crucible containing the liquid material according to the Czochralski method or by means of zone smelting. Now, however (see right side of Figure 6.6), the highly purified silicon is poured into the crucible, where it forms so-called "polyblocks," i.e., casted blocks of polycrystalline silicon with columnar structure. In the third step, the highly purified monocrystalline elements ("slabs") are cut with diamond saws. This can also be done with the polyblocks (see right side of Figure 6.6), in which case polyslices are obtained.

Another manufacturing path is the production of polyfoils by pouring, without the expensive step of sawing. The amount of energy given at the lower left in the figure corresponds to that for the production of 1 kg of monocrystalline silicon slabs and amounts to 2290 $kW_{el}h$. To recoup this amount of energy, the cells must be used for several decades, or virtually their entire useful lifetime. This underlines a problem of the classic method on the left side of the table, but the problem is solved by the process on the right. Thus, CdS–$Cu_{2-x}S$ cells need—despite their somewhat lesser efficiency (5%)—only one year to return the energy used in their manufacture, as will be described more fully in the following section.

A new method introduced by Wacker Chemitronic begins at the crystallization phase and uses a suitable temperature gradient in the melt after the pouring; a silicon slab obtained in this way shows an uninterrupted arrangement of monocrystalline sections of various orientations in the surface. The pouring process yields a large quantity of crystals per unit time and results in diminished costs for investment in raw materials and auxiliary substances needed, as well as for personnel. Compared with the conventional withdrawing of the crystal from the crucible by the Czochralski method, one obtains a crystallization rate at least 25 times faster than the classical one. The fact that the columnar structures of cells so obtained seem to give a high efficiency has encouraged the investigation of thicker foils by immediate pouring without using another material as substrate. There are considerable technological difficulties in doing this, and the original papers of Authier[5] should be consulted for full details. The following is only a brief summary of them:

1. High chemical reactivity of silicon, which wets the material of the die.
2. Anomalous volume changes on solidification, i.e., volume *increase* on solidification. Stresses and strains induced by the presence of impurities.

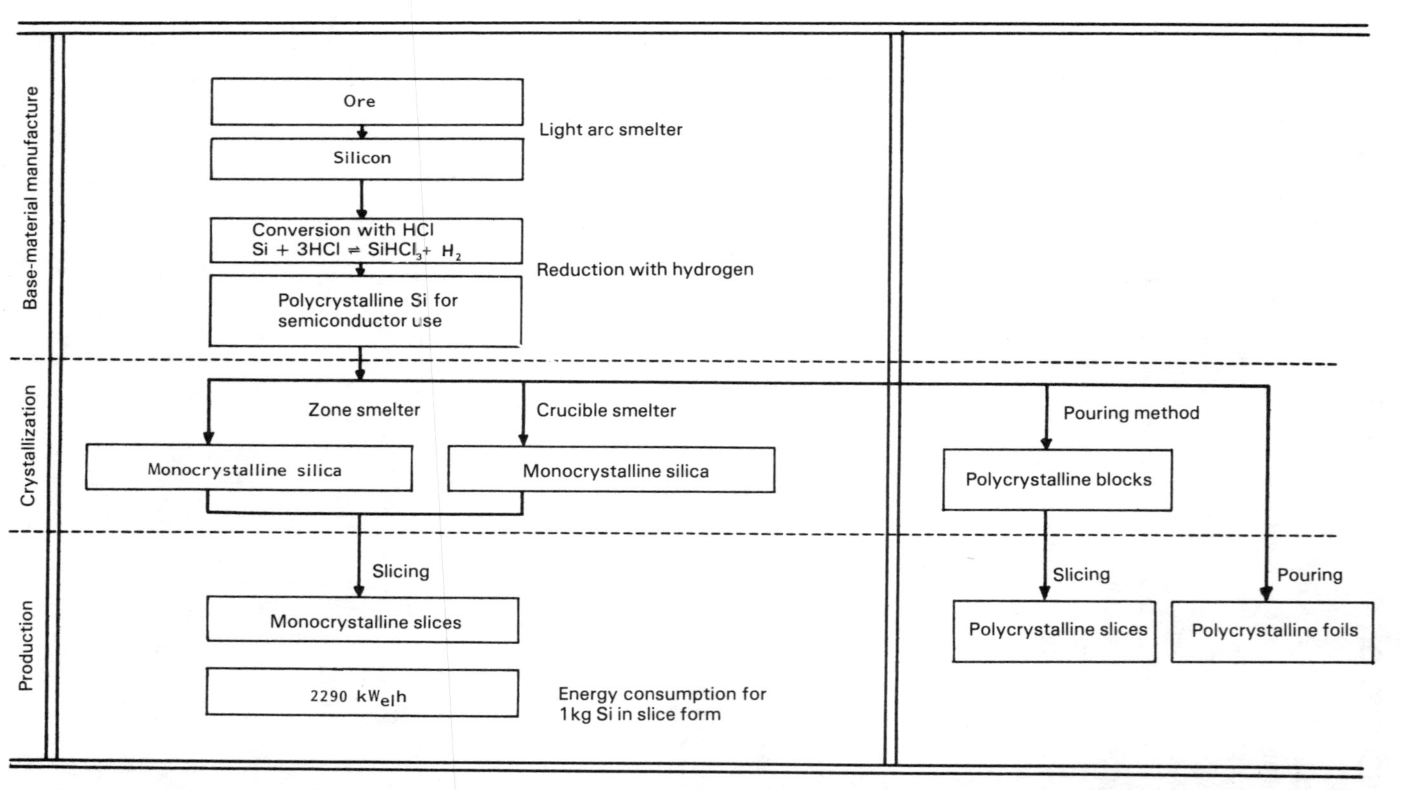

FIGURE 6.6. Diagram of the three-step production of standard high-purity monocrystalline Si (left) and columnar nonmonocrystalline Si (right). At bottom left is the total electricity consumption for the production of 1 kg monocrystalline disks; the enormous value of 2290 $kW_{el}h$ will be reduced using new production techniques that reduce the requirements in respect to temperature, purity, and monocrystallinity. From Authier.[5]

3. High melting heat (0.4 kcal/g, 5 times as high as for ice) and consequent extended melting times of Si-feed product; problems with the removal of the melting heat.
4. Solidification in a pronounced temperature gradient; thermally induced mechanical stresses.

It has been possible to produce cells with a size of about 11 cm^2, and Figure 6.7 gives an answer to the question about the distribution of the efficiency of conversion over the surface with the various crystallites.[12] The efficiency for air mass 0 (AM0) varies between about 8.5 and 10.3%, corresponding to a mean value for AM1 of 10%, which is equal to the intended goal of 10% efficiency. Of course, it is interesting to consider the introduction of these rather scattered experimental values into the usual current potential diagram. If one does this and also introduces the further parameter of the relevant grain size, one obtains

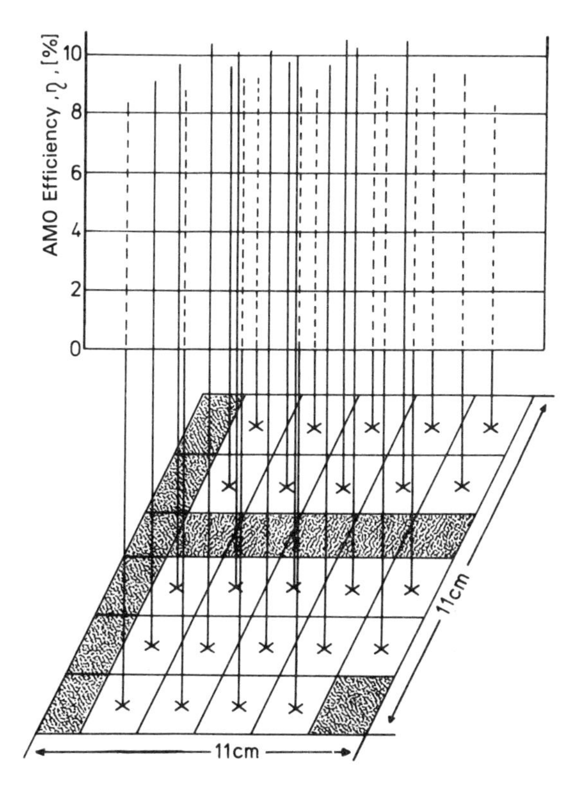

FIGURE 6.7. Distribution of the AM0 efficiency over parts of an 11 × 11 cm cell from nonmonocrystalline Si cells. The variation of the AM0 efficiencies between 8.5 and 10.3% is equal to an AM1 mean of 10% and corresponds to the minimal goal. From Fischer and Pschunder.[12]

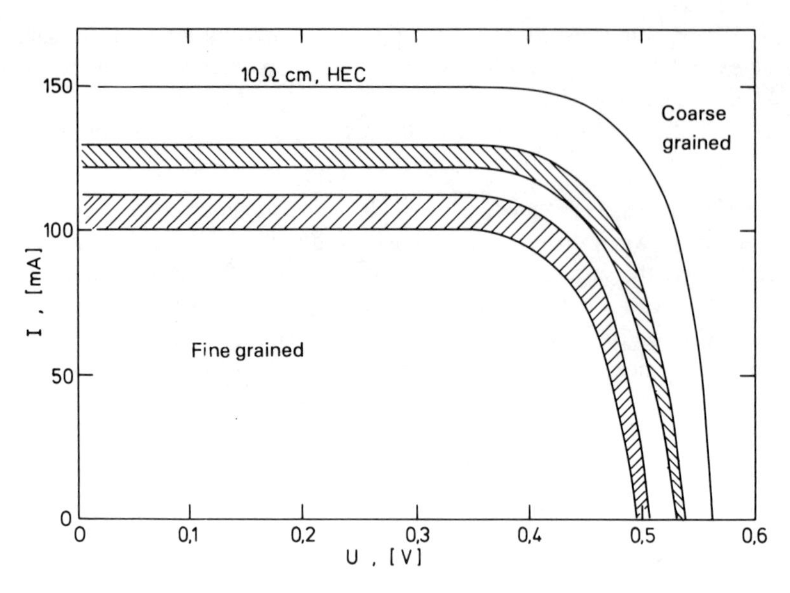

FIGURE 6.8. Experimental representation of the variation (as shown in Figure 6.7) of the conversion efficiency of a large-scale solar cell from nonmonocrystalline Si in a conventional current-charge diagram, in which the necessary points are ordered with growing grain size from bottom left to top right. At bottom is a band for the fine-grained and in the middle one for the coarse-grained areas. At the top, for comparison, is the i–U characteristic of a conventional monocrystalline high-efficiency cell (HEC) with 12% AM0 conversion efficiency. From Fischer and Pschunder.[12]

Figure 6.8, which gives a significant reduction of the scattering of the experimental points. At the top of the figure are the characteristics of the usual monocrystalline high-efficiency cell of the well-known optimized *p*-silicon Si of 10 ohm × cm specific resistance. Underneath is a narrow band of coarse-grained crystalline material and, beneath that, a broader series of results for very fine-grained silicon. The open circuit potential for the fine-grained material (5 ohm × cm) is 500 mV and for the coarse-grained, 540 mV. The respective values for the short circuit are 28.9 and 32.6 mA × cm^{-2}. The power density is 10.2 compared with 12.8 mW × cm^{-2}, from which one gets 7.5 and 9.4% for AM0 or 9 and 11.3% for AM1 conversion efficiencies. The fill factor is unexpectedly good at 70%. The high-efficiency cell shown at the top had 12% efficiency at AM0. The differences between the curves in Figure 6.8 can be explained quantitatively as being caused by the decrease of the effective charge collection by a reduction of the effective diffusion lengths arising from the smaller grain size.

The experts and scientists involved in the research and development programs funded by the Ministry of Science and Technology are quite reserved in their projections. However, they give the diagram in Figure 6.9 for the predicted

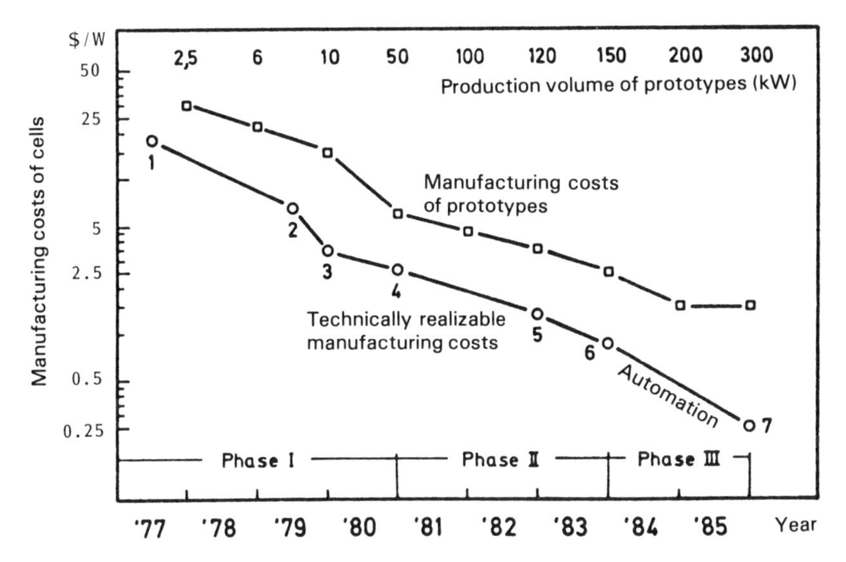

FIGURE 6.9. Milestones of the Ministry for Science and Technology research and development programs "Terrestrial Solar Generators" and expected volume of yearly "Pilot Production Quantity." Trial projection by Fischer.[4]

cell costs (shown on the ordinate at the left) and the production volume of prototypes in kilowatts as a function of Phase I (1977–1978), Phase II (until 1983), and Phase III (until 1986). The possible reduction of cost would arise from the difference between the upper curve for the prototype production costs and the lower curve for technically realizable production costs if automation were introduced. To some extent, this figure confirms the fears of experts in the field that a reduction of costs of silicon cells for terrestrial purposes is a scientifically very complex process and by no means just a simple consequence of a yearly rise in the numbers produced. Will final victory in the race for the best method to convert solar energy on earth to electricity go to nonmonocrystalline silicon layers? Or will it go to the corresponding polycrystalline evaporated thin-layer cells of the type $Cu_{2-x}S$–CdS?

6.4. LONG-LIVED HETEROJUNCTIONS FROM THIN-LAYER SOLAR CELLS CONSISTING OF CdS–$Cu_{2-x}S$

A look back at the schematic presentation of the three fundamentally possible configurations of barrier-layer photocells reminds us that among the semiconductor photocells, one of the alternatives to the so-called homojunctions of silicon is CdS–$Cu_{2-x}S$. This, of course, is a heterojunction. The fundamental expectation here is that one might get a much cheaper cell than the systems that at the moment yield the greatest efficiency, namely, silicon monocrystals. The reason is that, with cadmium sulfide cells, as they are generally called, one gives up the idea of a perfect crystal structure of *n* and *p* layers and thereby eliminates the major production cost factors and in return gets a specific investment cost ($/kW) which is an order of magnitude lower than that for high-purity monocrystals. Until very recently, it has not been possible to take advantage of this possibility because polycrystalline cells have had much too short a lifetime. However, significant progress has been made in recent years in lengthening the lifetime of such cells, and this work has been summarized by Martinuzzi *et al.*[13] and Pfisterer and Bloss[14]—for example, in the international Hamburg solar energy forum of 1978.

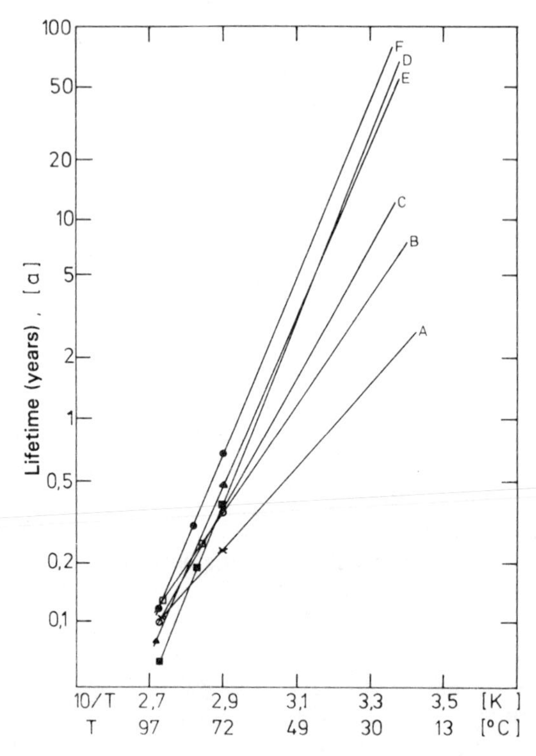

FIGURE 6.10. Influence of the surrounding gases (air, N_2, Ar) on the service life of CdS–$Cu_{2-x}S$ solar cells in the temperature range of 70–95°C, with continuous light and interrupted light radiation as parameters. (A) Air, continuous light; (B) air, interrupted light; (C) argon, continuous light; (D) argon, interrupted light; (E) nitrogen, interrupted light; (F) nitrogen, continuous light. From Partain and Sayed.[15]

6.5. SERVICE LIFE OF CADMIUM SULFIDE SOLAR CELLS

Progress in this area depends on diminishing the effect of the surrounding atmosphere, in particular O_2 and moist air, which interact with CdS while it is being irradiated.[15] To avoid these effects, the semiconductor is encapsulated between two glass plates and hermetically sealed, preferably in a nitrogen or argon atmosphere. Copper sheets serve as the agents for the passage of current. These measures have extended the lifetime of such cells to more than 10 years. Measurements of the degradation and plotting of the measurements against the log of the reciprocal of the absolute temperature (1/K) according to Arrhenius's law for reaction velocities in solid bodies[16] give rise to a linear extrapolation that tells us that the lifetime at 50°C will be about 20 years, and if the temperature is kept down to 25°C, about 100 years, although with the unproven assumption that exclusion of oxygen and moisture can be maintained for these long times. Figure 6.10 is such a plot by Partain and Sayed.[15] It is clear from this diagram that there are also differences in operation depending on whether the light is continuous or interrupted. One cause of the degradation rests on the migration and diffusion of ions, and this can be decreased near the *p–n* junction by doping. Table 6.1 shows the lifetimes deduced from the measurements that are mentioned above and illustrated in Figure 6.10.

6.6. PROCEDURES FOR THE PRODUCTION OF THIN-LAYER SOLAR CELLS FROM CdS–$Cu_{2-x}S$

Shirland,[17] working for the Clevite Corporation in Cleveland, Ohio, in the 1960s, achieved some very far-reaching advances on the way from discovery to commercial manufacture of practical barrier-layer solar cells, using

TABLE 6.1
Composite of the Extrapolated Results from Figure 6.10 Concerning Service Life of CdS–$Cu_{2-x}S$ Thin-Layer Cells[a]

Test conditions	At 25°	At 75°C
1. Air, continuous light	2.1 years	0.67 year
2. Air, interrupted light	5.4 years	1.2 years
3. Argon, continuous light	8.0 years	1.5 years
4. Argon, interrupted light	38.0 years	3.2 years
5. Nitrogen, continuous light	32.0 years	3.2 years
6. Nitrogen, interrupted light	60.0 years	5.1 years

[a]From Partain and Sayed.[15]

CdS–$Cu_{2-x}S$. Apart from the various geometric arrangements, the manufacturing processes differ mainly in the methods of production of photoelectrically active semiconductors CdS and Cu_2S, and here there are already many different possible combinations. CdS can be produced by sputtering or condensed from the gas phase, or it can be made out of CdS suspensions by silkscreen deposition and subsequent tempering. An alternative method is sintering CdS powder. Cu_2S can also be precipitated from the vapor or sputtered from an argon atmosphere; analogous to CdS, it can be produced by spraying or by means of pyrolytic reactions or deposited by combination with reactions such as the sulfiding of copper layers or reactive sputtering of copper in H_2S or in an argon–H_2S atmosphere.

Among the numerous procedures are a few that stand out in respect to cheapness, minimal weight, and lifetime. For example, plastic foils can be used as substrate or as a cover, because they weigh only 1 g/W and can be rolled up before the launching of a satellite equipped with them and then be unrolled into large areas when the vehicle is in orbit. All types of organic foils allow diffusion of water vapor and oxygen, which under the influence of solar irradiation give rise to a more or less rapid degradation of the crystal. However, if, for use on earth—where the cells are subject to atmospheric effects—one encapsulates the sensitive parts of this arrangement between two thin glass plates, one can achieve a desirable increase in lifetime to more than 10 years as well as a saving in weight by various constructions.

In the Federal Republic of Germany, the technological development of terrestrial CdS–$Cu_{2-x}S$ cells is being carried out with substantial and continuing research support by the Ministry of Science and Technology, and this money is going principally to the Institute for Physical Electronics of the Technical University in Stuttgart, the workers there being Pfisterer *et al.*[18] These workers use as the basis of their work the Clevite process.[19] They reported on their work at the Second International Solar Energy Conference in Hamburg in 1978 and compared the efficiency, costs, production yield, reproducibility, and lifetime of their cells with those produced in other laboratories. The substrate glass slides, $7 \times 7\ cm^2$ in area, are treated by well-known chemical processes of purification and then plated with chromium as an intermediate metal and finally with silver before making the contact to the cadmium sulfide layer by vapor-phase deposition. The layers are about 1 μm thick, and it is most important that they be protected from any contact with dust because this would affect the crystal growth of the following CdS condensation process. The central process in the manufacture is the CdS evaporation in high vacuum, and a great reduction in costs is obtained by making this process entirely automatic. The first furnaces used in this work had heating elements of tantalum, molybdenum, and tungsten. These elements have been replaced by elements of graphite, because graphite is not reactive with sulfur and therefore has a longer operational lifetime. In the pilot line in Stuttgart, three evaporation units were built; over these

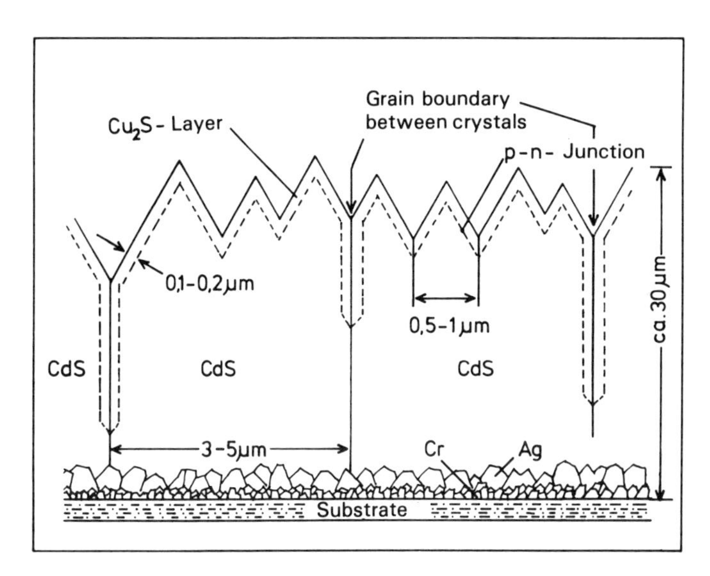

FIGURE 6.11. Schematic cross section through a $Cu_{2-x}S$–CdS thin-layer solar cell from a scanning electron-microscopic photograph. The cut shows grain boundary layers connecting from top to bottom; internally, the CdS crystals are ordered with respect to their c-axes, but not the other two coordinates. They are covered on top with prisms. On top of this, one sees the p-$Cu_{2-x}S$ layer, which is positive. The p–n junction with the indicated thickness is permeable enough for the sunlight, which penetrates from the top. From Pfisterer and Bloss.[14]

there was a turntable substrate plate carrying 12 sheets of glass, each $7 \times 7\ cm^2$. The arrangement was such that a mask covered the unused glass sheets, each sheet being exposed in turn to the operation. It was concluded from comprehensive optimization experiments that the following were the most successful parameters: The evaporation temperature had to be 1050°C, the substrate temperature 200°C, and the evaporation time per layer around 20 min, the distance between the evaporate and the substrate being 20 cm. In this process, the adhesive coefficient for CdS is rather high, but is quite low for cadmium, which is important in view of the unavoidably low yields. The CdS layers are n–type and contain about 10^{17}–10^{18} sulfur vacancies per square centimeter, which gives rise to a specific resistance about the same as that of the corresponding silicon layers.

The CdS layers are not really amorphous, but have hexagonal columns with pyramidal sections at the top with the c-axis perpendicular to the substrate (Figure 6.11). To avoid surface impurities and other additives to the surface as well as to reduce refraction to a minimum, the samples are etched at 60°C with 1 N hydrochloric acid and carefully rinsed. The Stuttgart workers produced the p-$Cu_{2-x}S$ layers by means of the very cheap but inherently less effective wet process, in which the CdS layer is briefly introduced into a CuCl solution acidified with HCl, whereupon the following substitution reaction occurs:

$$\text{CdS (crystalline)} + 2\text{Cu}^+ \rightarrow \text{Cu}_2\text{S (crystalline)} + \text{Cd}^{2+} + Q$$

Here, the optimal parameters have also been determined: The CuCl concentration should be about 10 g/liter; the bath temperature, about 90°C; the pH value of the solution in the bath, 4–5; and the reaction time 10 sec. The $Cu_{2-x}S$ layers that are obtained in this process are *p*-conducting and about 0.2–0.3 μm thick. Thus, they are sufficiently transparent, and in this situation, the density of the copper vacancies (the acceptors) is about 10^{19}–10^{20} cm^{-3}. This surface layer can be optimized further by a treatment devised by Bogus,[20] which consists of evaporating a further 50- to 100-Å-thick copper layer in high vacuum and then tempering this at 180°C in air until the optimum situation is reached. Time is not so critical here, but the situation is quite complicated because the system CuS has several stable Cu_xS phases, and only chalcocite gives a high performance. Another method according to Bogus, about equal in value to the first, is a follow-up treatment that heats the cells to incandescence in a hydrogen atmosphere. These procedures can give rise to considerable improvements; they increase the product yield and must be taken into account in contacts and in the encapsulation technology.

According to the process developed in Stuttgart, the latticelike front contact of the cells is produced on the 7×7 cm^2 glass slides, which in the subsequent encapsulation process are glued on as cover glasses. The cover glass is layered with hot sealing glue in such a way as to match the thickness of the copper lattice during the encapsulation process. The 35-μm-thick copper foil is glued using two heated laminar rollers, and the contact layer structure is introduced onto the copper foil by a pressure process involving etching. The etching is carried out in a spray etch machine using an alkaline etch for about 3 min, during which the etching perpendicular to the spray direction reaches about 5%. After dissolution of the etchant, the copper lattice is covered with a gold layer about 0.1 μm thick by an electrodeposition process and then the cells are glued together in a press at 2.5 atm and 160°C in vacuum.

It is extremely important, of course, to discuss the influence of the various technologies that have been described above on the efficiency as a function of the structure of the CdS and Cu_2S layers. Excellent discussions of this subject have been presented by Pfisterer and Bloss.[14,18] In the polycrystalline formations, there are grain boundaries in the proximity of which the all-important parameter of the diffusion length is greatly reduced. For this reason, the "topotactic" production procedures are to be preferred. In these procedures, the Cu_2S layers are produced in such a way that all the sulfur atoms are in place while all the cadmium atoms are replaced by two copper atoms in a way that corresponds to the equation above. From Figure 6.11, which is schematically drawn from a scanning electron-microscopic picture of a cross section, it can be seen that the *p–n* junction is pushed forward by the reaction from the surface and the grain boundaries into the center of the crystallite with the c-axis perpendicular to the substrate. This region is undisturbed, involving no reduction of the diffusion length for the charge carriers. This is quite different from the additive procedure described above, in which the *p–n* junction is coincident with the CdS surface. Here, the transition from the n to the p region is unavoidably disturbed, and the rate of recombination loss increases correspondingly. Another disadvantageous effect arising therefrom is that the grain boundaries of the CdS layers tend to form in the solar-irradiated sections of the *p–n* junctions, where the recombination losses are particularly high. Conversely, in the Bloss and Pfisterer wet process, the grain boundaries of the interrupted parts of the *p–n* junction lie particularly deep (2–5 μm) below the surface, i.e., in the dark, where they cannot reduce the quantum yield. In this way, one can understand the variation of the efficiency from about 4 to about 8%.

It is not easy to say what the commercial price would be unless one has some idea of the lowering of costs as a consequence of mass production. Pure material costs, including encapsulation, amount to about $10/m^2. However, the most important factor is the energy cost. The net cost estimates are related to the operational lifetime of the cell, and experiments on cells produced in the various ways must be used to obtain these figures, so that one can know how long it would take to get back the energy used in the production (30 $kW_{el}h$). Thus, in North Germany, the annual solar radiation is around 600 kWh per year, and if the efficiency of conversion is about 5%, one gets back about 30 $kW_{el}h$/year, so that the recovery time of the original energy costs is 30 kWh ÷ 30 kW/year = 1 year. This time is far shorter than for other converters, which take many years to return the energy used to make them. Continuation of developmental work on this type of photovoltaic cell would be very desirable.

6.7. SOLAR THERMAL GENERATING STATIONS WITH OPTICAL CONCENTRATORS

Solar generating stations that directly convert solar energy to electricity by photovoltaic cells would seem at first glance to be the right solution, but it would be necessary to lower the costs of investment by about 5 times, and although some further lowering is certainly likely, there is a possibility that even mass-produced cells would be too expensive. It is therefore reasonable to consider the more conventional approach of using *indirect* conversion with solar heat as the intermediate step. At first sight, this certainly appears to be the *less* desirable path because it introduces the Carnot factor in conversion of disordered to ordered energy, but when one compares the cost situation with the projected 25% efficiency for the very best photovoltaic cells, the advantages of indirect conversion seem to be worth reconsidering. Another aspect is that even in the use of photovoltaic cells, there are some aspects of the Carnot process, as has been shown previously under the discussion of quasi-Carnot processes. This arises from the fact that photovoltaic cells require cooling water, and the heat obtained therefrom can be regarded as part of the energy-conversion process of the solar cell. On the other hand, solar–thermal electricity generating stations would have an advantage over conventional ones in that they would not need any fuel or the corresponding expensive desulfurization processes, nor would they contribute to the problem of CO_2 in the atmosphere.

The pioneers of solar energy technology have always had two different ways of collecting solar energy and converting it into heat, and both have been well examined. The first way consists of a sufficient area of flat plate collectors in which water or a liquid of higher boiling point is used as a heat-transfer fluid, which is conducted to the heat exchanger. The second utilizes a single boiler with a hollow space absorber that is irradiated by a very large number of mirrors arranged so that they all reflect the sun's rays onto the boiler. The first alternative gives rise to "solar farms" or "distributor systems," and it has the advantage that flat collectors have already been manufactured, tried out in practical situations, and even mass produced, so that their price is now correspondingly lower. Further, there is no need for any additional equipment to rearrange (realign) the mirrors and, thanks to advances made in the last few years in selective coatings, it is possible to increase the evaporation temperature by about 50°C and thereby make the ideal Carnot efficiency about 10 points higher than it usually is (e.g., from 24.8 to 33.1%). On the other hand, heat collectors without any kind of optical concentration are probably best in small plants—e.g., solar plants which drive water pumps on farms—because in the large type of plants that would be suitable for urban use, the collector areas needed are so large and the amount of cooling water and steam conditioning so great that very considerable accumulated heat losses would reduce the overall efficiency of the plant. It seems at present much easier to cover the solar-irradiated area with a large number of flat mirrors about 1 m^2 in size; or, probably better, to cover it with slightly concave mirrors, the reflected rays of which (using computer control of the angle of the mirror according to the position of the sun) are always focused on a central furnace, which would be erected at the top of a tower several hundred meters high. Such systems are called "central receiver systems" or "power tower" generating stations. There is one difficulty with this method compared with flat-plate collectors in that for optical concentrater collectors, it is only *direct* sunlight and not *scattered* light that can be used. In Europe, this scattered (or cloudy) light is about half the sunlight available. Conversely, the theoretical value of the thermal efficiency factor climbs to about 38%.

Since the early flat-plate collectors were very inefficient and the steam engines that were available were low-pressure machines, the pioneers of solar energy generating stations used optical concentraters, parabolic or linear–parabolic mirrors. The power tower idea was first suggested by Barr in 1896, and in 1957, Baum *et al.*[21] built a model in southern Russia at a scale of 1 : 50 and studied it. The steady decline of oil prices, in constant dollars, made the approach seem uninteresting. Willsie and Boyle were pioneers of flat-plate collectors since 1902. They used a two-substance system whereby water was heated in the collector and in turn heated ammonia, ether, or sulfur dioxide, which provided the necessary higher pressure for the heat engine. A model of such a water–ammonia plant that was already commercially available in 1905 produced 15 kW and was used to irrigate the uncultivated land along the Colorado River; the water pumped was also used as cooling water for the machine. An interesting account of these early experiments is given in the monograph written by Meinel and Meinel.[22] The present status of solar generating stations in the United States was described by Köhne[23] in the first German solar energy forum in Hamburg in 1977. Other status reports were given at the International Solar Energy Forum in 1978 and the Seventeenth International COMPLES meeting in Tunis in 1977.[24] As a result of these studies, five European nations arrived at an accord whereby corresponding (high-cost) pilot plants, using both alternatives for the conversion of solar energy to electricity in the megawatt range were to be constructed and placed, for experimental purposes, in one of the southern countries of the group concerned.[25]

FIGURE 6.12. Photographic view of the helioelectric demonstration plant manufactured by Francia in Ilario near Genoa. One can see clearly the large number of parabolic heliostats that follow the sun and concentrate the rays on the absorber at the top of the tower.

6.8. ONE-MEGAWATT SOLAR POWER TOWER OF THE EUROPEAN ECONOMIC COMMUNITY

In 1975, the Commission of the European Economic Community (EEC) undertook as part of its research and development program the construction of a power tower to explore the possibilities in a demonstration setup, using a 1 MW prototype built by Francia Ilario near Genoa (Figure 6.12). The purpose of this work was to provide European industry with first-hand experience of the necessary hardware components at the system level for thermal solar power

TABLE 6.2
Technical Parameters of the One-Megawatt Solar Power Plant of the European Economic Community

General characteristics
- Experimental plant of the central receiver–multiheliostat type
  - Location: 37.5°N; 15.25°E (Sicily)
- Developmental starting point: Noon equinox and solarization of 1000 W/m^2
- Produced output: 1 MW_{el} in the existing network at the design-model location
  - Design parameter (amount of output at design-model location)
- Heliostat field: 4800 KW_{th} in the receiver, from a total mirror area of 7800 m^2

Heliostats
- Two types, steered by 2 axles, total precision ±4 mrad
- CETHEL type, about 52 m^2, cylindrical segments, 57% of the mirror fields
- MBB-type, about 23 m^2, 16 quadratic elements, 43% of the mirror fields

Tower receiver
- Cavity type with 4.5 × 4.5 m aperture and a height of 50 m, 110° inclination
  - Receiver outflow-steam condition: 510°C; 64 bar; 4700 kg/hr —5300 possible)

Steam process
- Turbine without heat exchanger, connected directly to receiver
- Output: 1200 kW_{mech} with steam at 450°C, 54 bar
- Supply water temperature at receiver inflow 36°C
- Cool water temperature maximum 25°C

Thermal storage
- Sufficient for about 30 min reduced electrical output
- Energy storage: Steam 300 kWh; Hitec[a] 78 kWh
- Equipment: Water pressure container (16 atm) for 7500 kg steam at 6 atm
- Two storage tanks, each holding 600 kg Hitec[a]
- Heat exchanger for 16 bar, 490°C and 240°C temperature

Electrical system
- Power production: Alternating current generator for minimum
  - 1100 kW with about 100 kW own consumption and 1000 kW for nonlocal consumption
- Transformers, emergency current unit
- Supply in the network: tuning into the public grid
- Steam process control system
- Order giving, operation, and controls combined in control center

[a] Hitec is an eutectic mixture of 53% KNO_3, 40% $NaNO_2$, and 7% $NaNO_3$.

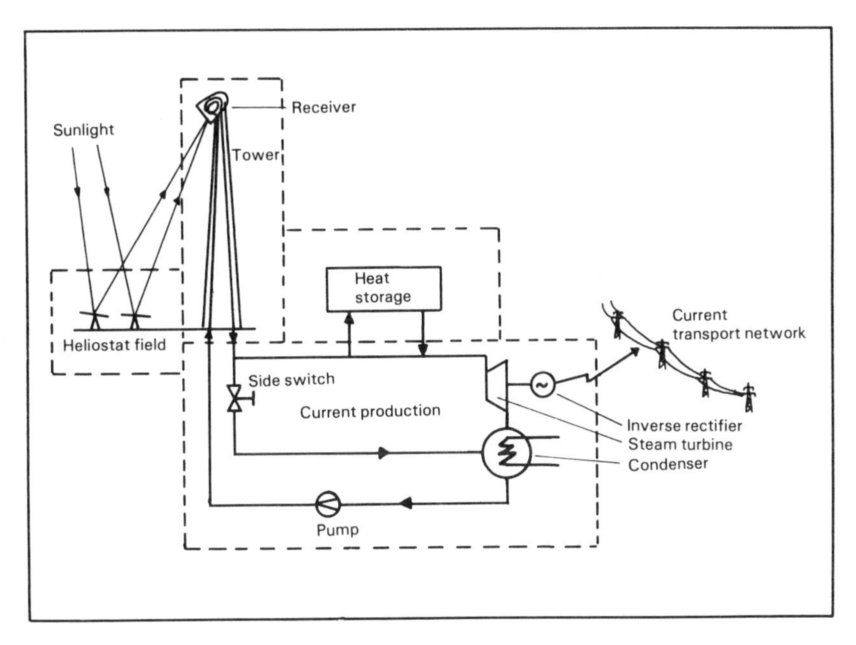

FIGURE 6.13. Block diagram of the EEC solar–thermal power plant for 1 MW_{el} output in Sicily. The quantitative data are in Table 6.2. From Hofmann and Gretz.[25]

station. Phase A, the systems study, was completed in 1976. It was followed by Phase B, a further systems study at a practical level, which set up a managerial system for the plant, planning, cost estimations as well as an engineering-level study with production and specifications details. Phase C comprises construction and installation, and Phase D will involve experimental work on the plant. The EEC Commission has budgeted a total cost for the project of $4.2 million. Of this money, half will be provided by the countries that have firms taking part—the Federal Republic of Germany, France, and Italy (Great Britain will take part only as an observer). Table 6.2 summarizes rather clearly the technical specifications (cf. also Figure 6.13).

Figure 6.13 is a block diagram of the entire plant, and Figure 6.14 is a schematic drawing of the heliostats. The surface area of each mirror is 23 m^2. Figure 6.15 is a sketch of the 50-m-high central tower that is being built by the Messerschmidt-Bölkow-Blohm Company. It can be seen that the receiver at the top of the tower is of the empty-cavity type. It will be recalled from basic physics that no black body is as black as a hole bored in it, because all the radiation reflected in the interior of the body is reabsorbed after multiple reflection. The prototype constructed by Ansoldo in Genoa at the suggestion of Francia is shown in Figure 6.16. There would probably be many difficulties with the materials of construction in such a design, because the receiver at the top of the tower works

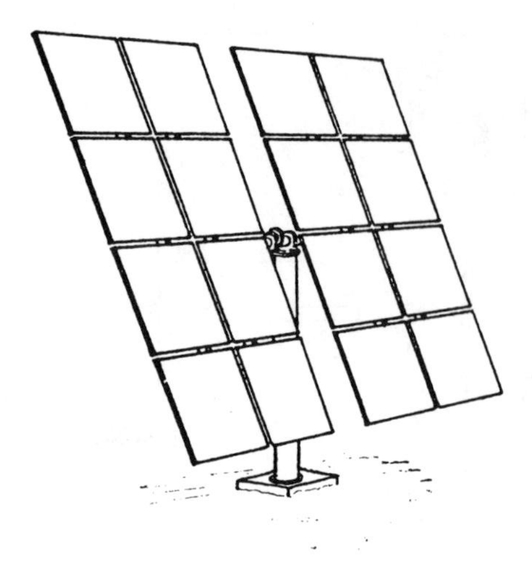

Technical data:

MBB:

Mirror area	23 m^2
Mirror elements	16 Elements, flat 3 mm thick glass 1.2 m × 1.2 m
Total height	5.5 m
Height of mirror construction	5.01 m
Width	5.67 m
Weight	≈ 1000 kg
Ground clearance	≈ 0.50 m
Above-ground clearance of axle	3.00 m

FIGURE 6.14. Schematic drawing of a Messerschmidt-Bölkow-Blohm heliostat system with a mirror area of 23 m^2.

at a maximum temperature of 600°C and would be subject to thermal and mechanical stresses, though it is supposed to have a considerable lifetime. Figure 6.17 is an artist's view of the heliostat field around the tower. Various studies relating to material properties of the components were described by Walton at the COMPLES Meeting in Tunis in 1977.[26]

6.9. 400 KW_{th} HIGH-TEMPERATURE SOLAR EXPERIMENTAL PLANT AT GEORGIA INSTITUTE OF TECHNOLOGY

Although the thermal performance of the Atlanta arrangement is only 15% of that of the EEC project described above, it is no less interesting because it

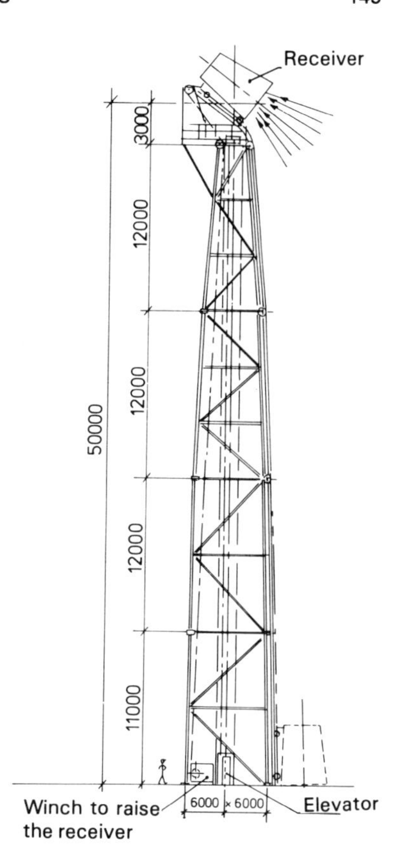

FIGURE 6.15. Construction sketch of the Messerschmidt-Bölkow-Blohm 50-m-high control tower with an empty-cavity-type receiver at the top.

rests on the experience obtained with the Department of Energy's experimental plant, the 5-MW_{th} experimental arrangement at the Sandia Laboratories in Albuquerque, New Mexico. Opened in May 1977, this plant produced 1.8 MW_{th} and reached 5 MW_{th} in 1978. A second plant was also funded by the U.S. Energy Research and Development Administration and planned for the 400-kW_{th} level based on the central 5-kW_{th} receiver plant method of the Georgia Institute of Technology in Atlanta. The arrangement comes from suggestions made by Francia along with those made by the Ansaldo Company.[27–30] The first experiments were begun in July 1977.

According to Walton and Harris,[26] the plant is provided with 550 circular mirrors, each 111 cm in diameter. The shape of the mirrors can be changed so that they are either flat or concave and are thus capable of varying their focal length, so that they can concentrate radiation energy on an experimental surface at an intensity between 25 and 200 W/cm^2. The principal goal of this installation

FIGURE 6.16. Prototype beam receiver constructed by the Ansoldo Company in Genoa. Photo Ansoldo.

is to produce steam at 600°C and 150 atm. In the course of these investigations, the following subjects are to be studied:

1. System for the heliostats, in particular the precision with which it is possible to focus the heat flux.
2. Functioning of the steam boiler, in respect to the influence of the honeycomb structure of Pyrex glass on the diminution of radiation losses, and the thermal efficiency of the performance.
3. Losses in energy storage in water and the pressures that will be produced by the thermal processes, as well as the conditions that arise from the steam in the boiler.
4. Setting up of a heliostat and the correlation between the distribution of the radiation flux and the frequency of adjustment of position.

The procedures for the use of the facility, which are open to some speculation, have been summarized and the types of experiments for which the plant is suitable discussed. The list of the various themes shows that a wide variety of experiments have to be done and data collected and verified before the power tower method can become accepted technology. In this context, the proceedings of the Hamburg solar energy forum of 1978 are relevant. In that forum, Barra *et al.*[31] reported on "Design problems of the hydraulic network in a linear collector power plant." Boese *et al.*[32] lectured on the subject of "Solar energy power plants using sodium as the primary working fluid, energy flux, and economics." Petrescu and co-workers lectured on "Experimental researches in the field of solar energy applications for the production of electrical energy," and Shishodia[33] described "A 5–10 kW solar energy power plant: Development of concept and analysis." DeGezelle[34] wrote on the readjustments of heliostats according to the sun's position.

FIGURE 6.17. View of the heliostat field and the central tower. Drawing: Messerschmidt-Bölkow-Blohm.

6.10. SOLAR POWER PLANTS IN THE UNITED STATES

The initiative of the EEC with its 1 MW_{el} experimental plant is all the more important because of the lead of several years that is held by the United States, as is well illustrated by a trip report written by Köhne in the first German solar energy forum of 1977. He describes the most important plants as well as the distribution systems and central receiver systems. We have already mentioned the solar energy test facility at the Sandia Laboratories in Albuquerque that produces the household heat needed for a large office building containing about 1100 m^2 of space. This plant contains several rows of linear–parabolic mirrors of about 800 m^2 mirror area oriented in an east–west direction and mounted on a common axis, around which they rotate, just as in the well-known arrangement of Perrot and Touchais[35] in Marseilles. The working fluid in the Albuquerque plant is Therminol 66 (a modified terphenyl). This liquid is heated to 315°C by using the solar source and then produces toluene vapor via a heat exchanger. It is this vapor that drives a 32 kW turbogenerator. The Sandia Laboratories have also built in Willard, New Mexico, a solar farm that has a 20-kW_{el} performance and is used for the irrigation of about 40 hectares of land. It consists of 14 rows of mirrors, mounted in a north–south direction, having a size of 1.44×31 m and a total area of 625 m^2. The absorbing surface is covered with a black chromium layer and is selectively black. A high-boiling-point oil is used as the heat-transfer fluid. The provision of the common horizontal axis, which takes the place, for the linear–parabolic mirrors, of the highly concentrating parabolic mirrors in the power-tower system, demands no great skill in construction and is carried out by the use of a 60-W motor driving a toothed-wheel worm drive. It is governed by a number of sensors that act when the sun's angle changes by 0.25° so that the total daily working time of the motor is less than 1 hr. Its power is sufficient to turn the reflecting side of the mirrors down in the case of very strong winds so that they are protected from sandstorms. This plant has great practical interest because there are over 160,000 irrigation sites in the United States, and it would also be of interest for developing countries where there is much sunlight but an insufficient infrastructure.

A third project described by Köhne is a solar farm near Phoenix, Arizona, built by the Batelle Memorial Institute. It has nine rows of cylindrical parabolic mirrors of about 510 m^2, and it is also employed for irrigation purposes. There are differences compared with the plant at Albuquerque; for example, there are counterweights on the back side of the mirrors, and this reduces the turning movement. The absorption tubes are turned with the mirrors. The working fluid is water, which is heated to 150°C and then heats Freon in a second circuit, the Freon operating a turbogenerator at 37 kW_{el}. This power is used to pump 2200 m^3 of water per hour from a 4-m-deep well. The mirrors are arranged in a sandwichlike construction. Aluminum sheets are mounted on an exactly parabolic honey-comb structure of aluminum, and these sheets have, on their reverse side, aluminized plastic foil as their reflecting surface.

6.10.1. More Recent High-Performance Solar Generating Stations in the United States

Among the more recent United States solar power stations, Köhne[23] names that of Barstow, California, having a 10 MW_{el} capability. This facility has been constructed under the general direction of the Sandia Laboratories. Various companies have been responsible for different components, some of which are shown in Figure 6.18 and 6.19. Boeing is responsible for the lightweight heliostats, which consist of plastic foils containing mirrors that are fixed over metal supports and are protected from environmental hazards by a hemispherical Tedlar tent. Martin Marietta proposed a design with side windows and Honeywell one with an opening at the bottom. McDonnell Douglas has proposed a cylindrical receiver containing a group of tubes. Heliostats contain between 30 and 40 m^2 of mirror surface and between 1700 and 2200 mirrors; the total area was designed

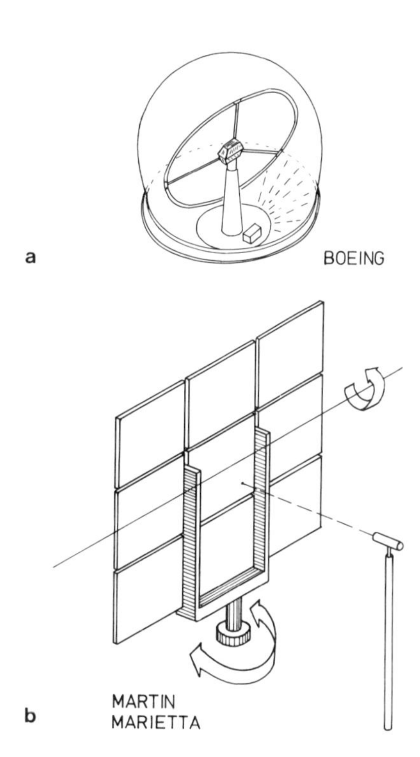

FIGURE 6.18. Different heliostat constructions of American firms. From Köhne.[23]

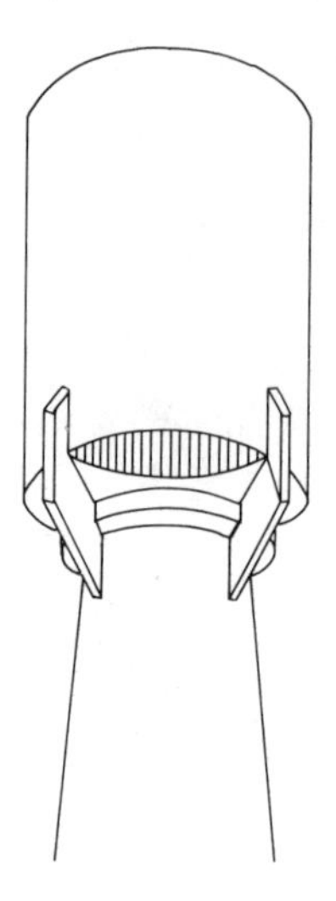

FIGURE 6.19. Cavity-receiver construction from Honeywell. From Köhne.[23]

to be 65,000 m^2. Tower heights designed were between 75 and 130 m. In all designs, the heat-transfer fluid was water, delivering steam at 510°C just as it does in modern coal-driven generating stations. The storage devices, which were meant to be effective for 6 hr, were designed to contain either oil, stones, or a eutectic liquid of 53% KNO_3, 40% $NaNO_2$, and 7% $NaNO_3$. The plan was that between 1980 and 1985, two further demonstration plants of 50 and 100 MW nominal performance would be produced, and these would give rise to the first commercial solar power stations between 100 and 300 MW_{el}, which would produce current that would compete with conventional power stations. Now, the southwest United States has something like 3600–4000 sunshine hours per year and actually offers optimal conditions. Thus, one should expect an expansion of such plants so that by the year 2020, about 25% of all American energy could be solar-derived. However, because of the size of the country, it would be absolutely necessary to distribute this energy by means of high-pressure hydrogen pipelines, rather than by sending it through high-tension cables.

At the time of writing, there is no detailed cost analysis for solar power stations, but some kind of rough analysis can be made based on the total cost of \$2 million for the 1-MW experimental station of the EEC, a cost that works out to \$2.0 $\times$ 10^3/kW. This figure includes the research and development costs. Were the construction to be replicated many times, i.e., were mass production to take place, the costs should drop considerably, let us estimate to one half, or \$1.0 $\times$ $10^3/kW_{el}$. This cost is only about double what Buch has estimated for the cost per kilowatt of a modern coal generating station with desulfurization

\$1150/kW. However, this comparison is a rather unfair one because in the case of the solar power station, it refers to a 1-MW plant, whereas for Buch's analysis, a 50-MW_{el} plant was the model, and of course, the cost goes down the increasing size of the plant (see Chapter 1). According to a more recent analysis given by Buch and Justi[36] of the coal plant using 8% interest and a 20-year lifetime (or 15% with a 5000 hr/yr use rate), the electricity cost in respect to the plant would be 2.85¢/kW_{el}h. In addition to this, one has to add the cost of the fuel, and if we take this as \$12/Gcal (1 Gcal = 1/860 MW_{th}h), then the fuel price of the electricity obtained from coal would be about 2.5¢/kW_{el}h. The total price of coal electricity would therefore be about 5.3¢/kW_{el}h, and this, then has to be compared with the situation that would obtain in New Mexico with 4000 hr/year of sunshine, which would produce a cost of 630/4000 or \$0.05/kW_{el}h, there being no fuel cost. This price is, of course, inversely proportional to the hours of sunshine, and were these to drop to, say, 2000 hr/year, then the cost would rise to 15.5¢/kW_{el}h—quite uncompetitive.*

The conclusion at this point is that the German figures somewhat confirm the estimates being made in the United States that in the fairly near future, solar electricity may be competitive with electricity from coal; this conclusion, however, is greatly dependent on the location of these solar power stations must be located in sites where maximum solar light is available and plenty of cooling water can also be obtained. These considerations give some indication that the power-tower concept—which was first known in the time of Archimedes[37]—may develop ahead of the other types of solar installations, e.g., the photovoltaic cell.

6.11. SOLAR SATELLITE POWER STATIONS†

One of the difficulties in the large-scale terrestrial use of solar energy is, of course, well known to be the energy density. Glaser came out in 1968 with the concept of a solar satellite power station (SSPS).[38,39] The design of an energy satellite to maintain such a station involves the concept of a stationary orbit at a height of about 40,000 km. The actual satellite would consist of about 50 km^2 of solar-cell area and a microwave transmission plant. The photovoltaic plant

* Translator's note: These estimates appear to originate from considerations relevant to about 1980. 1985 estimates of the Solarex Corporation for photovoltaic electricity for sufficienty large power plants suggest about \$0.08 $(kWh)^{-1}$. Estimates for solar–thermal plants depend on whether they refer to "one-off" projects or mass-produced heliostat buildings in a national plan carried out by corporations not dependent upon the economic dominance of the fossil fuel technology until fossil fuels have actually been exhausted. One-off costs seem to indicate around \$0.20 $(kWh)^{-1}$ in 1987 dollars.

† This section authored by Dr. P. W. Brennecke, Braunschweig.

would yield an electrical output in the region of 2000–15,000 MW, and this would then be transmitted in the form of microwaves to stationary, flat antennas on earth. After transmission line losses, there would be about 5000 MW available.[40] It is necessary to discuss this concept at a technical and economic level, because only then can one understand whether it will be a part of our energy future.

As already mentioned above, a satellite generating station consists of a large-area structure that itself supports thin-layer solar cells, a microwave generator, and microwave antennae at an earth station to take up the radiant energy. The necessary size of the satellite solar power station is determined to some extent by the energy-transmission system. For certain technical reasons, it is desirable to have a frequency of about 3.3 GHz. Thus, for this frequency, it is possible to have a rather efficient transmission of microwave radiation with very small atmospheric losses. The acceptable energy density on the ground is about 0.01 W/cm^2 (the usual value assumed in the United States), and Glaser and his collaborators have come to the conclusion that a unit between 2000 and 15,000 MW would act optimally in respect to the relationship between weight and performance.

If one accepts as the electricity source a solar cell–mirror combination that doubles the solar light intensity on the solar cells, then if the mirrors have an efficiency of transmission of 55%, and the photocells are 18% efficient, a collecting surface of 45 km^2 would have to have only 16 km^2 of solar cells. It must be observed, of course, that an efficiency of 18% for mass-produced photocells, although by no means impossible, is certainly an optimistic assumption. Although the theoretical upper limit for silicon photocells is 30%, the actual efficiency of cells that are coming onto the market at this time, the so-called "blue silicon cells," have an efficiency of 13.5%. The solar thermal alternative for space use must certainly be kept in mind, although at first it would seem to be less favorable from the technological point of view. However, it might be possible, for example, by combining a thermal process with a preceding series of low-temperature thermionic steps, to get about 35% efficiency; then, if the mirror efficiency were 85%, the overall efficiency for conversion of light to electricity would be 30%. Under these conditions, the collecting surface could be decreased to 19 km^2. It is certainly possible that the resulting smaller structure would justify the use of an energy-transfer technique somewhat more complicated than those that would be appropriate for the solar cell. Thus, the fact is that the photovoltaic cell, at the moment, does look to be more expensive to mass produce on a dollar-per-watt basis, than one of the thermal systems.

In any case, the problem of the mounting and the stability of the satellites—and we are talking about sizes about 1000 times greater than that of any satellite ever built—will certainly pose some problems that will have to be worked out. However, such problems are characteristic of the whole concept. It is not so much the technical possibility of the proposal, but really whether it can in practice be carried out. As far as the photovoltaic version is concerned, a satellite containing silicon cells and with a structure of synthetic-strengthened carbon fiber is basically possible.

According to present ideas, it seems that a solar satellite generation station should be best placed in a geostationary orbit. With this placement, a microwave energy-transmission system could indeed be practical and could serve the whole of middle Europe. Extraction of energy from the satellite would be carried out by a so-called "phase-array" antenna. In this array, a large number of antennas are arranged in a single plane. The phase of the signal would be regulated in such a way that a wavefront having a spherical surface would be formed in front of the antennas. The spherical midpoint would lie in the center of the receiving system. The energy transmission from space to earth is based on the following principles: According to the plans given, the nominal value for the control should be delivered by a light ray sent from the antenna array on earth. This earth-bound transmitter would then track the radiant energy to the region of its origin. Anything that disturbed or interfered with the main sender would have an effect on the radiation characteristics from the transmitting antennas, so that they would become diffuse and would not give rise to a concentrated degree of energy on the earth's surface. This method of energy transmission from space to earth would provide a simplified safety system, so that an immediate turning off of the energy input would be possible.

Amplification of the high frequency can be taken over by the so-called amplitron. A cold-cathode amplitron with a samarium–cobalt magnet and an efficiency of 80% has now reached a practical level of technical development. Assuming that there should be about 5 kW passing through each apparatus, 1.5 million amplitrons would be needed for the plant. The losses that would be sustained in the process of transmission by radiation through the atmosphere are less than 3%. The collecting area would have to be formed from a system of rectifying antennas, but the accuracy with which these can be positioned is not very great.[40] It has been suggested that the arrangement of the antennas with their electrical instrumentation be in the form of a foil, on which dipole and electronic elements would be printed. The attainable efficiency is about 70%, and this could probably be improved without breaking any really new ground in research.

To transmit space-produced electricity for reception on earth with the maximum microwave intensity of 0.03 W/cm^2 requires an area of antennas on earth of 43.5 km^2.

The question as to whether these concepts for a solar satellite generating station are realizable can be answered in two steps:

1. The first thing that has to be done is to calculate whether an energy satellite can be competitive with the other methods of providing energy. This comparison should give rise to clear plus and minus points and make a decision possible.

2. If these examinations show that the economics of the various systems are in the same ballpark, further criteria for the choice of one of these systems would have to be taken into account. Among the more general advantages of the solar satellite generating station are the complete absence of any pollution and its political independence.

The main priority is to estimate the time at which such an arrangement would become economically competitive. Of course, making such estimates needs a great number of details and the work required to get these details would be justified only if a positive result is expected. In all prognostications about the method, it must be taken into account that the present experience of space travel is only partly applicable to the concept of the solar satellite power station because the latter would involve size considerations so much greater than anything accomplished so far.

In planning, developing, and realizing satellite solar power stations, we are really looking into the future. However, if the various studies that are now being conducted do come out positively,[41] it seems probable that the building of such a power station will begin in this century.

REFERENCES

1. 1979 Photovoltaic Solar Energy Conference, April 23–26, (1979), Kongresshalle Berlin (West), organized by Commission of the European Communities and the IEEE.
2. *Workshop Proc., Photov. Conv. Solar Energ. for Terr. Appl.*, Oct. 25–27, (1974), NSF-RA-N 74-073.
3. Experimentalstudie zur Entwicklungsdefinition von terr. Solargeneratoren [Experimental studies on the Design and Development of Terrestrial Solar Energy Developments], AEG-Telefunken, Projekt des Bundesministerium für Forschung und Technologies, Bonn, Nr. 41885 A.
4. H. Fischer: Silicon solar cells from polycryst. material, *Proc. 1977 Photov. Sol. En. Conf.*, Luxemburg, Sept. 27–30 (1977); Solar cells based on nonsingle cryst. silicon, in: J. Treusch (ed.), *Festkörperprobleme* **18:**19–32, F. Vieweg and Sohn Verlagsgesellschaft, Braunsweig (1978).
5. B. A. Authier, DOS 25 08 83: Neuartige Siliziumkristalle and Verfahren zu ihrer Herstellung. [Novel kinds of Si crystals and processes for their productions]. Anm.: Wacker-Chemitronic; U.S. Appl. Ser. No. 652 359, Polycryst. Silicon with Columnar Structure, in: J. Treusch (ed.), *Festkörperprobleme* **18:**1 –17, F. Vieweg and Sohn Verlagsgesellschaf, Braunsweig, (1978).
6. H. J. Hovel: *Semiconductors and Semimetals*, Vol. XI, *Solar Cells*, Academic Press, New York (1975).
7. H. Fischer and W. Pschunder: Low cost solar cells based on a large area unconventional silicon, Conf. Rec. 12th PVSC Nov. 15–18 (1976), Baton Rouge, Louisiana.
8. S. I. Soclov and P. A. Iles: Grain boundary and impurity effects in low cost silicon solar cells, Conf. Rec. 11th PVSC, May 6–8, (1975), Scottsdale, Arizona.
9. W. Shockley: *Electrons and Holes in Semiconductors*, D. Van Nostrand Company, Toronto (1950).
10. H. Fischer and W. Pschunder: Impact of materiel and junction properties on silicon solar cell performance, Conf. Rec. 11th PVSC, May 6–8 (1975), Scottsdale, Arizona.

11. T. H. di Stefano and J. J. Cuomo: Reduction of grain boundary recombination in polycrystalline silicon solar cells, *Appl. Phys. Lett.* **30**(7):351 (1977).
12. H. Fischer and W. Pschunder: Low cost solar cells based on large area unconventional silicon, Techn. Ber. Nr. 5/76 der AEG-Telefunken, Fachbereich Halbleiter, Heilbronn, December 10, (1976).
13. S. Martinuzzi, A. Mostavan, F. Cabane-Brouty, J. Oualid, and J. Gervais: Properties and efficiency of Cu_2S-CdS sprayed layer photocells, Proc. 2nd Int. Sol. Forum, Hamburg (1978), Vol. 1, p. 385, Deutsche Gesellschaft für Sonnenenergie, Munich (1978).
14. F. Pfisterer and W. H. Bloss: Cu_2S-CdS Dünnschicht-Solarzellen [Thin-layer solar cells], Ibid, p. 369.
15. L. Partain and M. Sayed: Accelerated life test of CdS-Cu_2S solar cells, NSF Grant No. G 134 872, May (1973).
16. J. A. Hedvall: *Einführung in die Festkörperchemie* [*Introduction to Solid State Chemistry*], (Sammlung *Die Wissenschaft,* Vol. 106), Braunschweig (1952).
17. F. Shirland: 3rd Conf. On Large Scale Energy Converison for Terrestrial Use, Wilmington, Delaware, Oct. (1971).
18. F. Pfisterer, G. H. Hewig and W. H. Bloss: Conf. Rec. 11th IEEE Photov. Spec. Conf., Scottsdale, Arizona (1975), p. 460.
19. L. R. Shiozawa, F. Augustine, G. A. Sullivan, J. M. Smith, and W. R. Cook: Jr. Clevite Corp. Final Report, Contract AF 33(615)-5224 (1969).
20. K. Bogus and S. Mattes: Conf. Rec. 9th PVSC, Silver Springs, Maryland (1972), p. 106.
21. V. A. Baum, R. R. Apparase, and B. A. Garf: *Solar Energy* **1**:6 (1957).
22. A. B. Meinel and M. P. Meinel: *Applied Solar Energy,* Addison-Wesley, Manila, Philippines (1977).
23. R. Köhne: *Proc. d. 1. Deutschen Sonnenforums* [German Solar Energy Meeting], Hamburg (1977), Vol. 2, p. 283, Deutsche Gesellschaft für Sonnenenergie, Munich (1978).
24. J. D. Walton, S. H. Bomsar, C. T. Brown, and N. E. Poulos: *Rev. Int. Héliotechnique* (1978), 1st Semester, p. 29.
25. W. H. Hofmann and J. Gretz: *Proc. 2. Int. Solar Forum,* Hamburg (1978), Vol. 1, p. 185.
26. J. D. Walton and J. N. Harris: Development of a ceramic receiver for a Brayton cycle solar electr. power plant, Prepr. COMPLES Conf. Tunis (1977).
27. G. Francia: *Solar Energy* **12**(1):51 (1968).
28. G. Francia: The University of Genova solar furnace, Presented at the NSF Internat. Seminar on Large Scale Solar Energy Test Facilities, New Mexico State University, Las Cruces, Nov. 18–20, (1974).
29. G. Francia: The solar plant of S. Ilario-Genova-Nervi, *Ricerca* **36**(8): 779 (1966).
30. G. Francia: A new collector for solar radiant energy, Proc. Un. Nat. Conf. on New Sources of Energy, Rome (1961).
31. O. Barra, E. P. Caratelli, M. Conti, M. El Sawi and R. Visentin: Design problems of the hydraulic network in a linear collector power plant, *Proc. 2nd Int. Solar Forum,* Hamburg (1978), Vol. 1, p. 195.
32. F. K. Boese, W. Jansing, S. Kostrzewa, and D. Stahl: Solar central receiver power plant with sodium heat transfer, *Ibid.,* p. 209.
33. K. S. Shishodia: A 5–10 kW solar energy power plant, *Ibid.,* p. 265.
34. A. DeGezelle: *Ibid.,* p. 151.
35. M. Perrot and M. Touchais: l'Energie solaire et son avenir industriel. *Rev. Generale de Thermique*, Sept. (1964).
36. A. Buch to E. Justi: Personal communication, January (1979).
37. J. J. Hatzikakidis: From the history of solar applications: Archimedes' predecessors and successors, *Solar Energy* **1** (2,3) (1960).

37. J. J. Hatzikakidis: From the history of solar applications: Archimedes' predecessors and successors, *Solar Energy* **1** (2,3) (1960).
38. P. E. Glaser: Method and apparatus for converting solar radiation to electrical power, US Patent 3,781,647, Dec. 25 (1973).
39. P. E. Glaser, O. E. Maynard, J. Mackovciak, and E. L. Ralph: Feasibility study of a satellite solar power station, NASA CR-2357, NTIS N74-17784 (1974).
40. P. E. Glaser: *Physics Today* **30** (2):30–38 (1977).
41. D. Kassing: Solar power satellites—A review of the state of research, in: *Solar Energy '82*, Proceedings of a summer school held at Igls (Austria) on July 28–August 6 (1982), pp. 191–197.

CHAPTER 7

The Photolytic Production of Hydrogen

7.1. THE PRODUCTION OF HYDROGEN BY MEANS OF THE PHOTOCHEMICAL DECOMPOSITION OF WATER IN PLANTS*

Solar energy arrives on the earth at the rate of 170 trillion (10^{12}) kilowatts. It can be converted to useful energy not only along solar thermal or solar electric paths, but also along the solar chemical pathways. From a long-term point of view, this option could be the most important. Its goal is the photochemical decomposition of water to hydrogen and oxygen, after which the hydrogen could be collected and utilized in a hydrogen economy. It may well be that hydrogen can be produced from solar energy by indirect paths—namely, by first producing heat or electricity and then later utilizing these to produce hydrogen—yet the direct decomposition (photolysis) of water is the most immediate way and therefore fundamentally the best one.

Seen thermodynamically, the splitting of a mole of liquid water costs 57 kcal (237 kJ) in enthalpy. The reaction is strongly endothermic. However, these values are to be regarded as formal ones, because they do not take into account that the photochemical reaction usually gives rise to intermediates, and the energy of these exceeds the energy of the stable final products, H_2 and O_2. Thus, for example, if one considers the decomposition reaction $H_2O \rightarrow H^{\cdot} + OH^{\cdot}$— the formation of two radicals from one molecule of water—then this reaction needs no less than 123 kcal (516 kJ) per mole.

* This section authored by Professor E. Broda, Vienna.

An energy of 57 kcal per mole can be easily shown, by the use of Planck's formula $E = h\nu$, to be given (on a molar basis) by light having a wavelength in the region of 500 nm (green light). The amount of energy for this decomposition is on a molar basis (1 Einstein). In a quantum of violet or blue light, there would be a greater amount of energy, whereas the corresponding amount of light quanta from red or yellow light would contain a lesser amount of energy. Theoretically, therefore, the photolysis of 1 mole of water by 1 Einstein of light of short wavelength would be possible, whereas to obtain the necessary amount of light energy for the decomposition of 1 mole of water by means of light of longer wavelength, at least 2 photons would be needed. However, one has to take into account that the water molecule itself does not absorb in these ranges, so that it would be necessary to have a colored sensitizer (photocatalyst) present to absorb the energy and transfer it to water.

The few experiments that have so far been carried out with the aim of dissociating water photochemically have had little success. A stationary state always quickly establishes itself, and the net amounts of water decomposed are very small. Relatively favorable results—but they are not very good—were obtained with UV light, but the amount of this kind of light in sunlight (even in the outer reaches of the atmosphere) is small, and near the earth's surface, it is minimal.[1]

The reason for the lack of success in splitting water directly by the influence of light will be discussed later, but it can be said here that there is no natural law standing in the way of success. In a sense, nature has given us a strong indication that a solution to the problem of the photolysis of water is indeed possible. For three gigayears (3000 million years, i.e., 3000 megayears), plants have been carrying out photosynthesis. The amount of carbon that plants form yearly with the help of the energy of light entering the atmosphere is a hundred gigatons (100 billion tons). This is about 20 times the amount that is converted by our industrial civilization in the same time period into CO_2, or used in other ways.

The performance of plants is remarkable.[2,3] They use the entire range of light, from the violet to the red end of the spectrum. Their efficiency is essentially independent of the light intensity. The energy yield can—admittedly under optimal conditions in the laboratory—amount to about 33%; i.e., a full one third of the energy of the absorbed light can be utilized in the synthesis of biomass, and then a part of this can be reobtained in the form of heat by combustion of the material produced. Plants have solved the storage problem in a suitable way—a problem that is quite difficult to solve when energy is in the form of electricity—because biomass can be stored without any special difficulties. Biomass can be described by the formula CH_2O, the unit group in the carbohydrates. For example, the most frequently encountered carbohydrate is glucose, $C_6H_{12}O_6$. The carbohydrate group is represented as $[CH_2O]$, because otherwise it might be confused with formaldehyde, CH_2O.

To what extent is plant photosynthesis related to the splitting of water? An older concept—which was supported by the great biochemist Otto Warburg, among others—was that light was used directly by plants for the splitting of CO_2 in bound form. However, the Dutch microbiologist van Niel[4] showed in the 1930s that the primary action was water-splitting:

$$H_2O \rightarrow 2\,[H] + 0.5\,O_2$$

Here, [H] was a form of hydrogen that was not at first identified and was simply called "reduction power." After obtaining the [H], there is, according to van Niel, a second step, the reduction of CO_2 to biomass by means of [H]:

$$4[H] + CO_2 \rightarrow [CH_2O] + H_2O$$

This second step is independent of light; i.e., it depends on dark reactions, and was elucidated in the 1950s by Bassham and Calvin.[5]

It may be noted that in photosynthesis, it is not only the reduced substance called [H] that is formed but also the energy-rich substance adenosine triphosphate (ATP), and this substance is utilized in further decompositions (Figure 7.1). It is even possible that ATP is originally the only primary product of

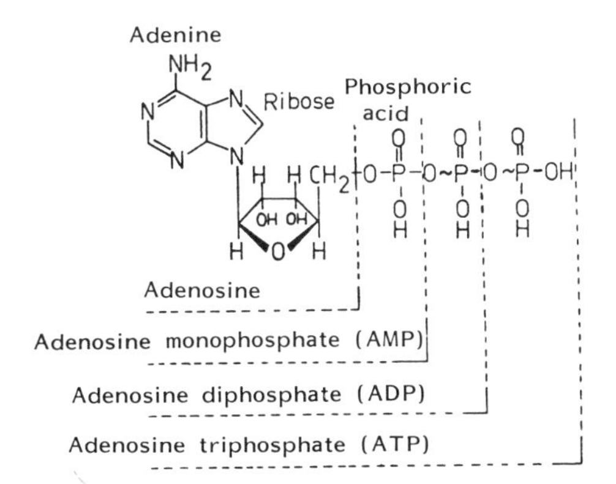

FIGURE 7.1. Structure of adenosine triphosphate (ATP) and its parts. The tails (~) are the "energy-rich" bonds. The dissociation of these bonds (with saturation of the products) gives a great amount of free enthalpy. ATP is shown in its nondissociated form. In reality, under physiological conditions at pH 7, several H atoms are dissociated. All plant and animal life uses this universal storage medium with fast, easy, and efficient convertibility to all sorts (mechanical, thermal, electrical, chemical, osmotic) of needed energy, which for example under hydrolytic dissociation of the final phosphoryl groups, and with the uptake of 1 mole H_2O according to $ATP + H_2O \rightarrow ADP + P$, produces adenosine diphosphate and inorganic phosphate. Under standard conditions, this delivers up to -7.3 kcal/mole, in animals, -12 kcal/mole, as energy for the aforementioned reactions, or takes up that energy in reversed phosphorolysis. One human, for example, produces and uses daily up to 100 kg ATP with a market value of $200,000. These conversions occur in every cell, so that is could be said that one person has 10^{14} decentralized energy-storage plants.

photosynthesis.[6] We are not concerned with this type of question at the moment, for ATP plays no role in technical photolysis.

Of course, no actual H_2 molecules are synthesized in plant photosynthesis under natural conditions. Plants themselves would not be helped by the evolution of hydrogen as a gas. For this reason, most plants do not contain the hydrogen-forming enzyme "hydrogenase." On the other hand, Hans Gaffron showed in 1942 that certain algae can produce limited amounts of hydrogen gas.

Under normal conditions, the reduction power represented above as [H] is in the form of a protein called "ferredoxin" in the reduced form. This is a well-defined substance that can be obtained in crystalline form and contains iron. When the iron is divalent, ferredoxin is strongly reducing, and the oxidation–reduction potential under standard conditions is about -0.4 V (Figure 7.2). This corresponds numerically to the redox potential of the hydrogen electrode in neutral solution; i.e., reduced ferredoxin has the same tendency to lose electrons as has a hydrogen molecule in a solution at pH $\approx$ 8. Thus, the photochemical–thermodynamic view of plants in photosynthesis is that their performance is equivalent to that of H_2. Under physiological conditions, reduced ferredoxin acts as the deliverer of electrons, i.e., as a reducing agent in the assimilation of CO_2 and the formation of biomass.

Now the production of biomass, looked at energetically, is difficult. If we think (schematically) of the overall process as

$$H_2O + CO_2 \rightarrow [CH_2O] + O_2$$

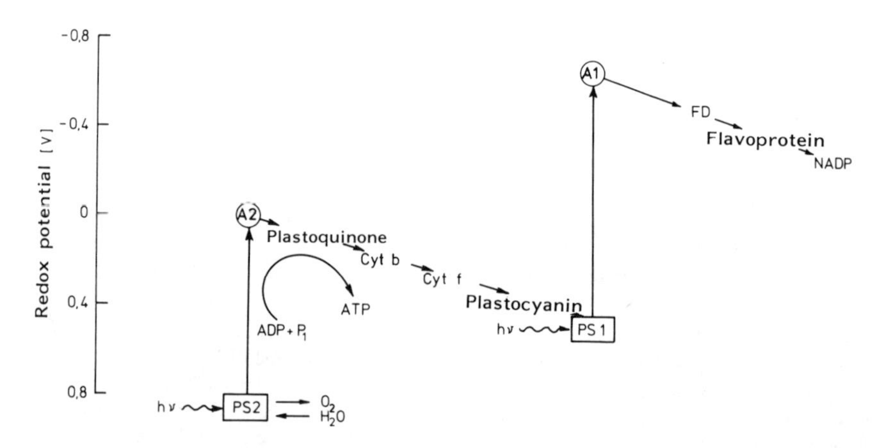

FIGURE 7.2. So-called Z scheme showing how electrons from water molecules are lifted "uphill" to an acceptor (A_2) with a redox potential of 0 V in the first light reaction of photosystem II (PSII). The electrons sink through a series of redox media, with discharge of energy, which is used for the synthesis of ATP from ADP and phosphate, to $+0.45$ V at photosystem I(PSI). In a second light-influenced reaction, the absorbed electrons are raised to -0.4 V to bind ferredoxin. From there, they flow downward, changing NADP into NADPH using protons obtained from the water. Abscissa: Reaction path.

then 112 kcal (472 kJ) per mole is necessary to drive the reaction from left to right. To make this process possible, a plant acts in two steps when it absorbs light. Two quanta are used successfully to transfer one and the same electron from water to ferredoxin, i.e., to reduce ferredoxin with water. The process takes place in an ordered way inside the structure of the solid, according to the Z scheme of Figure 7.2. Here, the two quanta are not applied to the electron simultaneously; rather, between the first and second absorption processes, a chain of dark reactions takes place by which the electron further increases its energy. The composition of this chain of reactions is known to a large extent through the work of Witt[7] and his colleagues.

Both the photochemical primary reactions of plants take place in so-called "reaction centers" that contain chlorophyll A (Figure 7.3). In contact with these centers are sensitizers called chlorophyll A, chlorophyll B, and the carotenoids. In the most primitive plants, the blue algae, there is no chlorophyll B, but a substance called phycocyanin. Sensitizers form so-called "antennae" through which light is absorbed and the available energy then given to the reaction centers. This is shown in Figure 7.4, an electron-microscopic photograph.

Plants, even the blue algae, work efficiently. Their photosynthetic apparatus is relatively complicated. One must assume, therefore, that plants themselves arose from somewhat simpler photosynthetic ancestors and probably from photosynthetic bacteria. At present, two such groups of bacteria are extant, the green and purple bacteria. They operate with a lesser efficiency than that of the main photosynthesis reaction of the higher plants, as can be seen from the fact that they do not have much "reducing power" [H] (e.g., they cannot photolyze water).

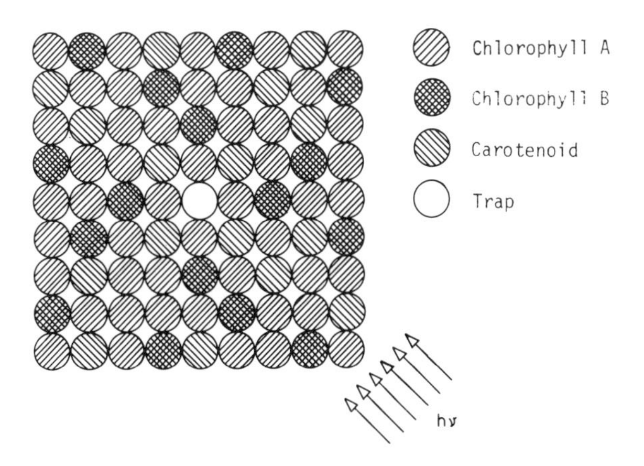

FIGURE 7.3. Structure of a photosynthetic system (PSI or PSII). The light-collecting "antenna molecules" consist of chlorophyll a, chlorophyll b, and about 200 other yellow pigment molecules called carotenoids. The quanta are transferred to a special molecule in chlorophyll a using the resonance process. This molecule is known as the reaction center or trap. Here, the sunlight energy collected by the antennas is converted to chemical energy according to the Z scheme.[3]

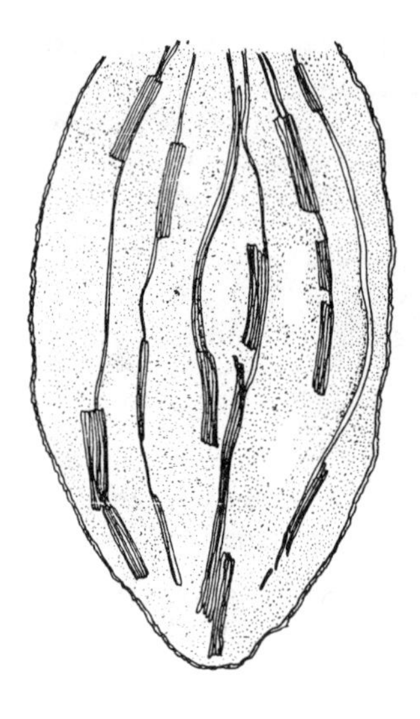

FIGURE 7.4. A partly schematized electron-microscopic picture of a chloroplast of a higher plant. The light reaction takes place in the lamellas, which are connected to form the stapels (grana). The products of the light reaction react further in the surrounding stroma, thanks to its ATP content, even in the dark. All lines in the picture symbolize membranes, which cause the vectorial product separation. The whole chloroplast is surrounded with a double membrane, which holds back the stroma, in which reactions also take place. The diameter of the chloroplasts is a few micrometers; the membranes are about 0.01 μm thick.

Indeed, they need as a source of electrons materials in which electrons are much less tightly bound, e.g., H_2S. In their photosynthesis, bacteria therefore do not release any oxygen, but release sulfur and other such substances. In this way, bacteria need only a single quantum of light to release an electron, not two quanta. Correspondingly, there are no photochemical following reactions. Although photosynthetic bacteria now play a much lesser role than plants in nature's overall scheme, they are nevertheless important objects of study for biochemists and biophysicists, because of their relative simplicity.[6]

Now, how did plants, and also the photosynthetic bacteria, solve the fundamental problem of photochemistry? The lack of success that research on this subject has had so far must be attributed to the fact that the primary products of photolysis, which are quite reactive, occur at a small distance apart and therefore probably meet easily. They then react with each other, and the resulting reaction probably gives rise to water. The absorbed energy is recovered as heat. To avoid such a degradation of energy, plants—and before them bacteria—developed the membrane principle. Light is absorbed inside assymetric membranes (Figure 7.4) that are capable of vectorial reactions. This means that the primary products of the photoreaction do not arise in a mixture but, according to their chemical natures, occur on one or the other side of the membrane. In this way, the back-reaction between the two sides is hindered and the products are able to exist as relatively stable intermediates. For example, in plant reactions, reaction step 1 produces O_2 on one side and reduced plastoquinone on the other side, and then in reaction step 2, after a series of dark reactions, oxidized plastocyanin and reduced ferredoxin (see Figure 7.2). Such vectorially working membranes have not yet been successfully constructed in laboratory experiments. However, except for electrochemical cells, no serious attempt has been made to do this because the problem was not really understood or formulated properly until quite recently.

It would be quite possible, in considering a solar chemical option similar to that of bacteria and plants, to apply the excellent mechanisms of photosynthesis and to utilize biomass or its products as a fuel. In some ways, this would be like returning to mankind's past, in which wood was the most important source of useful energy. The conversion of biomass into liquid or gaseous fuels (including hydrogen) that can be easily used can be carried out by pyrolysis (treatment at a higher temperature) or fermentation. For the synthesis of hydrogen, there is available a process that is basically similar to the so-called "re-forming" reaction, namely:

$$[CH_2O] + H_2O \rightarrow CO_2 + 2H_2$$

This reaction has free enthalpy of roughly zero, but by means of the absorption of the CO_2 that is available, it can be driven to the right; i.e., the gasification of biomass to form hydrogen is possible.

In fact, there has been throughout the years a persistent suggestion of "energy farming," i.e., the idea that plants could be grown with the specific aim of obtaining fuels on a large scale.[8] One of the possibilities that have been considered is poplar or, in tropical climates, sugar cane. Since hitherto only a small percentage of the available energy of photosynthesis has been used for the production of food, or for technical raw materials (fibers), one could really obtain much technically useful energy by plant-growing of this kind. The necessary procedures, which have been neglected during recent years because of the abundant supply of oil, could be carried out on the large amount of virtually free forest land that is available, and the economics would include the exploitation of decayed products from the forests. Thus, the rational use of straw, corncobs, branches, sawdust, and the organic material from municipal wastes is a pressing problem.

At present, a good deal of work is being carried out with "seminatural" systems. One can isolate from plants photosynthetic "organelles," the chloroplasts, combine them with enzymes which originate in bacteria, and thereby produce hydrogen from reduced ferredoxin and water. Experiments of this kind have been only partially successful so far. From the technological point of view, this mode of procedure is not all that promising, because the necessary biomaterial is rather special, and therefore costly, and easily susceptible to decomposition.[9]

From the long-term point of view, the principal task for solar chemical energy production would be the manufacture of synthetic systems. Their heart would presumably be the vectorially asymmetric membrane, which would incorporate photocatalysts and other necessary materials. Such membrane systems could first find application in tropical deserts, which receive so much sun, although in addition to these regions, there are large areas of more moderate climatic zones that could be considered. The hydrogen would then be transferred in gaseous form, or in the form of methanol or hydrocarbons, to user centers. Further into the next century, one could even consider floating islands on the oceans that would be devoted to energy production in this way.

The fact that water is necessary for these operations should cause no problem, even in deserts where the ambient temperature is very high, such as the Sahara. Thus, the amount needed in comparison with the amount for irrigation is very small. Plants lose 99.9% of the water they take up through transpiration and use a very tiny amount for actual photolysis. It is also conceivable that sea water, without preliminary desalination, could be directly applied to photolysis.

For the construction of vectorially working membranes, one can consider two ways, which would finally converge. On one hand, our considerable arsenal of physicochemical knowledge in the fields of thermodynamics, kinetics, photochemistry, electrochemistry, and structural chemistry is available for application to membrane synthesis. On the other hand, membranes from plants themselves should be studied much more vigorously. One should not necessarily try to imitate plants directly, but it might be possible to learn a good deal from them, and to allow the knowledge gained to lead us and perhaps inspire us.[10] Plants are not optimized with respect to the production of the maximum amount of biomass. Mechanisms of energy use in plants have to serve several other purposes, e.g., protection against unfavorable climatic conditions, manufacture of mineral substances and water, defense against herbivores, and other such purposes, as well as promulgation of the plant species. Apart from these factors, the mechanisms that exist in present-day plants have been influenced, and not always for the best by their development in the past and contain elements that are no longer of any use to them, just as city planning has to make do with conditions inherited from the past.

The construction of suitable asymmetric, vectorially working membranes is, in the opinion of the author, one of the greater tasks of contemporary physical chemistry. The solution of the problem could be one of the steps that give a different basis to the future of mankind.[10] The economic aspects could always be discussed after the synthetic vectorial membranes are available.

7.2. AN INTRODUCTION TO THE PRODUCTION OF HYDROGEN BY THE PHOTOCHEMICAL DECOMPOSITION OF WATER USING MONOMOLECULAR LAYERS*

It would be of great importance indeed if one could take the process by which plants use solar energy and convert it to chemical energy, and apply this to economic production. For example, one thinks of the possibilities of equipment

* This section authored by Professor H. W. Kuhn, Göttingen.

that would convert water economically into hydrogen and oxygen, or the carbon dioxide in the air into methane and oxygen. Production of an apparatus of this kind is a far-off goal, and only a few principles and preliminary experiments can be described in this section.

The complicated apparatus with which plants collect light and store energy is bound to a membrane only a few molecules thick (Figure 7.4). The light is absorbed, and there is then a separation of charges. It is assumed that electrons are pumped by light energy from one side of the membrane to the other. In this way, the energy is stored for a very short time as electrical energy, indeed, only until a chemical reaction causes the electrically charged membrane to discharge and to turn electrical energy into chemical energy again. Thus, the minimum energy loss occurs in this step, and it is important that the process be carried out very slowly, yet quickly enough that the membrane can be easily charged up again and be available for further light input. Under the normal conditions of nature, i.e., with normal solar irradiation, this means that these processes have to take place in about 0.01 sec.

Water is always dissociated to a small extent into H^+ and OH^-. A negative charge on the membrane is then used to reduce hydrogen ions and give rise to hydrogen. The positive charge is used in the oxidation of OH^- and gives rise to oxygen. The processes are connected with particular enzymes, i.e., with complicated structures.

The separation of charge by the membranes is the decisive trick by which the energy is stored long enough so that the associated chemical reactions are carried out sufficiently slowly and therefore run loss free.

Molecules of chlorophyll are dispersed over the whole membrane. A light quantum can be absorbed by one of these molecules. The molecule is excited, and the excitation energy is taken away over various chlorophyll molecules to the point where the apparatus able to carry out the charge separation is situated. There is a charge-separation apparatus for every 300 chlorophyll molecules that collect light in this way (see Figure 7.3). Then, the energy thus collected causes an electron to go from one side of the membrane to the other. At present, it is not known how this mechanism functions, and it must of course be known if the process is to be imitated artificially.

One can ask how such a process occurs in principle and then build an apparatus according to the same principle. A goal of this kind is important to the possible eventual application of the principle to the conversion of solar energy, independent of the correctness of the postulated mechanism in the biological system.

When the electron is taken from one side of the membrane to the other, it reaches a higher level because energy is added to it. The question arises as to how this occurs with the aid of light. The excitation energy is transported and arrives on a dye molecule situated at the point where the charge transport takes place. The excited electron of the dye molecule is taken through a potential energy barrier to the left (Figure 7.5) and then through one to the right, the

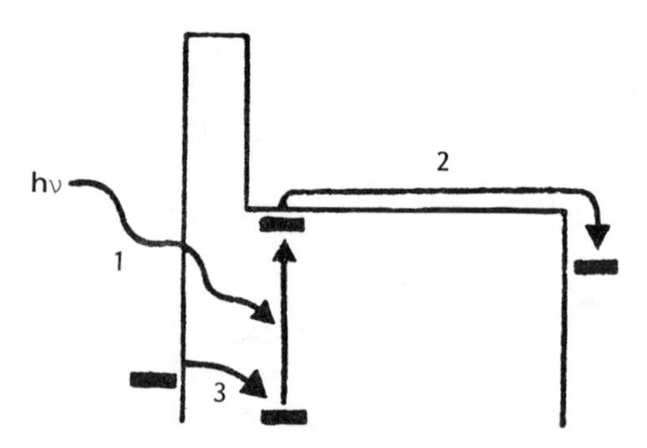

FIGURE 7.5. Potential profile for electron pump run by light.

energy of which reaches up to the desired level of the electron acceptors on the left. The problem with which one is now faced is how the electron in the deep level on the left can be taken to the dye molecule, i.e., how the dye molecule in its original state is decomposed back again so that the system is once more ready to take up a new light quantum. The high potential energy barrier prevents the back-reaction. However, if it is sufficiently thin, say only a couple of atoms thick, barrier penetration, which would be impossible according to classic theory, can take place according to quantum theory by means of tunneling. The electron has insufficient energy to climb the potential barrier, but it can go through the barrier, so long as the barrier is sufficiently thin and the lifetime sufficiently long.

To realize such an arrangement in practice, one would have to have certain molecules in an exactly planned way, arranged next to each other. In biology, many different molecules, arranged next to each other, are known to be used for the building of such apparatuses on a molecular scale. After mixing of these molecules, the appropriate components find each other in the process of diffusion; the organized systems build themselves up spontaneously. The problem consists in using this principle of self-organization to give rise to molecular aggregates artificially, with the intention that these will give rise to the conversion of light energy to chemical energy.

The chemist is confronted here with a new task. As with the earlier tasks, it consists in the synthesis of pure compounds. However, the synthesis of the individual compounds has to be planned out so that the individual synthesized molecules fit or match each other and—as a result of interaction among the molecular components—have a function as a whole. We want here to suggest pathways whereby this worthy goal could be attained. Molecules such as the soaps, which consist of a hydrophilic head and a hydrophobic carbohydrate chain, build aggregates as micelles in water or as a monomolecular layer on the water surface, the hydrophilic groups being within the water layer and the hydrocarbon chains being outside it, but mutually attracting each other (Figure 7.6). Now, if these molecules are dissolved in organic solvents, and if one introduces, for example, a glass plate (with a hydrophilic upper surface), a monomolecular layer is formed through adsorption. One can well imagine that so long as suitable molecules are present, this process could lead to the buildup of complex systems, e.g., the desired electron pump. However, it will not be easy to create molecules which lead to such complex systems in the desired way. It might be easier to attain the desired association by some kind of external action before one attempts to get directly to the desired goal.

Consequently, with the aim of creating the simplest functional systems from several interacting molecules in a planned way, one can envision placing suitable molecules of different kinds on the surface of water and pushing them together. This is a direct example of the process of exerting appropriate influence from the outside in the direction of constituting suitable systems from a mixture of molecules. Now, if one places the layer on some base plate, one can in this way take the various layers according to some prescribed plan and put one on top of the other. Here, one has a second possibility of some kind of steering of the right process from the outside. Monomolecular layers of fatty acids and the possibilities of making the layers one atop the other are well known (Agnes Pockles, 1891; Langmuir, 1917). One can see how an old process might be used for a new goal: the production of organized layers as functional systems.

We can consider in this respect first of all a system which shows the effect of the collection of light quanta and the conducting of excitation energy in a planned way. Like the soaps, dye molecules have a hydrophilic part, which is also the color-giving part of the system, and a hydrophobic part, two hydrocarbon chains. Using an appropriate device, one pushes these molecules together with a long-chain hydrocarbon on an aqueous surface. In this way, a compact monomolecular layer can be formed. The dye molecules are so precisely packed in this layer that they form a cooperative system that shows that light will be absorbed in a narrow spectral region. If we now introduce a small amount of dye II, its molecules will take up a certain fraction of the lattice positions of dye I (Figure 7.7). If the system is now irradiated with light that is absorbed by dye I, the excitation energy can be transferred to dye II. This is shown by the fact that dye II fluoresces with a characteristic color. Thus, after a molecule of dye I has absorbed a light quantum, the energy is then emitted from dye molecule

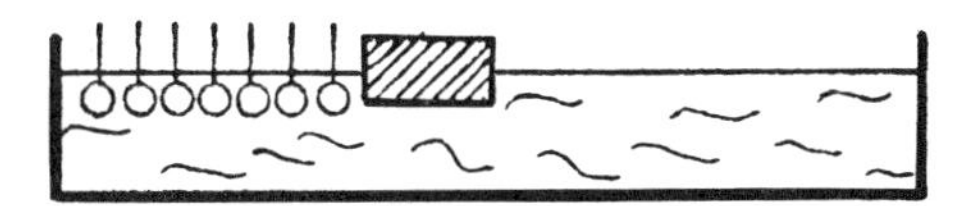

FIGURE 7.6. Formation of a monomolecular layer from a fatty acid on an aqueous surface.

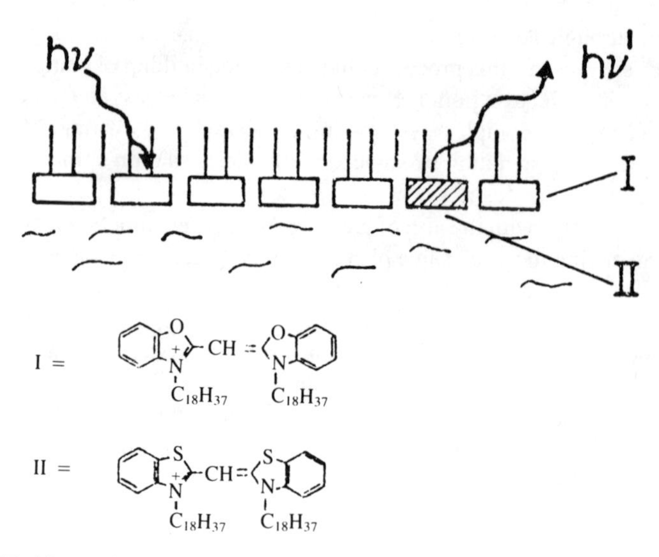

FIGURE 7.7. Monomolecular layer from dye energy donor (I) and energy acceptor (II). To obtain a compact layer, a long-chained hydrocarbon is added, which fills in the gaps between the dye molecules.

II, even though only 1 in 10,000 of the dye I molecules have been replaced by molecules of dye II. The effect of the cooperative carryover of energy is more pronounced here than in the photosynthetic apparatus of plants, in which only 1 in 300 molecules absorbs the primary excitation energy.

The effect of the transport of electrons from an excited dye molecule to an acceptor, and the blocking of this transition through a potential energy barrier, can be seen in Figure 7.8. The layer of dye molecules is covered, to the extent of about one half, with an inactive layer of a fatty acid, but the other half is covered with a layer of an electron acceptor. When light, in a frequency range absorbed by the dye, is incident on the system, one sees the fluorescence of the dye in the first half. The excited electron falls back into the ground state, and the energy would then be emitted as a light quantum. However, it is extinguished in the second part: the excited electron goes to the acceptor. If the two layers are then placed on opposite sides and brought into contact, then the dye and the acceptor are separated by the hydrocarbon. When the dye is excited, it fluoresces in both halves equally strongly, because the electron acceptor is blocked by the energy barrier and is therefore ineffective.

The gradual delivery of electrons through a high and narrow energy barrier is shown by the experiment in Figure 7.9. A dye molecule is excited and transmits an electron to an acceptor. The electron is supplied by a donor—leucomethylene blue. This follows from the fact the molecule is transformed into methylene blue, i.e., a blue color appears. In this way, gradually, by steadily improved

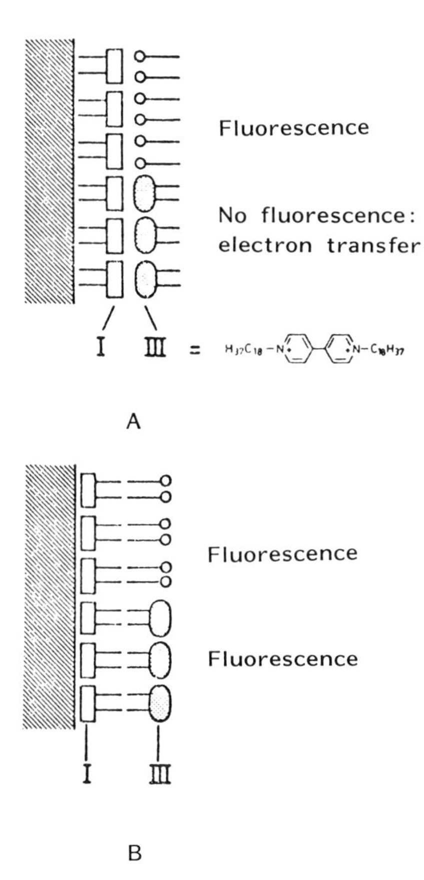

FIGURE 7.8. (A) Monomolecular layer of dye I, overlaid by a monomolecular layer of fatty acid (top) and electron acceptor III (bottom). (B) Electron donor and acceptor in contact (A), separated by carbon substituents.

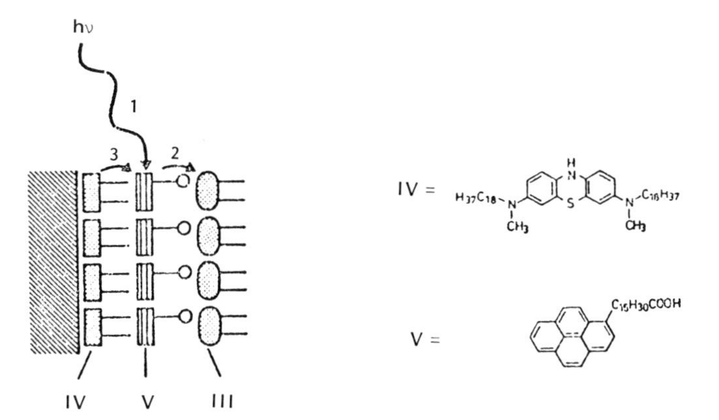

FIGURE 7.9. Layer formation of electron donor (IV), photocatalyzer (V), and electron acceptor (III).

coordination of the components, it may be possible to construct an effective electron pump and a mechanism to trigger the desired oxidation–reduction processes.

These presentations are meant to suggest how one might be able to build up simple organic systems from molecules into functional entities, but the technical realization of an arrangement for the production of hydrogen from solar energy in this fashion is still a long way off.[11–14]

7.3. PHOTOELECTROCHEMICAL PRODUCTION OF HYDROGEN*

In addition to photobiological processes and photochemical processes, the direct splitting of water can be carried out by means of natural sunlight with the help of photoelectrochemical systems.[15–17] These are largely semiconductor–electrolyte systems that are suitable for the production of electrical and chemical energy. For example, photoelectrochemical cells containing inorganic or organic semiconductors could be used for the direct production of hydrogen.[18,19]

Photoelectrochemical systems have certain similarities to purely solid-state systems and therefore will be given brief consideration. It is well known in solid-state physics that at junctions of semiconductor–metal interfaces, the equilibrium situation is characterized by the fact that there must be a common Fermi level throughout the whole system. At the interface there exists an electrical field. When the system is illuminated, electrons and holes are produced within the solid, and these are separated by the action of the electrical field. In the open-circuit condition, this process leads to a decrease of the electrical field; i.e., the energy bands within the solid are shifted and give rise to a corresponding photovoltage. Now, suppose this *p–n* junction is shorted through an external circuit. The position of the energy bands then remains unchanged on illumination. In this case, the excited electrons in the conduction band go into the interior of the *n* region and transfer through the external circuit to a *p* region, where they recombine with holes in the valency band. The photocurrent density is linearly dependent on the intensity of the incident radiation. The maximum photopotential is reached when the energy bands are flat to the surface. The maximum photovoltage depends on the original band bending, which is dependent on the position of the Fermi level at equilibrium.

In principle, the same kind of considerations may apply to the semiconductor–metal surface that is used in Schottky cells. Here, losses occur because the metal layer, which forms a barrier with the semiconductor, absorbs some of

* This section authored by Dr. P. W. Brennecke, Braunschweig.

the incident radiation. For this reason, the early investigators in this field decided to produce cells in which metal contacts could be replaced by an electrolyte.

If one brings a semiconductor into contact with an electrolyte, there forms at the phase boundary an excess of charge during the setting up of the equilibrium. This means there is a double layer and therefore, within the semiconductor, a space charge. When light is absorbed in the semiconductor, electron–hole pairs are produced within the semiconductor itself. They are immediately subject to the field associated with the band bending at the interface and therefore rapidly separated in opposite directions (thus reducing recombination). The electrical energy that is produced in the process of formation of the hole–electron pairs can be used for producing current in an external circuit, and eventually for the carrying out of electrochemical decomposition reactions, e.g., water-splitting, and thereby hydrogen production.

The original experiment in water-splitting with respect to photoelectrochemical reactions was carried out by Fujishima and Honda.[20] They initially observed water photolysis at the surface of an *n*-type semiconductor that was being irradiated with light (a 500-W xenon lamp). The photoanode they used first was a rutile crystal (n-TiO_2); the corresponding cathode was platinized platinum. The cell used by Fujishima and Honda is shown schematically in Figure 7.10. It consists of the usual arrangement of metal and semiconductor electrode; the cathode and anode compartments are separated by means of a glass frit. In the photoelectrochemical decomposition of water into hydrogen and oxygen, the redox potential of water/hydrogen and oxygen/water have to be taken into account, and these differ by 1.23 eV. The position of the Fermi level

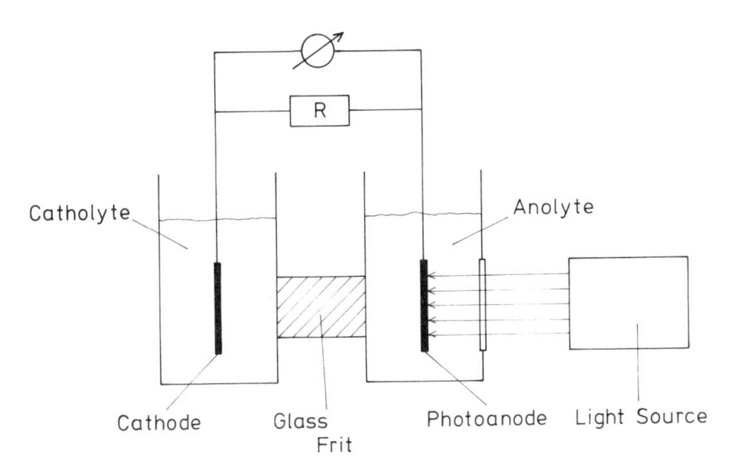

FIGURE 7.10. Schematic presentation of the photoelectrochemical cell used by Fujishima and Honda.[20]

in the electrolyte* is therefore determined by these two redox potentials and depends entirely on the oxygen and hydrogen concentrations. The equilibrium is supposed to be constant over the whole system. The electron–hole pairs that are produced by the excitation arising from illumination of the sample are separated by the electrical field. The holes migrate to the surface of the semiconductor electrode and the electrons to the interior of the metal electrode. When the electrodes are connected, the holes are used up on the semiconductor electrode by the following process:

$$H_2O + 2p^+ \rightarrow \tfrac{1}{2}O_2 + 2H^+$$

while the electrons flow around the external circuit to the metal electrode and promote the reaction

$$2H_2O + 2e^- \rightarrow H_2 + 2OH^-$$

on the cathode. Light absorption on the semiconductor therefore causes water decomposition.

In the choice of suitable semiconductors, some basic conditions should be observed. Thus, if only one semiconductor is to be irradiated, the difference between the bands in the semiconductor—the energy gap—must be greater than 1.23 eV. This condition is, of course, fulfilled by most of the semiconductors that are at present regarded as usable. Many are not usable because instead of oxygen evolution at the semiconductor–solution interface, a dissolution process (dissolution of the semiconductor material) takes place. It has been possible to suppress such dissolution processes only partially by the addition of other redox systems, so that researchers in this field have been forced to go toward the use of anodes in photoelectrochemical cells made of semiconducting materials of higher energy gap ($\approx$3 eV or more), which have the advantage that they are then stable with respect to dissolution. Good results have hitherto been obtained with oxide semiconductors such as TiO_2, ZrO_2, or Ta_2O_5, and their compounds. The yields are small because these oxide semiconductors have relatively large band gaps, with the consequence that they do not begin to absorb light until the extreme blue or the UV part of the spectrum is reached, and sunlight contains only a very small fraction of these wavelengths. With respect to the energy of the absorbed light quantum, say between 2.8 and 3.2 eV, only 1.23 eV is saved and stored in the form of hydrogen, and these light quanta, when absorbed, are being used only to the extent of about 50%.[21]

The development of photoelectrochemical processes and the technical realization of photoelectrochemical prototype cells have been active fields in the

* Translator's note: The concept of a Fermi level in an electrolyte may be difficult to accept because Fermi statistics apply to unbound electrons, but electrons in ions are bound.

last few years. They have given rise to experimental efficiencies for redox systems that are high, e.g., more than 10%, and to lifetimes of more than 1 year.[17]

In photoelectrochemical cells, two basically different cell types are to be distinguished: (1) electricity-producing regenerative cells and (2) fuel-producing nonregenerative cells. Current-producing cells use the electrical energy either directly (homogeneous) or as battery analogues, in which the energy is at first stored in two electrolytes (heterogeneous). Development of these systems has advanced particularly well in recent years. In the case of the use of crystalline semiconductors as electrodes, efficiencies of 10% and reaction stabilities of several months have been achieved. Of course, the state of the technical progress made with these cells is as yet by no means so great as that made with silicon solar cells, with which they are in competition. However, electrochemical cells may be longer-lived than photovoltaic cells because of their simpler construction and perhaps can be more cheaply produced.

With cells that produce materials that yield fuels (e.g., hydrogen), much lower efficiencies have been obtained. However, it seems reasonable to look toward the realization of a possible efficiency of about 20% for the production of hydrogen. The development of fuel-producing cells is more difficult than that of electricity-producing ones, because the integrated formation presents additional requirements for the use of materials and affects the considerations for the construction of the cell (separation and removal of the products). It is therefore not remarkable that the development of these cells is less advanced than that of the electricity-producing systems. However, water-splitting with an efficiency as high as 8% has already been achieved.

Apart from the simple semiconductor–electrolyte contacts, other constructions are possible:

1. Several semiconductor layers with different absorption characteristics can be put together so that solar energy is better absorbed (tandem systems).
2. The electrode can be stabilized by being covered with transparent corrosion-resistant films.
3. The light absorption of the electrode can be increased by the addition of sensitizers.
4. Instead of aqueous electrolytes, nonaqueous electrolytes can be used and the corrosion problem is thereby ameliorated.

At present, none of these four types of cells can be left out of consideration for the future. Because of their high possible efficiencies, tandem cells can be regarded as particularly promising. The development of cells with sensitizers must be looked at skeptically because, apart from the corrosion problems, one also has to deal with the problem of stability of the sensitizer. One part of the research needed here is basic research. Despite the low efficiency obtained so far, it is certainly desirable to undertake further basic studies so that a decision regarding photoelectrochemical processes can be well informed.[22] Apart from the several remaining problems in the mechanism of redox processes, there are also problems concerned with the properties of catalysts and sensitizers that have to be examined and solved.

REFERENCES

1. S. Claesson and L. Engström: *Solar Energy,* National Swedish Board for Energy Source Development, Stockholm (1977).
2. E. Rabinowitch and Govindjee: *Photosynthesis,* Pergamon Press, New York (1969).
3. N. K. Boardman and A. W. D. Larkum, Biological conversion of solar energy, in: H. Messel and S. T. Butler (eds.), *Solar Energy*, pp. 123–182, Pergamon Press, Oxford (1975).
4. C. B. van Niel: The present status of the comparative study of photosynthesis, *Annu. Rev. Plant Physiol.* **13:** 1 (1962).
5. J. A. Bassham and M. Calvin: *The Path of Carbon in Photosynthesis,* Prentice-Hall, Englewood Cliffs, New Jersey (1957).
6. E. Broda: *The Evolution of the Bioenergetic Processes,* Pergamon Press, Oxford (1978).
7. H. T. Witt: Biophysikalische Primärvorgänge in der Photosynthesemembrane [Biophysical primary process in photosynthesis membranes], *Naturwissenschaften* **6:**23 (1976).
8. M. Calvin: Photosynthesis is a source for energy and materials, *Photochemistry and Photobiology* **23:**425 (1976).
9. D. Hoffmann, R. Thauer, and A. Trebst: Photosynthetic hydrogen evolution by spinach chloroplasts coupled to a *Clostridium* hydrogenase, *Z. Naturforsch.* **32:**257 (1977); J. R. Benemann *et al.:* Energy Production by microbial photosynthesis, *Nature* **268:**19 (1977).
10. E. Broda: Grosstechnische Nutzbarmachung der Sonnenenergie durch Photochemie [Large scale use of solar energy by means of photochemistry], *Naturwiss. Rundschau* **28:**365 (1975).
11. H. Kuhn *et al.:* Manipulation in molecular dimensions, *Z. Physik. Chem. N. F.* **101:**337 (1976).
12. D. Möbius: *Ber. Bunsenges. Physik. Chem.* **82:**848 (1978).
13. H. Kuhn, D. Möbius, and H. Bücher, in: *Physical Methods of Chemistry,* Vol. 1, Pt. 3B (A. Weissberger and B. W. Rossiter, eds.), John Wiley & Sons, New York (1972).
14. H. Kuhn, in: *McGraw-Hill Yearbook of Science and Technology 1977,* McGraw-Hill, New York, p. 69 (1977).
15. R. Memming: *Electrochimica Acta* **25**(1)**:**77–78 (1980).
16. M. A. Butler and D. S. Ginley: *J. Mater. Sci.* **15**(1)**:**1–19 (1980).
17. E. Hamer, H. Möhwald, B. Obkircher, W. Schäfer, and B. Schröder: Möglichkeiten der verstärkten energetischen Nutzung photochemischer, photoelektrochemischer und biologischer Umwandlungsverfahren [Possibilities for increased conversion efficiencies in photochemical, photoelectrical, and biological conversion processes], BMFT-Forschungsbericht T 83-036, March (1983).
18. K. K. Rajeshwar, P. Singh, and J. Dubow: *Electrochimica Acta* **23**(11)**:**1117–1144 (1978).
19. A. J. Nozik: *Philos. Trans. R. Soc. London Ser. A* **295**(1414)**:**453–470 (1980).
20. A. Fujishima and K. Honda: *Nature* **228**(5358)**:**37/38 (1972).
21. H. Gerischer: *Umschau in Wissenschaft und Technik* **81**(14)**:**428–431 (1981).
22. Deutsche Forschungs- und Versuchsanstalt für Luft- und Raumfahrt: Wasserstoff als Sekundärenergieträger—Vorschlag für ein Forschungs–und Entwicklungsprogramm [Hydrogen as a secondary energy carrier—Proposal for a Research and Development Program], DFVLR-Mitteilung 81-10, Stuttgart (1981).

CHAPTER 8

The Electrolytic Process for the Production of Hydrogen

The basis of a modern hydrogen technology will be that process for the production of hydrogen which is the most favorable from both the energetic and economic points of view. A very adequate survey of the various methods of producing hydrogen has been given by Justi.[1]

Most of the hydrogen produced at present (about 80% of the world production) is made by petrochemical processes such as the catalytic re-forming process and the partial oxidation, or thermal and catalytic cracking, of hydrocarbons.[2] As far as the rest of the production is concerned, the next most important method is the gasification of coal and coke, which constitutes about 16% of the market. These processes ought to decline as a fraction of the whole during transfer to a hydrogen technology because they use fossil fuels, coal, oil, and natural gas, which should be conserved for use as raw materials in industry.

With respect to the large-scale manufacture of hydrogen, there remains the electrolytic process, which has a special position in this field, both with respect to its technical character and also in a qualitative sense, because the necessary raw material, water, is available in nearly unlimited amounts. With the use of the electrolytic method, it is possible to produce highly purified hydrogen and oxygen, duly separated, in an electrolyzer. Such a device is easy to construct, operate, and maintain.

Another interesting approach to the production of hydrogen, from both the energetic and technical points of view, is provided by cyclical chemical processes in which the necessary energy is provided by nuclear or solar heat.[3] With respect

This chapter authored by Dr. P. W. Brennecke, Braunschweig.

to the photolysis of water as well as biotechnical reactions leading to hydrogen production, which have been intensively examined only in recent years, we shall not discuss them further here, since they have already been discussed in Chapter 7.

Following a description of the thermodynamic basis of water decomposition, electrolytic processes will be described in this chapter, various ways to improve them will be discussed, and finally cyclical processes for the thermal production of hydrogen will be examined.

8.1. THERMODYNAMICS OF WATER DECOMPOSITION

The general meaning of the term "electrolysis" is the electrochemical decomposition of a substance that exists in electrolytic solution. With the passage of an electric current through the phase-boundary cathode-electrolyte and anode-electrolyte, a series of consecutive electrochemical reactions occur at each electrode; in the case of water electrolysis, it is not the electrolyte but the solvent that undergoes decomposition into its components, hydrogen and oxygen. In alkaline electrolyte, the overall reaction is

$$H_2O + \text{energy} \rightarrow H_2 + \tfrac{1}{2}O_2$$

Here, cathodic reduction (electron acceptance) of water molecules gives rise to negatively charged hydroxyl ions, with the evolution of hydrogen. Thus, the overall cathodic reaction is

$$2H_2O + 2e^- \rightarrow H_2 + 2OH^-$$

At the anode, the oxidation of OH^- ions (i.e., their giving up of electrons to the electrode) takes place to form water molecules and oxygen. The overall anodic reaction is

$$2OH^- \rightarrow \tfrac{1}{2}O_2 + H_2O + 2e^-$$

In the ideal case, the minimum energy that would have to be used in the adiabatic and isobaric processes for the splitting of water molecules would be equal to the reaction enthalpy, ΔH; this amount of energy would have to be introduced into the system in a purely electrical form. If the process is carried out isothermally and isobarically, then, according to the thermodynamic equation

$$\Delta H = \Delta G + T\Delta S$$

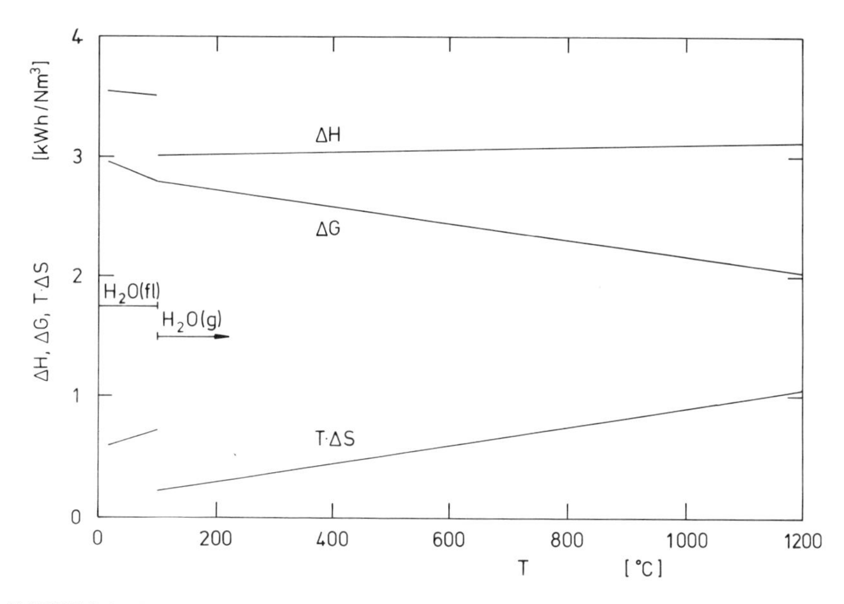

FIGURE 8.1. Temperature dependence of ΔH, ΔG, and $T\Delta S$ during electrolytic water decomposition.

$Q = T\Delta S$ has to be introduced from outside the system, in addition to the free energy necessary to make the reaction occur. In this case, the minimum, purely electrical amount of energy used is equal to the free energy, ΔG. If only ΔG is available in electrical form, the process takes place endothermally, because an additional amount of heat, $T\Delta S$ is required. Above the thermoneutral energy, ΔH, the process takes place exothermally, and excess heat is given off.

To gain some idea of the thermodynamic quantities that are involved in this reaction, the temperature dependence, calculated with the help of thermodynamic data, is given in Figure 8.1 for the temperature range between 25 and 1200°C.[4] In the temperature interval concerned, ΔH is almost constant up to the temperature of the boiling point, while the free energy and the quantity $T\Delta S$ vary to a far greater degree.

Because of the energetic relationships shown in the figure, it is sensible to carry out the process at the highest practical temperature so that a maximum amount of the necessary energy is derived from relatively cheap thermal energy.

For one complete unimolar reaction according to Faraday's First Law, the free energy is related to the reversible cell potential, ε, by the equation $\Delta G = nF\varepsilon$. The theoretical minimum potential for the decomposition of water follows, therefore, as $\varepsilon = \Delta G/nF(V)$; correspondingly, the electrical expression for the change in enthalpy in the reaction is $\varepsilon' = \Delta H/nF(V)$.

The temperature coefficient of the reversible cell potential is given by

$$\left(\frac{\partial \varepsilon}{\partial T}\right)_f = \frac{1}{nF}\left(\frac{\partial \Delta G}{\partial T}\right)_p = -\frac{\Delta S}{nF}$$

This equation shows that with increase in temperature, the cell potential is reduced. The value can be calculated by the use of the data of Barin and Knacke[4] for the temperatures between 25 and 100°C as -0.8 mV/K and $100°C \leqslant T \leqslant 1200°C$ as -0.2 mV/K. The temperature dependence of the cell potential can be obtained by integration of the aforestated relationship. In Figure 8.2, both ε and ε' are plotted for the range 25–1200°C. The dependence of ε on $\varepsilon(T)$ and of ε' on $\varepsilon'(T)$ corresponds to the trends that have already been illustrated in Figure 8.1; while ε' in the temperature interval considered varies little up to the point at which one has to take into account the latent heat of evaporation, ε varies at a considerably greater rate. The value of the reversible decomposition potential for water under standard conditions comes out to be $\varepsilon_0 = 1.229$ V. Correspondingly, $\varepsilon'_0 = 1.481$ V. Knowledge of this quantity allows determination of the minimum *electrical* work needed for the deposition

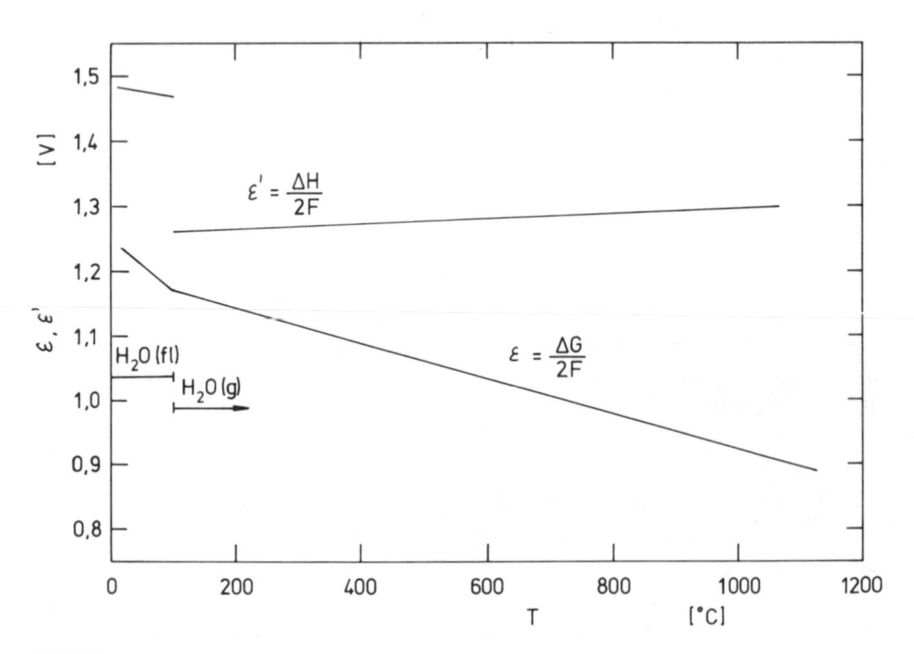

FIGURE 8.2. Temperature dependence of the thermodynamically calculated minimum decomposition potential ε (representing ΔG) and ε' (representing ΔH).

of 1 Nm^3 of hydrogen and 0.5 Nm^3 of oxygen. It also gives rise to the definition of the efficiency of electrolysis, η. In general, one has for the necessary energy quantity, W_e, the quantity $U_K NnF$ (in kWh), where U_{Kl} is the working potential of the electrolytic cell, N is the number of moles of hydrogen, and nF is the amount of electricity needed for the deposition of 1 mole of hydrogen. If, for U_{Kl}, one substitutes ε (or, correspondingly, ε'), the minimum electrical energy needed, W_e^{min}, follows from the equation and at $T = 25°C$ amounts to 2.94 (or, correspondingly, 3.54) kWh. If the temperature is increased to 100°C, this is reduced to 2.79 (or, correspondingly, 3.51) kWh. The electrochemical efficiency of electrolysis (cf. Figure 8.2) is then calculated from the ratio of the energies theoretically necessary in this ideal calculation to those that have to be used, that is:

$$\eta = W_e^{min}/W_e^{exp} = \varepsilon/U\ (\%)$$

8.2. CONSTRUCTION OF PRACTICAL WATER ELECTROLYZERS

A schematic of an electrolytic cell is given in Figure 8.3. The electrodes which are connected to the direct current source are immersed in the electrolytic solution, usually an aqueous KOH solution, which fills the cell. To carry out the electrolytic process at constant electrolyte concentration, water has to be added to the cell at a steady rate. The cathode and the anode are separated by a membrane that allows the passage of ions, but not of gases. The gases produced are taken from the cathode and anode compartments for purification and storage. Since the anodic and cathodic partial reactions are influenced by the type, concentration, and temperature of the electrolyte, and also by the shape and size of the cell, there are considerable deviations from the theoretical ideal potential and the actual potential that must be applied to obtain a given current. The difference between the actual potential necessary to drive the cell at a given rate and the ideal thermodynamic potential is called the "polarization" or "overpotential."

A hydrogen electrolysis plant such as would be needed for large-scale manufacture consists of a rectifier for the production of direct current, the actual cell block, and the necessary subsidiary parts of the plant, such as the container for the fresh water supply, various devices for appropriate electrolyte preparation, gas coolers and washers, catalysts for hydrogen and oxygen purification, and drying and storage apparatus.

There are two types of hydrogen electrolyzers, unipolar and bipolar. In a unipolar tank cell, there is a series of electrolytic cells in which one or several anode–cathode pairs connected in parallel are immersed. In Figure 8.4, this type of cell construction is shown schematically.[5] Such an arrangement is easily

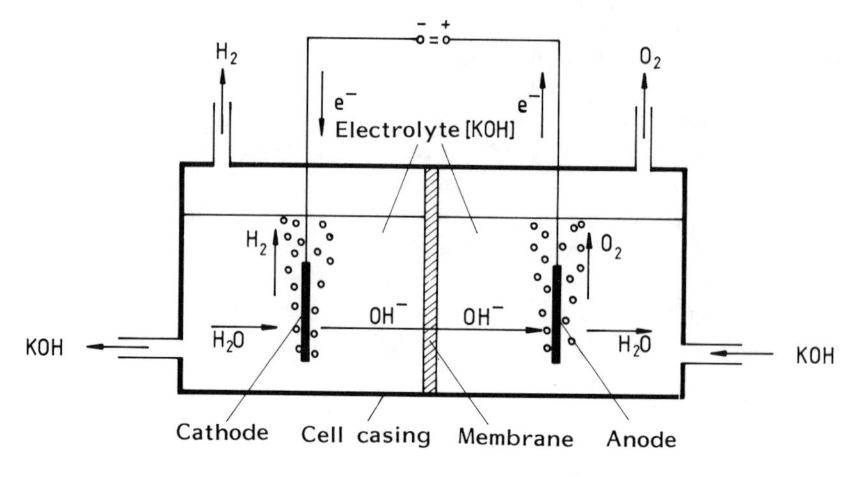

FIGURE 8.3. Schematic diagram of an alkaline water electrolyzer cell and simplified representation of the reaction process.

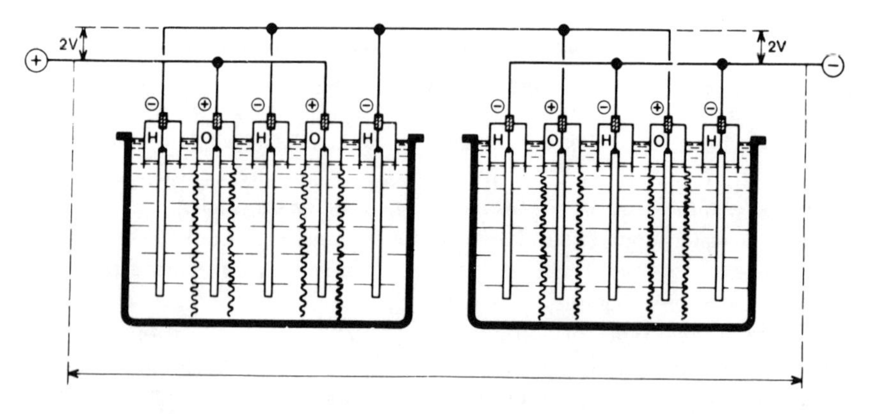

FIGURE 8.4. Schematic diagram of a unipolar electrolyzer. From Stuart.[5]

constructed, and the failure of an individual cell does not lead to an interruption of the current. On the other hand, such cells require considerable space, low operational temperature (60°C), and relatively small current densities (maximum 70 mA × cm^{-2}). Because of these factors as well as their high total current and total energy use, they are now considered obsolete.[6,7]

Most water electrolyzers manufactured at present are made up of bipolar cells, consisting of numerous single cells fitted together; the arrangement is similar to that of the filter press principle and consists of several compact cell blocks. In these bipolar cells, one side of each electrode acts as an anode and the other side as a cathode, although there is no connection to a current source.

Atmospheric or normal-pressure electrolyzers work between 10 and 70 mbar above atmospheric pressure.[8–10] As the schematic diagram in Figure 8.5 shows, the anodes, made of perforated pore-free nickel-plated iron or steel sheets, and the cathodes, made of mild steel, are mounted as outer electrodes on a middle electrode. The outer electrodes are roughened either mechanically by sandblasting or activated chemically by various methods to increase their surface area, the objective being to reduce the polarization potentials. The vertical arrangement of the electrodes facilitates circulation of the electrolyte simply by means of the

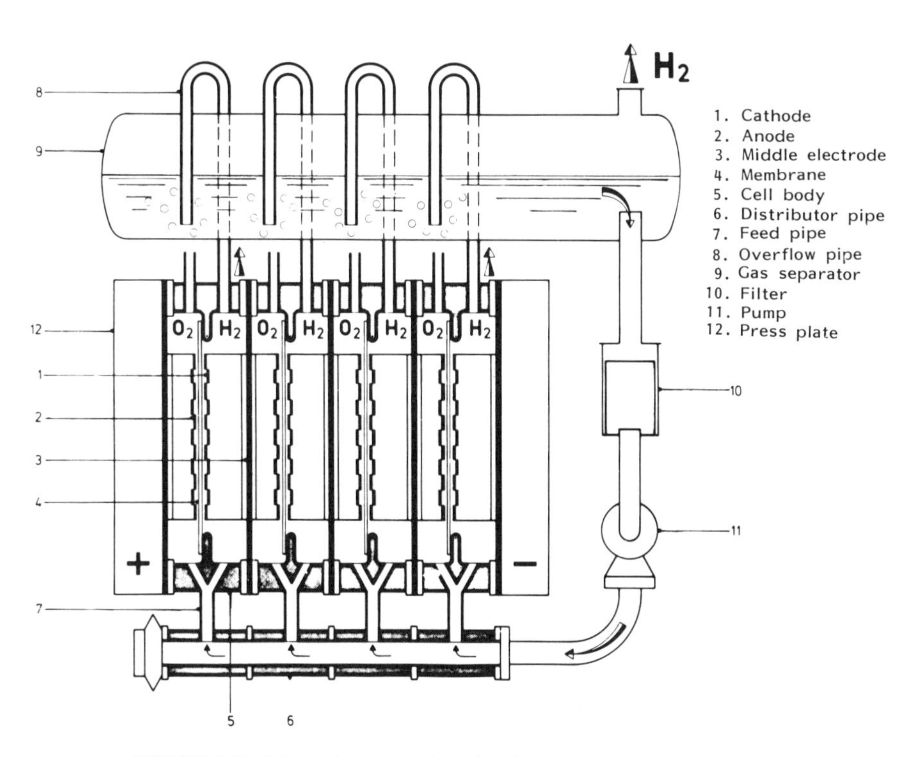

FIGURE 8.5. Schematic construction of a bipolar atmospheric electrolyzer.

resultant rising stream of gas bubbles. The anode and cathode compartments are separated by the introduction of a pure asbestos membrane into the cell. The perforated outer electrodes are kept as close to the membrane as possible to avoid or reduce large ohmic potential differences that might arise in the electrolyte because of the rising stream of gas bubbles. The deposited gases are thus allowed to reach the current-free region between the outer and middle electrodes by passing through the perforations in the electrode sheets. They finally reach the gas–water vapor–electrolyte mixture in the gas-separating and -cooling vessel above the cell block. The electrolysed gases are dried in a purification vessel connected to the cell, compressed, and stored. Depending on the required performance and the current density applied, the cell voltage in such normal-pressure electrolyzers varies between 1.88 and 2.27 V. The specific energy use per 1 Nm^3 of hydrogen and 0.5 Nm^3 of oxygen produced is about 4.5–5.45 kWh.

A corresponding process developed by Zedansky and Lonza involves working at a higher pressure, approximately 30 bar, to reduce the primary energy use.[6,11,12] As the pressure is increased, the decomposition potential (and hence the energy necessary) for the electrolyte actually *increases;* this increase is more than compensated, however, by the effect of the increased pressure on the gas bubbles, which are much reduced in size, thereby decreasing the ohmic potential difference to which they give rise to the electrolyte. Moreover, it is possible to increase the operating temperature to about 90°C without disturbing the formation of gas bubbles in the neighborhood of the electrodes. Thus, by operating under pressure, it is possible to reduce the specific energy used in a water electrolyzer by about 20% or to work at the same cell potential and obtain a higher current density or operating capacity.

Pressure electrolyzers are made according to the filter press principle, as are normal-pressure cells. The middle electrode is stamped out as a thin-sheet completely nickel-plated, and an annular nickel-plated and activated steel wire mesh is applied as an outer electrode. Figure 8.6 shows such an electrolyzer. Pressure electrolyzers operate at cell potentials up to a maximum of 1.95 V and their specific energy use for the production of 1 Nm^3 H_2 and 0.5 Nm^3 O_2 is between 4.3 and 4.6 kWh, including the gas compression energy.

The largest electrolyzer plants in the world have been built for the nitrogen industry to produce hydrogen for the synthesis of ammonia, which is used in the manufacture of artificial fertilizers. These large plants, for which very low-cost hydroelectric energy is available, are to be found in Egypt, India, Norway, and Peru. The largest electrolyzer plant at present, which is fitted with bipolar plates, is a normal pressure electrolyzer in Norway, with a daily production of 1,434,000 Nm^3 of hydrogen; another very large plant is the one at the Aswan Dam in Egypt, which has a daily production of 894,000 Nm^3H_2.[13] In the Egyptian plant as well as in the Norwegian one, the Norsk-Hydro, heavy water is also a by-product of the electrolytic process. All these plants have proved themselves in long-term operation and are characterized by simple servicing, high operational

FIGURE 8.6. Pressure electrolyzer according to Zdansky-Lonza. Photo: Lurgi GmbH, Frankfurt.

safety, and excellent performance conforming to their expected economic characteristics.

However, water electrolysis is intrinsically a costly process, due to the large amount of electrical energy needed (the high price of electrical energy is due to the low Carnot efficiency in the conversion of heat to mechanical energy, a maximum 35–40% in present power stations). Consequently, there has been no intensive building of large electrolytic plants because of the cost attributable to this low overall efficiency. In the concept of a future solar–hydrogen economy, the necessary electricity would be produced from renewable energy sources.

Such a source might be high-temperature collectors,[14,15] producing high-temperature steam to drive low-pressure turbines, which would produce the necessary current. Alternatively, current could be obtained directly from solar cells. Another eventual possibility is the use of wind- or tide-driven generating stations.

8.3. FUTURE POSSIBILITIES AND NEW-TYPE ELECTROLYZERS

Water electrolyzers for the production of hydrogen and oxygen use much more energy than thermodynamic calculations would indicate. An accurate analysis of the specific energy needs shows that the additional component of energy use is overwhelmingly attributable to losses in potential that occur in the form of cathodic and anodic polarization as well as to the ohmic potential drops in the electrolyte. The current efficiency of almost all conventional electrolyzers is more than 98%; however, the goal of improvement in electrolytic technology is the reduction of the necessary applied potential for a given current density and thereby of the electrical energy use. By way of modifying present systems, one might consider using different electrode materials and structures, improving the catalysis of the electrode reactions, changing cell construction and operating parameters, or even introducing radically new electrolytic procedures.

Since the undesirable polarizations that arise at the cathode and the anode lead to an increase in the applied potential, it is desirable to use as electrodes only metals or alloys that have minimal overpotentials (of hydrogen and of oxygen, respectively). Furthermore, there is the problem of corrosion, particularly at the anode, where, with reasonably high current densities, there will be a very positive potential region that obviously increases the possibilities of breakdown. This problem sharply limits the number of substances that can be considered as anode materials. According to present knowledge, the optimal electrode material is nickel, which has good electrolytic properties and is also resistant to hot, highly concentrated potassium hydroxide.[16,17]

Apart from the question of working materials, the electrode structure has to be considered. It is well known from fuel-cell technology that polarization voltages can be reduced if one uses highly porous electrodes with a large internal surface area.[18] In the structure of such electrodes, the catalytically active surface area is much greater than the apparent geometric external surface area, so that the actual current density is substantially less than that calculated from the specific external surface area; by means of these increases of internal area, the overpotential can often be reduced significantly. The advantages of porous electrodes can be illustrated in production prototypes made by the Allis-Chalmers Company[19] as well as the ELOFLUX water electrolysis cell, which is described in the next section. The schematic structure of the Allis-Chalmers cell is shown in Figure

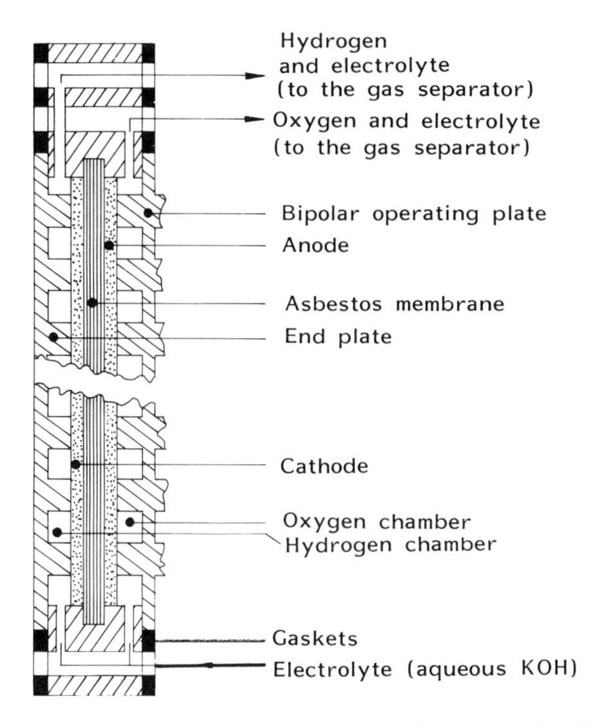

FIGURE 8.7. Schematic construction of the Allis-Chalmers water electrolytic cell.

8.7. The electrodes are fashioned of nickel without any catalyst and are arranged according to the filter press principle. They are circuited in a bipolar scheme and are separated by an asbestos paper matrix that is 0.76 mm thick. The electrolytic gases are piped from the back of the cell and flow together with the electrolyte through the spaces in the external cell structure to the gas-separation chambers. At a cell current density of 300 mA × cm^{-2}, the measured cell potential at 93°C is 1.8 V, so that production of 1 Nm^3 of hydrogen and 0.5 Nm^3 of oxygen requires 4.3 kWh. Although a fully commercialized version of this apparatus has not yet been developed because of the unsolved problems with corrosion and the consequent questions concerning durability, the simple cell construction and the small energy need show that the proper use of porous electrodes results in a decided improvement in the water electrolyzer.

According to Vuilleumier and Braun,[8] the value of the necessary cell voltage is influenced strongly by ohmic potential drops. Different cell constructions and changed operating conditions could therefore be advantageous. In the construction of an electrolytic cell, the interelectrode distances are kept to a minimum, although the presence of concentration polarization has to be taken into account.

Apart from this, it is necessary to avoid potential losses due to gas bubbles in the interelectrode spaces; therefore, the gases that are being evolved are better removed from the back of the electrode rather than from the front.

It is necessary now to look into more future-oriented concepts of the further development of water electrolysis technology—processes that avoid the use of ordinary membranes and instead utilize solid electrolytes consisting of stabilized ZrO_2 ceramics or synthetic ion-exchange membranes. In this way, corrosive alkaline or acid electrolytes can be avoided, and a particular advantage is the constant electrolyte concentration. There are two such advanced concepts in electrolysis: the solid-polymer electrolyte (SPE) process and the high-temperature steam process.

The General Electric Company developed the SPE process utilizing a special sulfonic acid polymer as the solid electrolyte. This electrolyte has physical properties that are somewhat similar to those of Teflon, a high hydrogen-ion conductivity, and sufficient stability with respect to oxygen.[20] As shown in Figure 8.8, the anode (covered with catalytic-active alloys) and the cathode (1–5 mg black platinum/cm^{-2}) are placed directly on the exchanger, which is about 0.5 mm thick. Distilled water vapor circulates through the cell and at the anode is decomposed into hydrogen ions and oxygen. The hydrogen ions then migrate via the fixed SO_3^- groups in the polymer to the cathode and are discharged there, forming hydrogen. At a temperature of $T = 93°C$ and $i = 300 \text{ mA} \times \text{cm}^{-2}$, the cell potential is 1.55 V. The necessary energy use for 1 Nm^3 H_2 and 0.5 Nm^3 O_2 is only 3.7 kWh. Whether the SPE process ever finds large-scale application depends on the introduction of modifications that would allow the avoidance of the expensive noble-metal catalyst as well as the solution of gasket problems.

The second process to be considered here is the high-temperature steam electrolytic process. As can be seen in Figure 8.1, a reduction in the cost of energy needed in electrolysis could be attained if it were possible to increase the thermal component of the energy used in water decomposition.[21] At a temperature of about 1000°C, the required energies are about $\Delta H = 3.1 \text{ kWh/Nm}^3$,

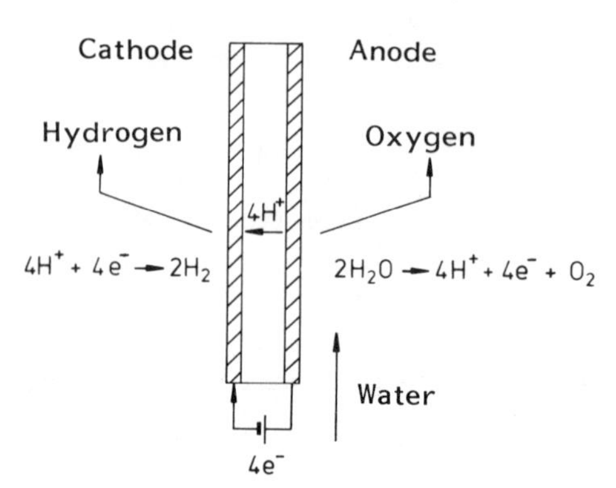

FIGURE 8.8. Principle of SPE water electrolysis.

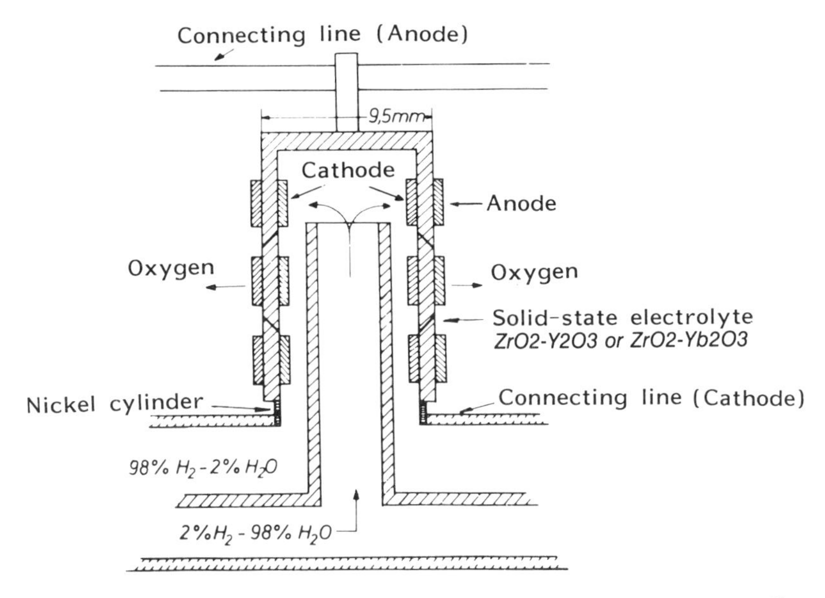

FIGURE 8.9. Principles of construction of a high-temperature steam electrolytic cell.[22]

$\Delta G = 2.2$ kWh/Nm^3, and $T\Delta S = 0.89$ kWh/Nm^3, while the reversible potential for the steam decomposition decreases to 0.92 V. At the high temperatures involved here, the kinetic losses concerned with overpotential at the electrodes are almost zero, so that high-performance current densities can be obtained. The decomposition of the hot water vapor is best carried out on a solid electrolyte in which the basic material is ZrO_2 stabilized by the presence of Y_2O. The arrangement is schematically shown in Figure 8.9, in which the electrodes in an annular form are arranged around a ceramic cylinder consisting of solid electrolyte.[22] The water vapor undergoes the reaction $H_2O + 2e^- \rightarrow H_2 + O^{2-}$ and the steam becomes a hydrogen–water vapor mixture, while the oxygen migrates in the ionized form through the solid electrolyte to the anode. The energy requirement for 1 Nm^3 H_2 and 0.5 Nm^3 O_2 is less than 3 kWh,[23] but there is still much to do to get the process to work in a practical way. Thus, there are considerable problems due to corrosion of the ceramic electrolyte. One must also carefully choose the working electrode materials; further, the handling of oxygen at 1000°C is not easy, and one must question the ready availability of cheap high-temperature heat.

8.4. ELOFLUX WATER ELECTROLYSIS CELL

The developmental problems associated with new methods of electrolysis described in the last chapter are quite considerable, and hitherto there has been no breakthrough that would make these methods immediately practical. Since electrolytic decomposition of water is the reverse of the fuel-cell reaction, in which hydrogen and oxygen combine to make water, Brennecke[17] attempted to improve the water electrolysis technique, utilizing the experience gathered in the examination of alkaline low-temperature fuel cells. From fuel-cell technology, it is well known that polarization potentials can be reduced considerably by the use of highly porous electrodes with large internal surface areas. In electrodes that are structured in this way, the catalytically reactive surface is greater than the geometric surface, so that the surface density for a specific electrode surface area is much reduced and the overpotentials are made correspondingly smaller. The best electrodes to develop would be the porous electrodes that originated in the work of Justi *et al.*[24] and were developed according to the double skeletal catalyzer principle (sometimes abbreviated: DSK principle). DSK electrodes consist of a very stable metallic supporting lattice in which the highly disordered active catalyst is embedded. The high porosity needed is not obtained by the use of sintering, but by a process used in powder metallurgy, the simultaneous hot pressing of metal powder mixtures followed by the dissolving out of a salt filler portion within the structure. The structure of such an electrode is illustrated in Figure 8.10. The pores produced by the salt filler are shown in black, the supporting lattice in white, and the catalyst grains in gray.

In the DSK electrode, two pore systems are available, one within the other. The finer-pore system is controlled by the microporosity of the actual grains, while the larger-pore system is formed by the wedge-shaped regions between the actual supporting lattice and the catalyst grains as well as by the macropores that are produced by the dissolving of the salt filler. The large electrode surface area gives rise to a reduction of overpotential in the phase boundaries between electrodes and electrolyte. This in turn causes a lessening of the energy need or, for the same cell potential, an increase in the current density. This fact is shown in Figure 8.11 in which the potentials for a nickel anode evolving oxygen at 80°C are plotted as a function of porosity. It can be seen that with growing porosity, the overpotentials are sharply reduced. For example, they are reduced by about 375 mV for a current density of $i = 350 \text{ mA} \times \text{cm}^{-2}$ when the porosity varies from 0 (nickel sheet) to 83.5% (hot pressed carbonyl nickel electrode).

Justi *et al.*[24] have introduced porous anodes and cathodes into fuel cells, in which the use of highly active, nonnoble, and relatively cheap Raney nickel catalyst gives rise to a considerable increase in efficiency. Normal Raney nickel is an alloy with about 50 weight-% nickel and 50 weight-% aluminum, the aluminum component being dissolved out after the hot pressing. The result is a nickel structure that is highly disordered and somewhat spongy in appearance, with a large internal surface area ($\approx$100 m^2/g) and high porosity. Use of a Raney nickel mixed catalyst, the activity of which is increased by the addition of 2 weight-% titantium, leads to particularly low overpotentials.[17]

In the pores of the gas-diffusion electrodes that are used in hydrogen–oxygen

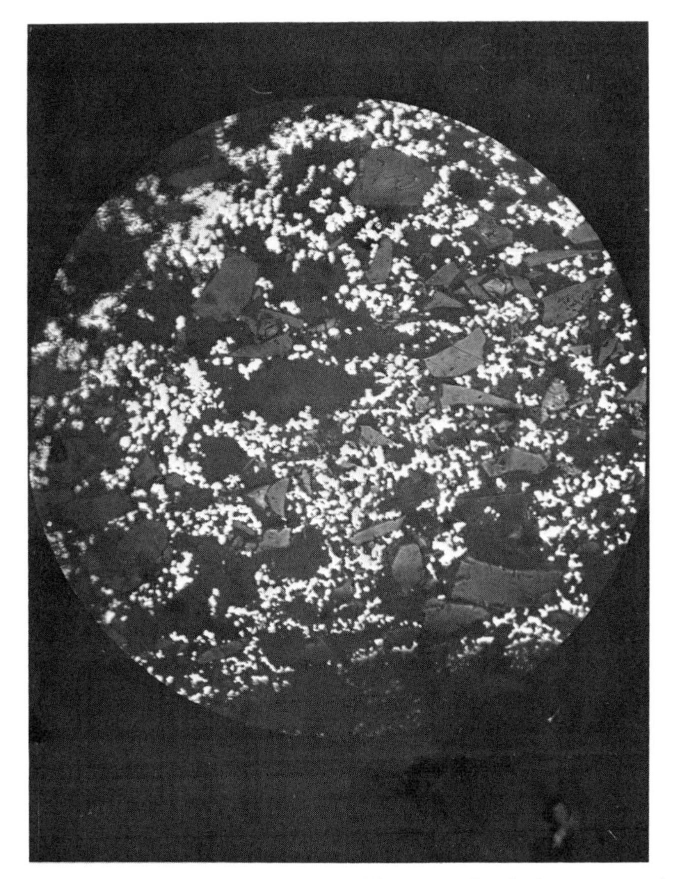

FIGURE 8.10. Photograph of a DSK electrode (white: supporting lattice; gray: catalyst grains; black: pores produced by salt filler). $\times 500$.

fuel cells, there are considerable concentration polarizations at high current densities. These polarizations influence the current distribution and the limiting current density that can be expected from the cells. To avoid these undesirable concentration differences between the interior of the electrodes and the free electrolyte consisting of potassium hydroxide, there must be free electrolyte flowing through the porous anodes and cathodes.[25] This is particularly easy in the case of DSK electrodes because a slow, continuous flow of electrolyte through the cell can be applied without difficulties by means of the hydrostatic pressure. The application of this flow principle has led to a process called the ELOFLUX arrangement.[26]

The construction of a water electrolysis cell[27] on the ELOFLUX principle is based on the idea that the two pore systems are in contact; because of the

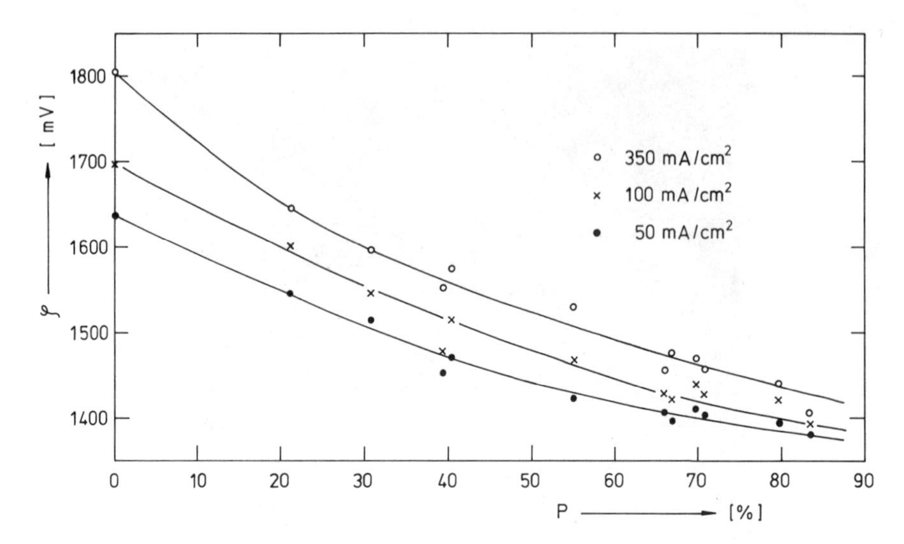

FIGURE 8.11. Influence of the porosity of a hot pressed nickel electrode on the potentials produced by anodic O_2 formation at 80°C.

high capillary pressure, the fine-pore system is filled with electrolyte, and the large-pore system is filled with hydrogen or oxygen being evolved. According to Figure 8.12, there are no free electrolyte or gas spaces, but there are alternating porous anodes and cathodes with fine-pore asbestos membranes between them to allow for the separation of the gases, leading to electrical isolation in a compact cell block. To provide a tight seal around the rim, the electrodes are pressed into stainless steel rings that are resistent to alkaline electrolyte. The flow of electrolyte occurs axially through large-pored separators in the cell. Even at high cell loadings, there is little effect if the gas pressure blows through the electrolyte and empties the pores. As the evolved electrolysis gases are removed to the boundaries of the cell, the gas and the electrolyte streams meet within the electrodes.

The ELOFLUX water electrolysis system has many advantages:

1. Due to the compact construction, the cell volume is small and it can be arranged as a pressure electrolysis cell.
2. In anodes and cathodes made according to the DSK principle, it is possible to build in Raney nickel catalysts.
3. The thickness of the asbestos paper membranes controls the spacing of the individual electrodes and thereby the ohmic potential drops.
4. The axial electrolyte streaming eliminates most of the concentration polarization at high current densities.

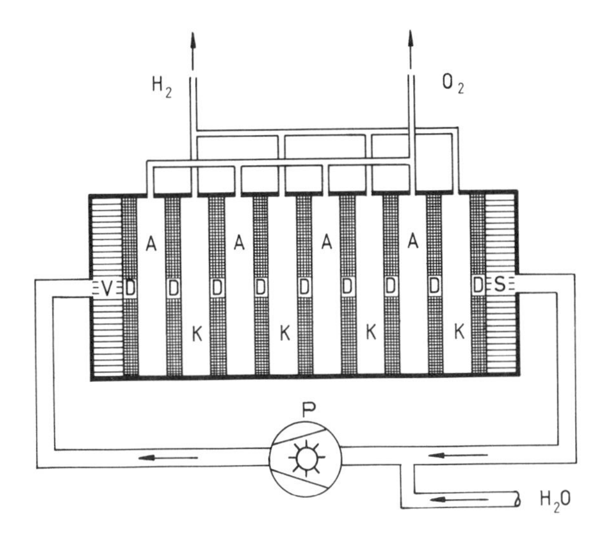

FIGURE 8.12. Schematic construction of the ELOFLUX water electolysis cell. (A) Anode; (K) cathode; (D) asbestos paper membrane; (V) electrolyte separator; (S) electrolyte collector; (P) electrolyte pump.

5. It is easy to convert the electrolysis device into a three-electrode storage cell that can work both as an electrolyzer and as a fuel cell.[30]

Due to these characteristics, the ELOFLUX cell represents one of the better advances toward an improved electrolytic cell with increased performance.

The anodes and cathodes needed in the construction of an ELOFLUX stack are made of a mixture of carbonyl nickel and Raney nickel with 2 weight-% titanium and a sodium carbonate filler; the porosity of the electrodes is about 80%. At a flow pressure driving the electrolyte through the electrode of about 145 mbar, the maximal flow per square centimeter of electrode cross-sectional surface is 0.6 cm^3 KOH/hr. The energy use for this small flow is negligible. Figure 8.13 shows the current–voltage curve (......) for an ELOFLUX cell working at 90°C and with an excess gas pressure of 0.5 bar. It may be compared with the corresponding curves for various other electrolysis cells,[31] both old and new. This figure shows that electrolytic hydrogen production with an ELOFLUX cell takes place with a relatively small primary energy use. At 90°C and 200 mA $\times$ cm^{-2}, the applied potential is 1.57 V, whereas in conventional normal-pressure cells, the potential varies between 1.88 and 2.27 V.

In Figure 8.14, the electrolysis efficiencies, both η and η', are given, and they are correspondingly high. If one plots the value of η, i.e., the efficiency related to the free energy ΔG (for 90°C, ε = 1.76 V), (thus relating it to the electrical energy) then the efficiency is about 80%, which is considerably more

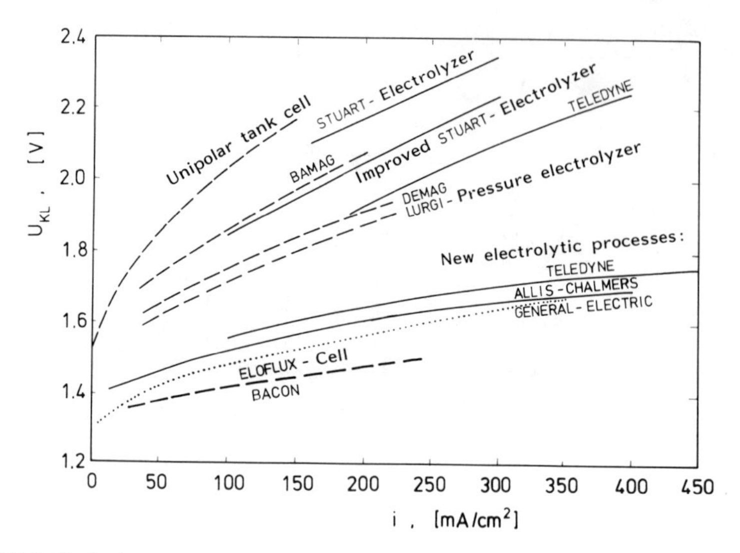

FIGURE 8-13. Stationary current characteristic curve of an ELOFLUX water electrolysis cell in comparison with technical electrolyzers and other new developments. From Gregory *et al.*[31]

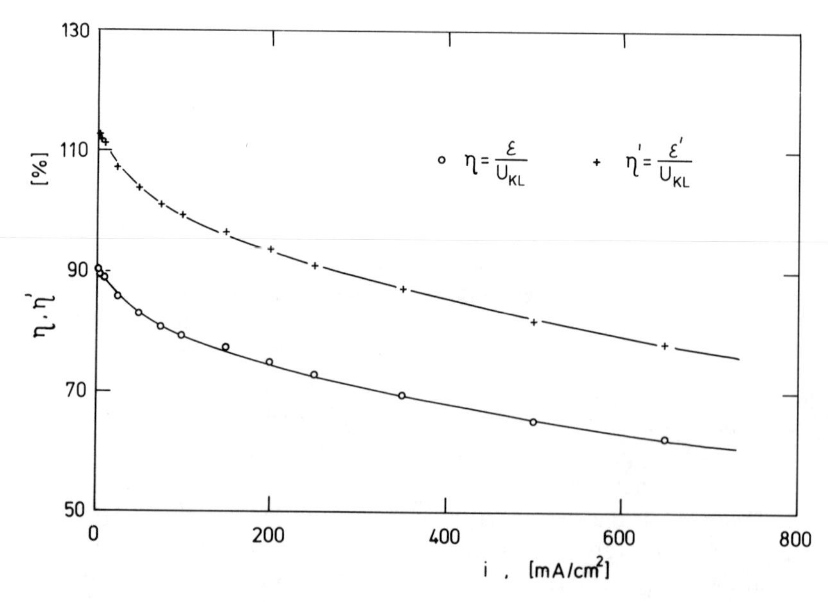

FIGURE 8.14. Dependence of electrolytic efficiency η (related to ΔG) and η' (related to ΔH) of the ELOFLUX water electrolysis cell, operated at 80°C, on the cell current density.

than the 60–70% that is characteristic of most technical electrolyzers.[22] On the other hand, if one relates the efficiency to the enthalpy, ΔH, and uses 90°C and the corresponding thermoneutral voltage of 1.471 V, the efficiency may exceed 100% with respect to the electrical energy. The additional energy $T\Delta S = \Delta H - \Delta G$ is provided either by heating in the cell by means of the I^2R losses or by externally applied heat (otherwise the cell tends to cool).

With respect to energy use, in electrolyzers working at atmospheric pressure, the specific use for the production of 1 Nm^3 H_2 and 0.5 Nm^3 O_2 is 4.5–5.45 kWh, while with the Zedansky–Lonza pressure electrolyzer, it is about 4.3–4.6 kWh. However, in the case of the ELOFLUX electrolysis cell, there is a smaller energy requirement. For example, at 90°C and 200 mA $\times$ cm^{-2}, it is necessary to utilize only 3.75 kWh. These figures can be compared with the theoretical minimum energy needs, referred respectively to the ΔH and ΔG quantities of 3.51 kWh (for the ΔH at 90°C) and 2.81 kWh (for the ΔG at 90°C).

8.5. THERMOCHEMICAL PROCESSES

Many publications have appeared in recent years concerning an alternative to electrolytic hydrogen production, i.e., thermochemical processes, by which—with no detour to make electrical energy—hydrogen can be obtained from water using only process heat.[3,32,33] Although none of these thermochemical processes has been fully developed, they certainly appear to be promising.

No significant direct thermal dissociation of water molecules occurs until the temperature reaches 2000–3000°C; since this is above the practical temperature at which processes can be carried out (because of material difficulties), direct thermal dissociation can be discarded. On the other hand, it is possible to run water decomposition in a closed cycle if the overall processes are split up into various partial reactions, in the course of which only the water that is to be decomposed is added and hydrogen and oxygen are removed, the intermediate products being cycled with minimal loss. Corresponding to the thermodynamic relationship $\Delta H = \Delta G + T\Delta S$, it would be best to run reactions that have a positive entropy at high temperatures, while reactions that have a negative ΔS should be run at the lowest possible temperatures. The change in free energy of the whole process would thereby be minimized. Cyclical processes that are considered in this regard generally consist of three or four reaction steps. If the higher temperature is reduced, the number of partial reactions increases; and with each increasing step, the thermal efficiency of the overall process is reduced.[3] In this way, according to the various suggestions that have been made recently, thermal energy from high-temperature reactions or from solar energy plants could be utilized for the direct production of hydrogen and oxygen.

With the use of computer calculations of the thermodynamic data for various chemical steps, it is possible to set up various cyclical processes and, according to what substances are being used, arrange them into families of, for example, such metal salts as halides or sulfates. The best-known families are represented by iron/chlorine, iron/sulfur, and manganese/sulfur. Among the more classic processes that have been developed is that of deBeni and Marchetti.[34] It is called the Mark I Cyclical Process, and the reactants with which it begins are $CaBr_2$ and Hg:

$$CaBr_2 + 2H_2O \rightarrow (CaOH)_2 + 2HBr \qquad 730°C$$

$$Hg + 2HBr \rightarrow HgBr_2 + H_2 \qquad 250°C$$

$$HgBr_2 + Ca(OH)_2 \rightarrow CaBr_2 + H_2O + HgO \qquad 200°C$$

$$HgO \rightarrow H + \tfrac{1}{2}O_2 \qquad 600°C$$

The Mark I Cycle is the most developed and energetically the most favorable, when compared with many of the other cycles that have been discussed and that consist of several steps.

A typical representative of the iron/chlorine family of cycles is the following four-step process[35,36]:

$$3FeCl_2 + 4H_2O \rightarrow Fe_3O_4 + 6HCl + H_2 \qquad 930°C$$

$$Fe_3O_4 + 8HCl \rightarrow FeCl_2 + 2FeCl_3 + 4H_2O \qquad 380°C$$

$$2FeCl_3 \rightarrow 2FeCl_2 + Cl_2 \qquad 330°C$$

$$Cl_2 + H_2O \rightarrow 2HCl + \tfrac{1}{2}O_2 \qquad 1030°C$$

In this reaction sequence, there are several difficulties.

One is the coupling of the necessary process heat, both in the internal heat exchange and in the separation of the gaseous reaction products. Another is the properties of iron(II) chlorides, which are liquid at about 680°C. These difficulties are in addition to the material problems associated with the continuous running of reactors at these temperatures. For example, the actual running of the step $2FeCl_3 \rightarrow 2FeCl_2 + Cl_2$ at 330°C has proved to be rather disappointing (there being side reactions).

At present, it is not possible to judge the practicality of these processes because a la-ge amount of data has to be collected with respect to the reaction kinetics of each single reaction, and an examination must be made of whether they work in combination.[32] With respect to cycles that are related to the iron/chlorine family, one of the main difficulties is presented by the hydrolysis of $FeCl_2$ and the thermal decomposition of $FeCl_3$.[37] Apart from this, there are at present no commercially practicable high-temperature reactions or solar power stations that give the necessary thermal energy at a level of about 1000°C. Another fundamental question that must be asked concerns the material needs and very considerable corrosion problems that would arise; among the most critical are the safety risks associated with working at these high temperatures with such corrosive and highly toxic substances as Cl_2, CO, SO_2, and Hg. Since some of these dangerous gases could be released from the process and give rise to considerable environmental hazards, thermochemical cyclical processes do not rate highly from the safety viewpoint.

8.6. CONCLUSIONS

Compared with all the other process possibilities, water electrolysis is by far the best-known and comparatively simplest method of hydrogen production. Although the electrolyzers that exist at present have proved themselves in long service to give economical operation within the performance expected, the hydrogen they produce is relatively costly because of the use of expensive (Carnot-dependent) electrical energy in producing it. Thus, the water electrolyzer will be a real alternative for the direct production of hydrogen when electrical energy can be produced from renewable sources (solar plants) at a sufficiently cost-favorable rate.

To reduce the primary energy needs for water electrolysis sufficiently, the main aim of current research on water electrolysis should be to carry out constructive measures for the introduction of more suitable electrode materials, particularly those containing catalysts. In addition, an effort should also be made to improve conditions of operation. Development of the thermochemical cyclical processes should also be continued, and the first experimental station should be set up as a research operation to determine whether the direct use of nuclear or solar process heat can given rise to the realization of multistep processes at a technical level.

REFERENCES

1. E. Justi, in: U. Bossel (ed.), 1. Deutsches Sonnenforum [German Solar-energy Forum], Tagungsbericht der 5. Tagung der Deutschen Gesellschaft für Sonnenenergie e.V. (DGS) [Report of the 5th Meeting of the Society for Solar Energy], Hamburg 26–28. 9. 1977, Vol. II, pp. 517–549, DGS, Munich (1977).
2. G. Kaske, *Chemie-Ingenieur-Technik* **48**(2):95–99 (1976).
3. H. Hofmann, *Chemie-Ingenieur-Technik* **48**(2):87–91 (1976).

4. I. Barin and O. Knacke: *Thermochemical Properties of Inorganic Substances*, pp. 316, 322, 323, 584, Springer Verlag, Berlin/Heidelberg/New York, and Verlag Stahleisen, Düsseldorf (1973).
5. A. K. Stuart: Modern electrolyser technology in industry, Paper presented at the American Chemical Society Symposium on Non-Fossil Fuels, Boston, April 9–14 (1972).
6. E. A. Chapman, *Chem. Process Eng.* **46**(8):387–393 (1965).
7. Bundesministerium für Forschung und Technologie (publisher): *Programmstudie: Auf dem Wege zu neuen Energiesystemen [On the Way to New Energy Systems]*, Part 3, pp. 7–8, Bonn (1975).
8. H. Vuilleumier and M. Braun, *Rev. Gen. Electr.* **85**(6):534–536 (1976).
9. Lurgi GmbH, Lurgi Schnellinformation T 1190/3.76, Frankfurt a. M. (1976).
10. Davy-Bamag GmbH, Technische Information 6.2, Butzbach (undated).
11. Lurgi GmbH, Lurgi Schnellinformation T 1073/5.73, Frankfurt a. M. (1973).
12. H. Wuellenweber, *Rev. Gen. Electr.* **85**(6):537–541 (1976).
13. M. S. Casper (ed.): Hydrogen Manufacture by Electrolysis, Thermal Decomposition and Unusual Techniques, S. 177/8, Noyes Data Corporation, Park Ridge, Illinois (1978).
14. E. Justi, in: *Akademie der Wissenschaften und der Literatur, Mainz (1949–1974)*, pp. 41–53, Franz Steiner Verlag, Wiesbaden (1974).
15. J. Hofmann and J. Gretz, in: A Derichsweiler and H. Krinninger (eds.), Tagungsbericht [Daily report] 2. Internationales Sonnenforum [International Solar-Energy Forum], Hamburg 12.–14.7. 1978, Vol. I, pp. 185–193, Deutsche Gesellschaft für Sonnenenergie e.V., Munich (1978).
16. M. H. Miles: *J. Electroanalyt. Chem. Interfacial Electrochem.* **60**:89–95 (1975).
17. P. Brennecke: Thesis, TU Braunschweig (1978).
18. F. Justi and A. Winsel: *Kalte Verbrennung-Fuel Cells*, [Cold Combustion—Fuel Cells], F. Steiner, Verlag, Wiesbaden (1962).
19. R. L. Costa and P. G. Grimes: *Chem. Eng. Progr.* **63**(4):56–58 (1967).
20. L. J. Nuttal, A. P. Fickett, and W. A. Titterington, in: T. N. Veziroglu (ed.), Hydrogen Energy, Proceedings of The Hydrogen Economy Miami Energy Conference (THEME), Miami Beach, March 18–20 (1974), Part A, pp. 441–455, New York (1976).
21. W. Dönitz, R. Schmidberger, E. Steinheil, and R. Streicher, in: T. N. Veziroglu and W. Seifritz (eds.), Hydrogen Energy System, Proceedings of the 2nd World Hydrogen Energy Conference, Zürich, August 21–24, 1978, Vol. 1, pp. 403–421, Pergamon Press, Oxford/New York/ Toronto/Sydney/Paris/Frankfurt (1978).
22. A. Gann: Über die Herstellung von Wasserstoff durch, Wasserelektrolyse [On the production of hydrogen by water electrolysis] Deutsche Luft-und Rourmfahrt, DLR-Mitteilung 74–39, Stuttgart (1974)
23. W. Baukal, H. Döbrich, and W. Kuhn: *Chemie-Ingenieur-Technik* **48**(2):132–133 (1976).
24. E. Justi, W. Scheibe, and A. Winsel: DBP 1 019 361 (1954).
25. R. Wendtland: Thesis, TH Braunschweig (1966).
26. R. Wendtland and A. Winsel: DBP 1 496 241 (1965).
27. P. Brennecke, H. Ewe, and E. Justi: Heissgepresste poröse Elektrode und Verfahren zu ihrer Herstellung [Hot-pressed porous electrodes: Processes for their production], DPA P 29 03 407 vom 30.1. 1979, DBP 29 03 407 vom 15. 12. 1983.
28. P. Brennecke, H. Ewe, and E. Justi, in: A Derichsweiler and H. Krinninger (eds.), Tagungsbericht 2. Internationales Sonnenforum [Report on the Second Solar-Energy Forum], Hamburg 12–14. 7. (1978), Vol. II, pp. 247–263, Deutsche Gesellschaft für Sonnenenergie e.V., Munich (1978).
29. P. Brennecke and H. Ewe: *Chemie-Ingenieur-Technik* **52**(5):426–428 (1980).
30. H. Ewe and E. Justi: VDI-Berichte, No. 223, pp. 108–117 (1974).
31. D. P. Gregory *et al.*: A Hydrogen-Energy System, American Gas Association/Institute for Gas Technology, Chicago (1973).

32. Bundesministerium für Forschung und Technologie (publisher): *Programmstudie auf dem Wege zu neuen Energiesystemen [Program for the Study of New Energy Systems]*, Part 3, pp. 45–105, Bonn (1975).
33. T. N. Veziroglu and W. Seifritz (eds.): Hydrogen Energy System, Proceedings of the 2nd World Hydrogen Energy Conference, Zürich, August 21–24, 1978, Vol. 2, Pergamon Press, Oxford/New York/Toronto/Sydney/Paris/Frankfurt (1978).
34. G. deBeni and C. Marchetti: *Euro Spectra* **9**(2):46–50 (1970).
35. K. F. Knoche, H. Cremer, and G. Steinborn: *Int. J. Hydrogen Energy* **1**(1):23–32 (1976).
36. K. F. Knoche: *Chemie-Ingenieur-Technik* **49**(3):238–242 (1977).
37. D. van Velzen and H. Langenkamp: *Int. J. Hydrogen Energy* **3**(4):419–429 (1978).

CHAPTER 9

The Transmission of Energy over Large Distances

The vast energy sources of the future will originate far away from where they are used; because of climatic considerations, solar and OTEC (ocean thermal energy conversion) power stations will have to be located in relatively circumscribed areas of the world. The world's remaining coal reserves are largely in Siberia and Alaska, and nuclear power plants cannot be put in cities because of the danger they present and because of their enormous requirements for cooling water.

Apart from the direct transport of coal (and over great distances this may become uneconomical because coal quality is likely to decline with time), there are three basic processes to be considered:

1. Direct transmission of electricity through power grids or through underground or underwater cables.
2. Microwave transmission via satellites.
3. Transport of hydrogen gas through pipelines in a way analogous to that by which natural gas is already transported.

In this chapter, these three methods for the transmission of energy over great distances and in large quantities will be studied comparatively.

9.1. DIRECT TRANSMISSION OF ELECTRICAL ENERGY

A typical voltage used for transmitting electrical power over large distances is about 600 kV with present technology. Even at this rather high potential, the loss of energy during transmission through a 4000-km transmission line consisting of copper would be about 50%. If the operating voltage could be increased, the loss would be diminished, but there is a limit to the amount by which the potential can be increased because of corona discharge and strong electromagnetic influences on the environment. At present, experiments with 1000 kV are being made, and substantially higher potentials seem to be out of reach.

Conduction losses can also be decreased by reducing the resistance of the cables. However, wiring of greater thickness and weight is very costly. Cooling the cable by means of liquid air (83K) lowers the resistance of Cu or Al by about 15 or 11%, respectively (cryogenic cables). However, the liquefaction of air is an inefficient process. It requires about 1 kWh/kg, and the efficiency of the cooling process is only 0.80 kWh/kg. It is interesting, however, to consider supercooling of the conductor with the aim of completely abolishing the resistance, i.e., cooling it down to temperatures at which superconductivity intervenes.[1–4] The early superconductors were effective only in the range below 10K, and they had to be kept cool by means of liquid helium, below 4.2K. According to the Carnot efficiency formula, the production of 1 cold calorie in the cooling process at 4K requires the use, in an ideal calculation, of 300K/4K = 75 calories of energy at an operating temperature of 300K. In reality, however, the amount required is more like 1000 calories. Furthermore, world supplies of helium are being exhausted; in the United States, no research on superconduction technologies that need liquid helium is being done because of these factors. In 1942, Aschermann *et al.*[5] discovered the nonstoichiometric superconductor NbN, which functions in boiling hydrogen at 16K and this led to the first demonstration of the use of superconducting magnets and superconductivity in a public presentation.

Although research on superconductors is currently well funded, particularly for those viable at higher temperatures, and important scientific meetings are taking place to discuss them, the best superconductors still need a temperature lower than 100K. Another difficulty is the contact between a superconductor and a normal conductor because of the enormous electrical and thermal current that arise in the latter. Finally, there are the practical difficulties connected with the cooling of cables over long distances, the ease of breakdown on the system, and its high costs.*

9.2. TRANSMISSION THROUGH DIRECTED MICROWAVE RADIATION

A concept formulated by Gläser[7] has already been mentioned, the concept being to build satellites that would carry solar cells and microwave transmission

* Translator's note: Discoveries made in 1987 make the possibilities of superconductors of liquid N_2 (rather than liquid H_2) temperatures seem likely. Even if room temperature superconductors become possible, their expense may prohibit their use for large items or transmission.

stations. By means of the latter, solar energy received by orbiting satellites would be beamed to earth. It would also be possible to use such satellites to receive solar energy beamed up from earth and transmit it over very great distances to other countries.

9.3. TRANSMISSION BY MEANS OF HYDROGEN

Recalling the introductory remarks made in this book, and before going on to the more detailed considerations, we think it helpful to summarize the advantages of hydrogen over conventional technologies of fuel distribution. It offers the cheapest method for the transmission of energy, it is convenient for a multitude of applications, avoids the difficulties of the distribution of current in urban centers, and water; the only product of its use as a fuel, is entirely non-polluting. Figure 9.1 gives the investment costs of electricity as a function of distance in comparison to natural gas and hydrogen. According to these data, energy transmission by means of hydrogen will cost \$207.50/kW$_{th}$ (1972 figures) for the

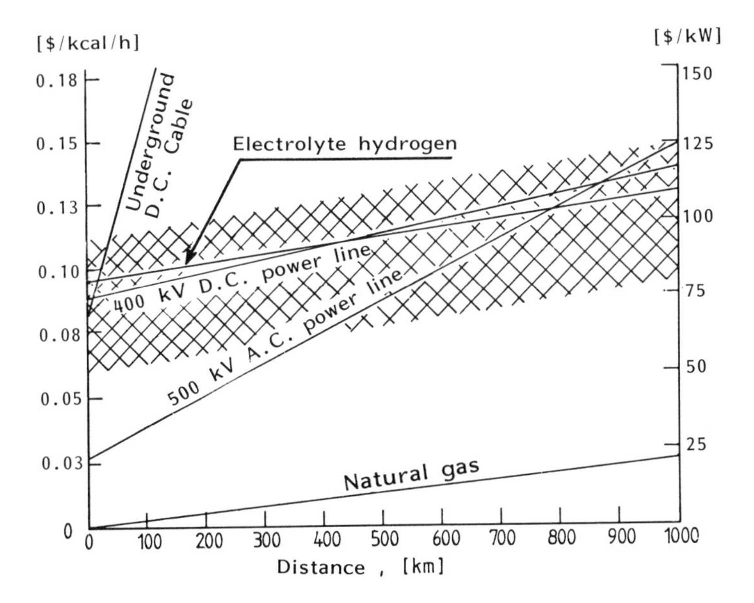

FIGURE 9.1. Comparative presentation of the investment costs (in \$/kW) as a function of the distance between the places of production and consumption, for underground cable, 400 kV D.C. power line, 500 kV A.C. power line, natural gas pipeline of 762 mm net diameter, and H_2 pipeline of 762 mm net diameter. The cross-hatched area above and below the H_2 curve represents the range of variation of the degree of H_2 electrolysis. From Gregory *et al.*[8]

TABLE 9.1
Transmission Costs of Electrical Energy[a]

Data	Estimated price (in dollars)
Average costs of five 500 kV lines, erected since 1969, including costs for right-of-way, excluding costs for one urban power line	161,500/km
Costs of D.C. power line at 65% of the costs of a comparable A.C. power line	105,000/km
Typical output capacity of a 500 kV power line	900,000/kW
Typical costs of an A.C. transformer station	15.45/kW
Typical costs of a D.C. transformer station	60.00/kW
Average costs of a 500 kW power line	0.18/kW × km
Average costs of a D.C. power line	0.11/kW × km
Costs of two A.C. transformer stations	32.00/kW
Costs of two D.C. transformer stations	120.00/kW
Cost ratio of underground and aboveground power transportation	10 : 1 to 30 : 1
Costs of a ground line	1.79/kW × km
Total costs of a power line and transformer station for 320 km	0.138/kW × km
Total costs of a ground line and transformer station for 320 km	2.09/kW × km

[a]From Gregory *et al.*[8] Exchange rate for 1986: $1 U.S. = 2.00 DM. Price increases not considered.

first 1000 miles (1609 km), but after this only $35.50/kW$_{th}$. Hydrogen is somewhat more expensive to transport than methane because of the high cost of the electrolysis of water to produce the water, while there is no equivalent of such costs for methane. The cost of the transmission of hydrogen becomes less than that by a 400 kV D.C. line at 650 km and less than that by a 500 kV A.C. line at 900 km. If one considers underground cable distribution, hydrogen is cheaper even at 50 km.[8]

TABLE 9.2
Specific Cost Comparison for Energy Supply by Electricity, CH_4, or H_2[a]

	Electricity	CH_4	H_2
Production	1.45[b]	0.093[c]	1.6–1.75[d]
Transport	0.33	0.11[e]	0.29
Distribution	0.88	0.15	0.19
TOTAL	2.66	0.35	2.08–2.2

[a] From Meyerhoff.[9] Prices in ¢/kWh$_{th}$ (as of 1986).
[b] Equivalent to 1.45¢/kWh$_{th}$.
[c] Prices for 1986 are higher.
[d] Assumed purchase price for electricity: 1.45¢/kWh$_{th}$.
[e] Assumed pipe for H_2 at 1.65¢/kWh$_{th}$ used to power the compressor in an optimal pipeline, compared with natural gas at 0.15¢/kWh$_{th}$.

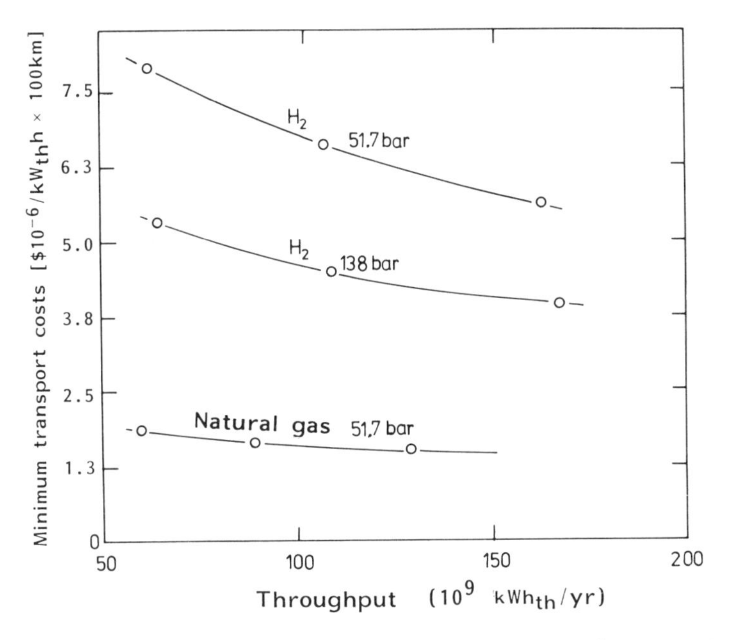

FIGURE 9.2. Comparative presentation of the energy-transport costs (in $/kW$_{th}$ × 100 km) for CH_4 pipelines compared to H_2 pipelines (middle and top) as a function of the throughput (in kWh$_{th}$/yr). The H_2 compression costs are 1.65¢/kWh$_{th}$. Parameter: gas pressure. Note the decrease in transportation costs with pressure and throughput. From Gregory *et al.*[8]

9.4. DIFFERENCES BETWEEN PIPELINE NETWORKS FOR NATURAL GAS AND FOR HYDROGEN

The maximum heating value of hydrogen (H_u = 28.57 kcal/kg) is high, thanks to its low molecular weight. The heating value per unit weight is $2\frac{1}{2}$ times higher than that of methane (H_u = 11.30 kcal/kg). On the other hand, precisely because of the low density of hydrogen (ρ_n = 0.09 kg/m^3), the heating value per unit volume (H_u = 2570 kcal/Nm^3)* is 3.31 times *lower* than that of methane (8550 kcal/Nm^3).[10] Thus, hydrogen is less advantageous than methane with respect to the economics of transmission through pipelines. Because of its minimum density and viscosity, however, hydrogen is a very easily flowing medium, and at an equal pressure ratio (p_1/p_2 = 1.35) and pipe diameter (1 m), the flow velocity is 2.68 times greater than that of methane, so that the transported heat for hydrogen (for a given pressure gradient) is only 9.9% lower than that of methane. It is certainly true that, as Gregory *et al.*[8] state, the required compressor performance for hydrogen is 3.3 times greater than that for methane. The re-

* 1 Nm^3 = 1 m^3 (normal cubic meter) at T = 0°C, and P = 1 bar.

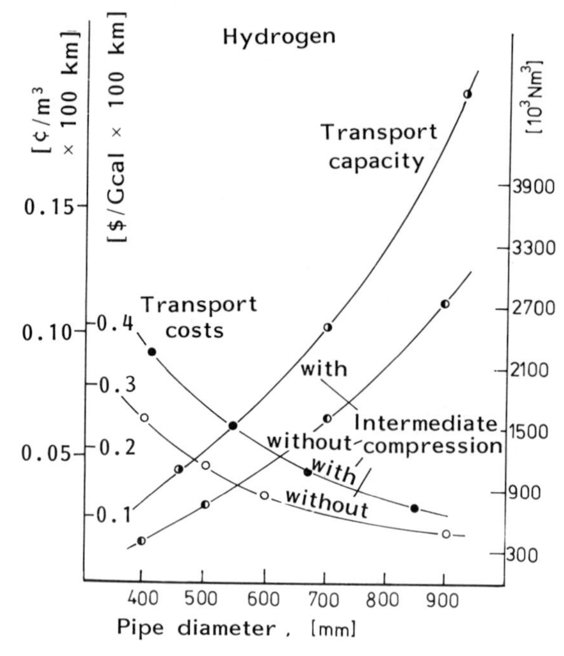

FIGURE 9.3. Costs of energy transport by means of H_2 pressurized gas (in ¢/m^3 × 100 km) as a function of the pipe diameter (in mm) without (lower curve) or with intermediate compression $p_2/p_1 = 1.35$ with $p_i = 67.5$ bar. Intermediate compression decreases the transportation costs with decreasing pipe diameter. The right-hand ordinate shows the increasing transportation capacity with increasing pipe diameter using intermediate compression (in 10^3 Nm^3 H_2).[11]

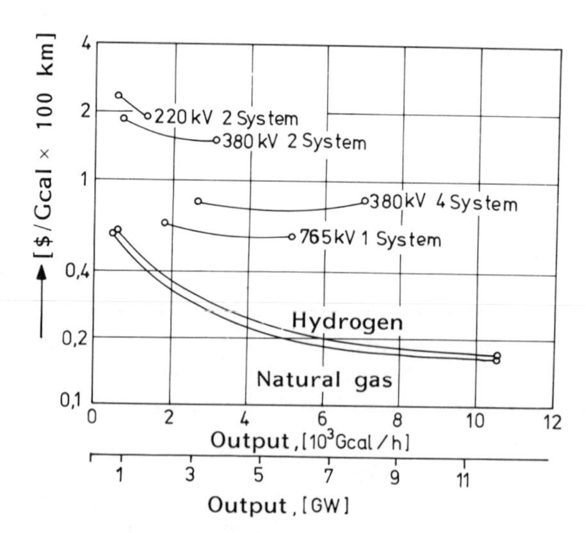

FIGURE 9.4. Comparison of energy-transport costs (in \$/Gcal × 100 km) as a function of the output in 10^3/ Gcal/hr (upper abscissa) or in GW (lower abscissa) for electrical power lines (above ground) with 220 kV and 380 kV in two systems, 380 kV in four systems, 765 kV in one system, contrasting H_2 and CH_4 in optimized pressurized gas pipes. Taken into consideration here are only the power lines or pressurized-pipe investments without costs for switching ends, transformers, or compression, which can be calculated when the transport distance is known. From Thissen.[12]

lationships are shown in Figure 9.2, in which transmission costs are given as a function of throughput. There are interdependent relationships among the pipeline diameter, the pressure ratio, and the throughput; further, the introduction of compressor stations gives rise to the expectation of further optimization. Also, there is a relationship between investment costs C (\$) and the pipeline diameter D (m) that is given by the equation $C = C_0 \times D^n$, where n is 1.09 (for values of D between about 0.5 and 1 m). There is a corresponding formula for the flow velocity, V(Nm³/hr), and diameter, given by the equation $V = V_0 \times D^{2.5}$, so that the specific investment costs C/V can be reduced by a small increase of the pipeline diameter, D, as can be seen in Figure 9.3.[11] Figure 9.4 compares the energy transmission costs in high-voltage lines (excluding switching units) with those of transporting methane or hydrogen through pipes (excluding the condensation costs) as a function of output. Gas transport is clearly cheaper. Figure 9.5 is one of the figures of Gregory and colleagues, calculated for transmission

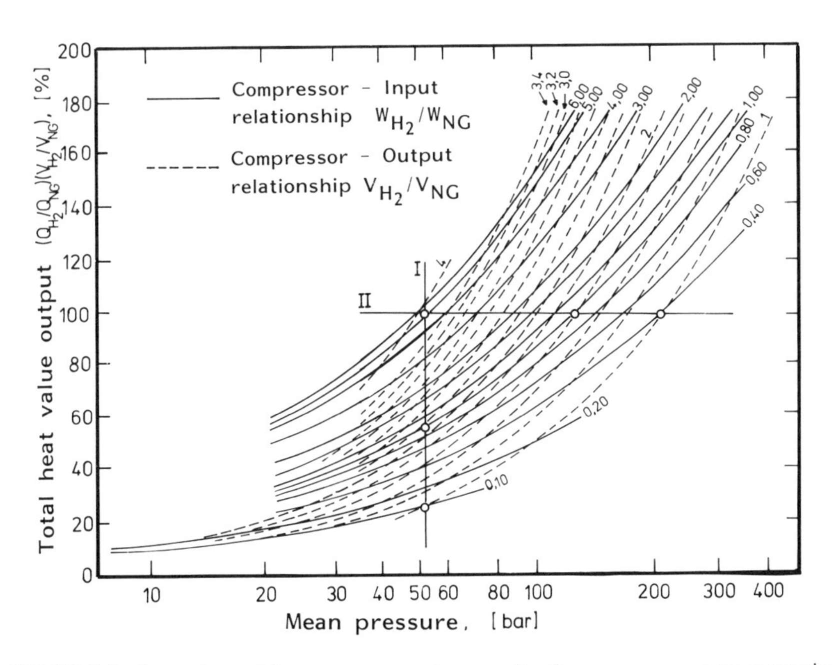

FIGURE 9.5. Comparison of the energy-transporting capacity, the necessary compressor capacity, and the drive power of an H_2 pipeline at different pressures with that of natural gas at 57.1 bar and 15.6°C and a distance between compressors equivalent to the pressure relationship of $\pi_{NG} = 1.3 : 1$. Basic assumptions are: (1) H_2 pressure is variable; (2) pipes are of similar quality; (3) gas flow is totally turbulent; (4) volumetric and thermodynamic degrees of efficiency are the same; (5) fuel use and use of auxiliary energy are the same; (6) reference conditions for natural gas having mean value of 51.7 bar at 15.6°C and pressure relationship of 1.3; (7) mean H_2 temperature of 15.6°C. Ordinate: total heat value output $(Q_{H_2}/Q_{NG})(V_{H_2}/V_{NG})$ in %). Abscissa: mean pressure (in bar). From Gregory *et al.*[8]

and performance needs of compressors for both hydrogen and methane as a function of their dependence on the working pressure.

9.5. OPERATING HYDROGEN PIPELINES AND OTHER NETWORKS

Largely because of the pioneering work of the German chemical industry, the Federal Republic of Germany has some 50 years' practical experience in the construction and operation of high-pressure hydrogen pipelines in which, to date, there has been no record of accidents. The chemical company Hüls (CWH), being skilled in supplying its factories and workshops with 22 different reactive gases, has an 875-km-long pipeline network, of which 208 km is connected with German petrochemical factories which partly use the hydrogen and partly supply more to the pipeline (Figure 9.6). This series of pipes is the longest hydrogen transport and storage network in the world, and Hüls probably represents the greatest source of experience in this technology. For example, in thermal cracking of hydrocarbons in the electrical arc and in chloralkali electrolysis, large quantities of hydrogen are released, and only a part of this material is used by CWH. Consequently, transporting the unused hydrogen to distant users sufficiently cheaply and safely becomes an important objective. The hydrogen comes out of the arc at a pressure of 15 bar, so that hydrogen from the chlorine electrolysis plants has to be compressed to this value. As early as 1938, a hydrogen pipeline of 14 km was run to the hydride works in Gelsenkirchen-Scholven (VEBA-Chemie) and a further pipeline was laid over 9 km to Gelsenkirchen-Horst (Gelsenberg AG). In 1954, there was a systematic extension and rebuilding of the hydrogen pipeline with the idea of transporting further amounts of hydrogen to numerous users, particularly those in the Rhine-Ruhr area. Extension of the manufacturing facilities at Bayer AG in Leverkusen and Krefeld-Uerdingen also gave rise to the production of hydrogen, and these companies became both producers and users. Correspondingly, Esso in Cologne also obtained a connection so that it could both give to and take from the pipeline network. During the entire 30 years of operation of these facilities, no material damage through hydrogen has been observed. This is important with respect to the often-feared danger of diffusion of hydrogen through the grain boundaries of the pipeline materials and the possibility of decarbonization through the reaction with carbides. The experience gained is in agreement with the conclusion that hydrogen embrittlement through grain boundaries occurs only at higher pressures and temperatures.

The pipes used are made of normal steel, and have a maximum nominal internal diameter of 400 mm. They are covered with bitumen insulation which also has a glass layer. In recent years, synthetic insulators have been used. In addition to this, all the pipes are protected from corrosion from outside sources

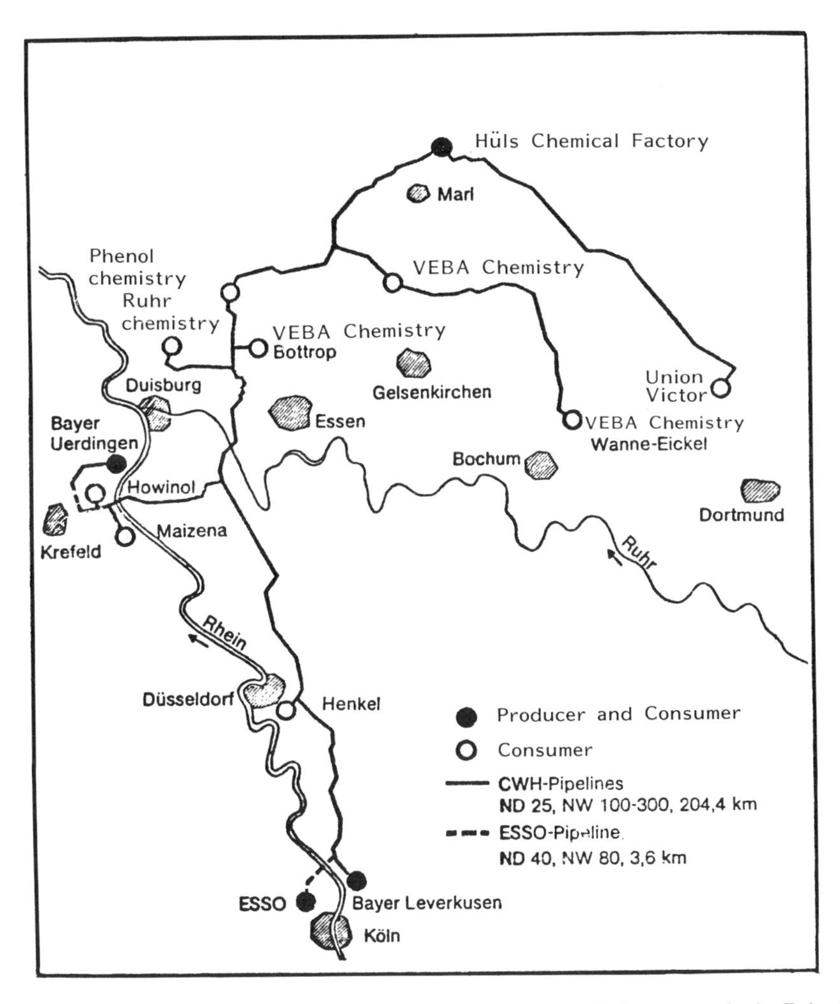

FIGURE 9.6. Map of the integrated hydrogen network sites of the Hüls company in the Federal Republic of Germany. From Isting and Thier.[13]

through cathodic protection. The pipeline network is laid about 1 m deep and has only a 10-m-wide protective zone around it (in which no buildings are allowed). Pressure gas lines in areas of mining operations demand particular precautionary measures in construction. Due to mining operations, there are earth movements (a consequence of various stresses and strains); these, of course, lead to changes in the pipeline lengths, which must be allowed for by the inclusion of sections that permit some degree of extension. The places in which these expansion sections have to be introduced are determined during the pipe-laying operation.

The operation of a pipeline network for hydrogen is not very different from that of a network for natural gas. Control of the entire network is carried out from a central command station, which is manned around the clock and is in telephone contact with the participants in the network, both producers and consumers (Figure 9.7). The amounts, pressures and temperatures are measured, communicated to a central management building, and automatically registered there. In the management building, the layout of the entire network is displayed above the measuring instruments, so that remote control can be monitored. This quantitative control, which has been going on for several decades, has given rise to a remarkable record of reliability. Since the beginning of the operation, no hydrogen has been lost; the leak rate from the pipes is within the design limits.

The objectives for the use of hydrogen by some of the users provide strict requirements with respect to purity; e.g., CO is a poison and may be present only in amounts of a few parts per million. Oxygen and water vapor can also produce some negative effects. Checking the purity is a task for the users of the pipeline. Limits are set, and if these are exceeded, further introduction of hydrogen is halted.

FIGURE 9.7. Central control unit of the H_2 pressure pipe system of the Hüls company in the western Germany industrial area.

In summary, the pipeline method is the most economical and the safest method of transporting hydrogen, and there have as yet been no difficulties relating to the lack of the use of special steels. The yearly hydrogen throughput is about 312×10^6 Nm^3.[13]

9.6. DISTRIBUTION OF HYDROGEN IN TRANSPORTABLE STEEL CYLINDERS*

After energy is transported in the form of electricity over long distances in high-tension lines, the next step is usually to distribute the power over ordinary lines in a stepwise fashion. Correspondingly, in a hydrogen economy, something analogous would be done. In the Federal Republic of Germany, this has already been started in the pioneer work of Messer-Griesheim and Co., who have several years' experience with the technique. Thus, for distances too small for it to be economical to build and operate a steel pipeline, due to as yet insufficient amounts of hydrogen in operation, it is necessary to distribute and deliver to individual customers medium to large amounts of hydrogen from a hydrogen production point or central pipeline collecting point, in steel cylinders under pressure, using specially modified trucks for delivery.

The best-known example of the conventional steel container is the ordinary steel cylinder with a volume of 40 liters, a working pressure of 150 bar, an empty weight of 70 kg, and a ratio between contents and packaging between 0.6 : 70 and 1 : 110. A remarkable example of progress during the last decade has been the introduction of lighter-weight cylinders made of chrome–molybdenum steel and having an upper pressure limit of 755 N/mm^2 compared with the 390 N/mm^2 of ordinary carbon steel. The new cylinders contain, at 50 liter volume and 200 bar pressure, about 10 Nm^3 H_2, and since they weigh only 63 kg, their weight ratio is about 40% better at 1 : 70 than that of ordinary steel. If the usage is over 500 Nm^3/month, it is better to make use of a cluster of cylinders. Such a cluster is made up of a certain number of individual cylinders bound together and mounted on a movable vehicle. The cylinders are connected with high-pressure tubing, so that they can be filled and emptied through a single valve. It is usual to use 12–28 cylinders, which, at 200 bar pressure, would contain 120–280 Nm^3 H_2. The ratio of the full to the empty weight is just as unfavorable as with individual cylinders, but there are some cost savings in using a cluster of cylinders, among them being simplification and acceleration in recharging the cylinders. Given the growing problem of deliveries and the increased importance of hydrogen in industry, improving the weight ratio of chrome–molybdenum steels is very important for the development of specialized transport vessels for both rail and road use. In this area, one should distinguish between

* This section authored by Prof. H. H. Ewe, Hamburg, and Prof. E. W. Justi, Braunschweig.

trailers and semitrailers for use in road transport and cylinder cars intended for rail transport. With the trailers, about 100 single cylinders, each of 50 liters, are bound together in groups of three and four and then connected through high-pressure tubing and mounted on the trailer. In this way, about 1000 Nm^3 H_2 can be stored. The economical level is reached at a usage rate between 5 and 25 kNm^3 month; when greater amounts are to be delivered, it is more appropriate to use a semitrailer, which would have a carrying capacity of 2.7–4.2 kNm^3. Several of the clusters are securely mounted on a truck (Figure 9.8).

If the consumer has access to a rail line, it is worthwhile to send the hydrogen by this means of transport, so long as the amount to be sent is more than 5 kNm^3/month. A comparison of Figure 9.9 with road transport shows that there are only five very large cylinders per rail car; these five cylinders weigh more in aggregate than do the many small cylinders used in road transport, but do not further increase the carrier capacity of 4.2 kNm^3. The reason that such very much larger and heavier cylinders can be used for rail transport is that the allowable load is higher than it is on the road and the weight has no adverse effect on the freight rates.

The best procedure for delivery in stationary containers has not been decided, and in fact two procedures are in use: Either the hydrogen is allowed to fill the customer's container or a cascade process, in which containers are grouped according to their internal hydrogen pressure is used to distribute the gas. In the latter case (cf. Figure 9.10), stationary storage groups are available with different

FIGURE 9.8. Semitrailer truck with 60 slender pressure containers, made of chrome–molybdenum steel, securely mounted and bound together. Total capacity: 4200 Nm^3 H_2 at a pressure of 200 bar. Constructed by Messer-Griesheim GmbH in 1973. From Kipker.[14]

FIGURE 9.9. Rail car with five high-capacity containers, made of chrome–molybdenum steel, securely mounted and bound together, each containing 800 Nm^3 H_2. Four-axle model with five containers = 4000 Nm^3. Working pressure: 200 atm. It is essential to take good advantage of the entire length of the long car. Constructed by Messer-Griesheim GmbH in 1973. From Kipker.[14]

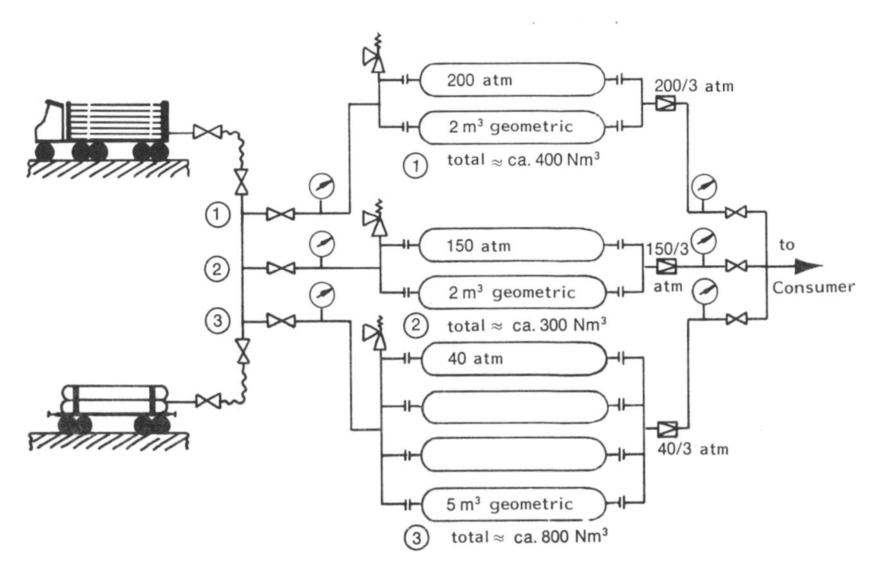

FIGURE 9.10. Block diagram for the filling of a semitrailer truck or rail car with chrome–molybdenum steel containers for H_2 by the cascade procedure. From Kipker[14] and Neuroth.[15]

working pressures, e.g., 150 and 40 bar. The 200-bar containers (consisting of 60 individual members) on the transport vehicle are distributed in several groups, after which first the containers at higher pressures and then those at lower pressures are filled gradually. This procedure allows the transport vehicle to be emptied efficiently, though the time consumed by the entire procedure is greater than it is with one-step filling of a container at a medium pressure. The more complex filling process is important in practice, because it is only when the contents of the specialized transport vehicle can be transferred at high pressure and quickly—and further when several deliveries per day are possible—that the cost of the delivered hydrogen can be brought down to acceptable levels.

9.7. ENERGY STORAGE AND TRANSPORT WITH LIQUID AND SLUSH HYDROGEN

Up to now, we have been discussing the Hydrogen Economy in terms of gaseous hydrogen, and we have shown that for long distances, those greater than some 700 km, a gaseous method of transport is cheaper than sending the energy through an electrical grid or by a satellite. If the hydrogen is liquefied first, the cost difference is smaller (e.g., if current costs 1.12¢/kWh, then liquefaction costs 0.41–0.52¢/Nm^3). Apart from this, there are difficulties in keeping the system at sufficiently low temperatures[16] (Figure 9.11). Nevertheless, there are situations in which the transport of liquid hydrogen through pipes will find application:

1. There could be situations in which hydrogen has already been liquefied, e.g., when it is shipped by tanker. Under these circumstances, it might

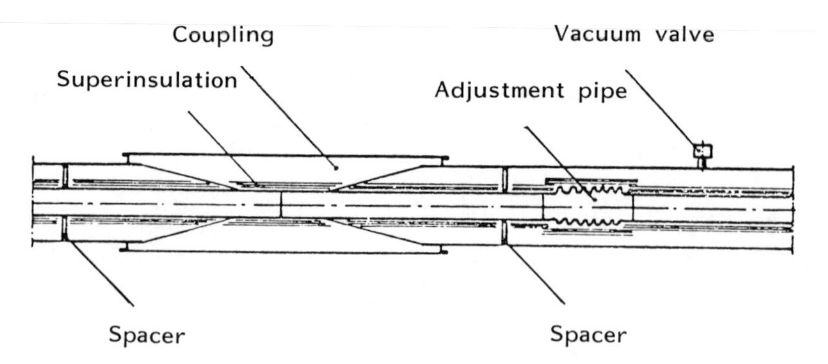

FIGURE 9.11. Schematic cross section of a collapsible vacuum-insulated double-walled pipeline for liquid and slush hydrogen. From Schraewer.[16]

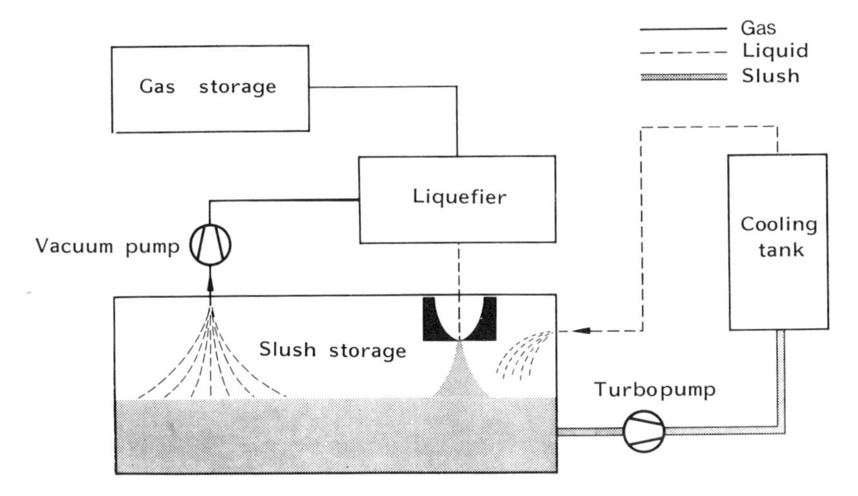

FIGURE 9.12. Block diagram of a liquid hydrogen storage system with pulse operation for the periodic production of slush hydrogen at approximately 14K (= −259°C) for the flow cooling of superconducting cable. Procedure from Messer-Griesheim GmbH from Schraewer.[17]

be practical to pipe the hydrogen through extremely low-temperature refrigerated pipes, without allowing it to gasify, and store it in tanks.

2. If large-scale superconductors having a sufficiently high critical temperature (i.e., >20K) were to be developed at an acceptable cost, it might be possible to combine the piping of liquid hydrogen with electrical transmission through superconducting cables, and this combination might be sufficiently cost-effective to be attractive. On one hand, liquid helium is too expensive for superconducting cables, and on the other hand, using liquid hydrogen as a coolant, the lowest critical temperature that can now be reached is $T_s = 20.4K$. In this type of situation, extremely low-temperature refrigeration can be achieved with so-called "slush hydrogen," which is a mixture of gaseous liquid and frozen hydrogen at equilibrium at the freezing temperature of about 40K, just as one obtains an equilibrium mixture of water and ice. A fundamental investigation has been carried out by Schraewer[17] (Figure 9.12) at Messer-Greisheim in Frankfurt with the support of the BMFT (German Federal Ministry of Research and Technology).

9.8. CONCLUSIONS

1. The direct transport of electrical energy by underground cable is very expensive. As for sending electricity through a high-tension grid, utilizing the

maximal voltage of 700 kV at present available, hydrogen transmission becomes cheaper at about 700 km. Transport by superconductors seems to be technically unfeasible. Even if it were to become possible, it would probably be uneconomical.

2. Microwave transmission to satellites in geosynchronous orbit and subsequent beaming down from these satellites to distant usage centers could be useful for the distribution of solar energy, the latter being first gathered in isolated and sparsely populated desert areas and at sea. This would be cheaper than large orbiting satellite platforms collecting solar radiation in space and beaming it directly to usage centers.

3. There could be situations that would involve preliminarily liquefying hydrogen (but this would take up a substantial fraction of the energy available) and then transporting the liquid hydrogen in a container ship specially constructed for the purpose. Alternatively, extreme low-temperature refrigerated pipes could be used, followed by storage of the liquid hydrogen in large containers as is now done at Cape Kennedy with hydrogen for the fueling of space vehicles (see Fig. 11.8).

4. Slush hydrogen is an equilibrium mixture of crystalline, liquid, and gaseous hydrogen at the triple point temperature of 14K. Slush hydrogen could be pumped and combined with cooled superconducting cables to form a system that might well be economical in some circumstances.

REFERENCES

1. R. P. Feynmann, R. B. Leighton, and M. Sands: *Lectures in Physics,* Addison-Wesley, New York (1964).
2. W. F. Gauster, D. C. Freemann, and H. M. Long, Proc. World Power Conf. Paper 56, p. 1954 (1964).
3. P. A. Klaud: Austrian Patent. 256, 956, Nov. 11 (1967).
4. G. Bogner and F. Schmid: *Naturwissenschaften* **57:**414 (1970); *Elektrotechische Zeitschrift* **A 92:**740 (1971).
5. G. Aschermann, E. Friederich, E. Justi, and J. Kramer: *Phys. Z.* **42:**(21/22): 349–360 (1941).
6. W. Buckel: *Supraleitung* [*Superconductivity*], Verlag Chemie, Weinheim (1972), p. 219.
7. P. Gläser: The case of solar energy, Conf. on Energy and Humanity, Queen Mary Univ. Coll. London, *Astronaut. Aeronaut.* **60** (August 1973).
8. D. P. Gregory, assisted by P. J. Andrews, R. J. Dufour, R. H. Elkins, W. J. D. Esher, R. B. Foster, G. M. Long, J. Wurm and G. G. Yie: *A Hydrogen Energy System,* prep. for the Am. Gas Assoc. by the Inst. of Gas Technol. Chicago (1972).
9. R. A. Meyerhoff, *Cryogenics* **4**:91 (1971).
10. E. Justi: *Spezifische Wärme, Entropie, Enthalpie und Dissoziation technicher Gase* [*Specific Heat, Entropy, Enthalpy, and the Dissociation of Technically Important Gases*], Verlag Julius Springer, Berlin (1938).
11. Bundesministerium für Forschung und Technologie: *Einsatzmöglichkeiten neuer Energiesysteme* [*Opportunities for the introduction of new energy systems*], Vol. 3, *Wasserstoff*, Bonn (1975).
12. G. Thissen: VDI-Tagung, Integrierte Energieversorgung [Integrated Energy Handling], Hamburg (1974).

13. Cr. Isting and B. Thier: Erfahrungen bein Wasserstofftransport, *Lichtbogen* **24:**16 (1975), No. 176, Hüls, CWH.
14. R. Kipker: *Gas Wärme Int.* **22** (6) (1973).
15. W. Neuroth: *Gas Aktuell* 905505 9/7063.
16. R. Schraewer: The hydrogen energy concept, ISPRA Courses. No. 14, CEC Joint Res. Center, Ispra (1975).
17. R. Schraewer: *Gas Aktuell* No. 7, 13 (1974).

CHAPTER 10

The Transmission of Hydrogen in High-Pressure Pipelines and the Storage of Hydrogen in Pipes

The comparative capital and operating costs for the transmission of hydrogen over great distances and those for the transmission of the same amount of energy through high-tension power lines are so important to the objective assessment of the transition to a hydrogen economy that they should be subjected to inquiry not only by academic scientists, but also by the planning office of a uniquely qualified company. For this reason, Justi requested Messer-Griesheim GmbH, a Düsseldorf company concerned with the transmission of gases, to undertake a detailed project concerning the investment and operating costs of a hydrogen pipeline 2000 km long and having an annual throughput of 1×10^{10} Nm^3 (= 10 GNm^3/year). This chapter, which discusses a project for transporting hydrogen over great distances and storing it in the pipeline, is not science fiction, but an account of a proposed design by the most experienced German company in this area.

This chapter authored by Dr. P. W. Brennecke, Braunschweig, Professor E. W. Justi, Braunschweig, and J. Kleinwächter, Lörrach.

10.1 CALCULATIONS FOR A 2150-KILOMETER HYDROGEN PIPELINE AND DISTRIBUTION NETWORK WITH A CAPACITY OF 10^{10} Nm^3 H_2 PER YEAR AND WITH THREE PRESSURE STATIONS USED FOR TRANSMISSION AT 100, 60, AND 40 ATMOSPHERES

10.1.1. Amount to be Transported

According to the COMPLES commission for solar irradiation and the environment—under the direction of Professors P. Blanco and L. Azcarraga, Madrid—it has been established in work over many years that southern Spain and southern Italy are absolutely the best places in Europe for solar-energy power plants to be established. The particular places where they thought the installations could be made with special advantage are on the Mediterranean coast between Malaga and Almeria and in Huelva near Gibraltar.

Messer-Griesheim was therefore asked to draw up a detailed plan for a solar–thermal hydrogen power plant that would operate approximately 8700 hr per year, requiring 10^{10} Nm^3 H_2 per year. The plant would consume 1.15 million Nm^3 H^2/ hr. Assuming 3640 hr of sunshine per year (10 hr per day), 2.75 million Nm^3 of H_2 must be produced/hr of sunshine and compressed to 100 bar initial pressure. Such a power station was to supply 10% of the electrical power needed by the Federal Republic of West Germany in 1974.

10.1.2. Investment

10.1.2.1. Pipeline Costs

The hydrogen pipeline from Gibraltar to Karlsruhe as measured on a map would be 2150 km long and would consist of steel tubes of nominal internal diameter 1000 mm, and outer diameter 1100 mm. These tubes would contain gas at pressures of 100, 60, and 40 atm in successive segments.

10.1.2.2. Additional Costs

Total cost for tubing in pipeline (1974 figures)	= \$0.920 billion
Administrative Buildings, Operating Supervision Facilities, and Others	= \$0.005 billion
	\$0.925 billion

10.1.2.3. Salaries

4 Management personnel @ \$60,000/yr	= \$240,000/yr
10 Staff and technicians @ \$35,000/yr	= \$350,000/yr
20 Repair and security personnel @ \$20,000/yr	= \$400,000/yr
	≈ \$1,000,000/yr

10.1.2.4. Capital Investment

The life of the storage pipes is estimated at a reasonable mean of 25 years.

4%	Depreciation
1.5%	Repairs and security
1.5%	Taxes and insurance
7%	(without profit and calculated interest)

7% of $0.925 billion rounded off = $64.75 million ≈ $65 million/yr

10.1.2.5. Operating Costs (Supervisory Costs)

Capital investment according to Section 10.1.2.4 = $65 million/yr
Salaries according to Section 10.1.2.3 (rounded off) = $1 million/yr
= $66 million/yr

$66 million/yr per 10^{10} Nm^3 H_2 (without profit or calculated interest) = 0.66¢/Nm^3

With 4% interest on $0.925 billion there are additional costs of $37 million, which are obtained from the total volume transported yearly:

$$\frac{\$37 \text{ million/yr}}{10^{10} \text{ Nm}^3 \text{ H}_2\text{/yr}} = 0.37¢/\text{Nm}^3 \text{ H}_2 \qquad = 0.37¢/\text{Nm}^3 \text{ H}_2$$

To this is added a further 8% provision for unforeseen costs:

8% × $0.925 billion = $74 million

Dividing this amount by the total volume transported yearly:

$$\frac{\$74 \text{ million/yr}}{10^{10} \text{ Nm}^3 \text{ H}_2\text{/yr}} = 0.74¢/\text{Nm}^3 \text{ H}_2$$

Sum of the operating costs and the costs for financing and other technical costs = 1.77¢/Nm^3 H_2

10.1.3. Calculated Details of the Compressor Station

10.1.3.1. Transported Amount

2.75×10^6 Nm^3 H_2/hr = 2.75 million Nm^3 H_2/hr
Initial pressure: atmospheric pressure; final pressure: 100 atm
Prior compression with 13 axial and 7 radial compressors with intermediate cooling to 1–25 atm. Later compression with piston machines, with intermediate cooling, to 25–100 atm.
Required drive output for an ideal gas: 2.75 million Nm^3/hr = 0.20 kWh/Nm^3 = 550 MW

Additional expenditure for energy consumption for closed cold-water system	= 25 MW
Total energy expenditure	= 575 MW

10.1.3.2. Investment for the Compressor Station (1974 Figures)

Compressors, including installation of 35 machines	= $150 million
Cooler; pipelines to the machines, including installation	= $ 75 million
Motors or turbines and electrical switches	= $ 80 million
Buildings, foundations, and auxiliary buildings, including grounds and development of gas, water, and electric lines	= $ 75 million
Cold-water supply, back-cooling system, and water pipeline system	= $ 55 million
Unforeseen expenses	= $ 15 million
TOTAL	$450 million

10.1.3.3. Salaries (see also Section 10.1.2.3)

4 Management personnel @ $60,000/yr	= $240,000
20 Machinists (5 men × 4 shifts) @ $20,000/yr	= $400,000
10 Repair personnel @ $20,000	= $200,000
10 Staff and technicians @ $35,000/yr	= $350,000
	$1,190,000
Rounded off:	$1,200,000

10.1.3.4. Capital Investment

Compressor station, life 10–20 years, reasonable mean 15 years, including buildings and foundations

7% Devaluation
3% Repairs and security
1.5% Taxes and insurance
11.5% (without profit and calculated interest)

11.5% of $450 million = $51.75 million/yr

Rounded off: = $52 million

10.1.3.5. Energy Costs

Energy consumption including cold-water supply according to Section 10.1.3.1	= 575 MW

Cost of electricity assumed as 1¢/kWh:

$$\frac{575 \text{ MW} \times 1¢/\text{Nm}^3 \times 1000 \times 8700 \text{ hr/yr}}{10^{10} \text{ Nm}^3/\text{yr}}$$

Rounded off: $= 0.5¢/\text{Nm}^3$

Additional operating materials (grease, chemicals, etc.):

$$\frac{\$5 \text{ million/yr}}{10^{10} \text{ Nm}^3/\text{yr}} = 0.05¢/\text{Nm}^3 \qquad = 0.05¢/\text{Nm}^3$$

10.1.3.6. Total Compression Costs at 1–100 Atmospheres

TOTAL $= \underline{0.55¢/\text{Nm}^3}$

Capital investment according to Section 10.1.3.4

$$\frac{\$52 \text{ million/yr}}{10^{10} \text{ Nm}^3/\text{yr}} \qquad = 0.52¢/\text{Nm}^3$$

Salaries according to Section 10.1.3.3

$$\frac{\$12 \text{ million/yr}}{10^{10} \text{ Nm}^3/\text{yr}} \qquad = 0.012¢/\text{Nm}^3$$

Energy costs according to 10.1.3.5 $= \underline{0.55¢/\text{Nm}^3}$

Compression costs (without profit and calculated interest) $= 1.082¢/\text{Nm}^3$

With 4% calculated interest from \$450 million $= \$18$ million $= 0.18¢/\text{Nm}^3$

With 8% profit from \$450 million in addition $= \$36$ million $= 0.36¢/\text{Nm}^3$

Compression costs, including profit and calculated interest $= \underline{\underline{1.6¢/\text{Nm}^3}}$

10.1.4. Detailed Calculations for the Hydrogen Pipeline from Huelva to Karlsruhe for Transport of 10^{10} Nm^3 Hydrogen per Year

10.1.4.1. Length of the Pipeline

The length of the hydrogen pipeline from Huelva to Karlsruhe was measured taking into account the characteristics of mountain terrain from a land map with a scale of 1 : 2,500,000, but without taking height differences into account, and came to 2150 km.

10.1.4.2. Amount to be Transported

The yearly needs for hydrogen transmission were calculated to be 10^{10} Nm^3 H_2 per year, which would amount to about 10% of the electrical power needed in the Federal Republic of Germany in 1974. This yearly requirement would correspond to a delivery of hydrogen at an hourly rate of 1.15×10^6 Nm^3 H_2/hr. Obviously, hydrogen electrolysis using solar energy can be carried out only while the sun is shining, and for southern Europe, one can assume an hourly average of 10 hr of sunshine per day, but this increases to 12 hr per day in Huelva, where the necessary amount of cooling water is also readily available.

This site has two considerable advantages that were first pointed out by Justi on the basis of meteorological observations made by a Spanish commission of COMPLES.[1] There are areas available that are of little value from the agricultural point of view, and therefore cheap, but that are at the same time excellent receivers of solar energy—a rather rare situation.

Since hydrogen production would take place only during the day, the amounts to be delivered at night would have to be reserved by storage in the pipeline network from the day's production.

To optimize the transportation of hydrogen, plans call for the pipeline to be divided into three sections by two compressor stations. Later, we will discuss an alternate plan which does not involve compressor stations.

Section I: 725 km from Huelva or Gibraltar in the direction of Karlsruhe to pressure station A at a nominal pressure of 100 bar.

Section II: 725 km from pressure station A in the direction of Karlsruhe to pressure station B at a nominal pressure of 64 bar.

Section III: 700 km from pressure station B to Karlsruhe at a nominal pressure of 40 bar.

The calculation for Section III assumes that 1.15 million Nm^3 H_2 are transported per hour. Section II should store one third of the nighttime demand (1.15 million Nm^3/hr $\times$ 12 hr/night $\times$ 0.33 = 4.6 million Nm^3/12 hr of night $\rightarrow$ 0.38 million Nm^3/night hour) during the 12-hr daytime.

The calculation in Section I, therefore, must consider the following hourly flow of hydrogen:

Hourly volume delivered in Karlsruhe:	1.15×10^6 Nm^3/hr
Hourly volume stored in Section II:	0.3833×10^6 Nm^3/hr
Hourly volume flowing through Section I:	1.5333×10^6 Nm^3/hr

Thus, two thirds of the Section I should store during the 12 hr of daytime/nightly demand (1.15 million Nm^3/hr $\times$ 12 hr/night $\times$ 0.66 = 9.2 million Nm^3/12 hr of night $\rightarrow$ 0.77 million Nm^3/night hour).

10.1.4.3. Pipe Material

According to the German standard DIN 2170, the thickness of the pipeline wall S, for long-distance transport in pipelines, must be at least 0.01 × pipeline diameter. Since we have chosen pipes of internal diameter 1000 mm, the minimum wall thickness must be at least 10 mm. Because the pipes are to be made of DIN 17172 steel, according to API-Norm X52, there would be a yield point of 37 kg/mm^2, and X70 would have to have a yield point of 49 kg/mm^2.

The pipes chosen were welded steel pipes of the type DIN 2458, which means:

100 bar pipe 1016 × 20 mm from steel × 70

64 bar pipe 1016 × 12.5 mm from steel × 70

40 bar pipe 1016 × 10 mm from steel × 52

10.1.4.4. Recalculation of the Wall Thickness of the Steel Pipe

The following symbols are used in the recalculation of the wall thickness:

S = Safety cofactor = 1.7

V = Welding seam factor = 1

C_1 = Additional cost of deviation in the thickness of the pipe wall

C_2 = Additional cost for corrosion

K = Stretch limit of materials

$S = S_0 + C_1 + C_2$ (wall thickness)

$$S_0 = \frac{d \times p \times S}{200 \times V \times K}$$

$$S_{0_{100\ \text{bar}}} = \frac{1016 \times 100 \times 1.7}{200 \times 1 \times 49} = 17.62. \text{ Selected value: } S = 20 \text{ mm}$$

$$S_{0_{60\ \text{bar}}} = \frac{1016 \times 64 \times 1.7}{200 \times 1 \times 49} = 11.28. \text{ Selected value: } S = 12.5 \text{ mm}$$

$$S_{0_{40\ \text{bar}}} = \frac{1016 \times 40 \times 1.7}{200 \times 1 \times 3.7} = 9.33. \text{ Selected value: } S = 10 \text{ mm}$$

10.1.4.5. Recalculation of the Pipe Diameter and Optimization of the Operating Costs

A gas pipeline must be optimized to achieve minimal transportation costs at a given rate of flow. Such a calculation is mathematically possible because the costs are the various subtotals of energy expenditure for the compression, inclusive of investment and upkeep of the unit, as well as the capital requirement for the pipeline. A minimum is reached because the expenditure of energy decreases, but the capital requirement increases, with increase in diameter of the pipe, as shown in Figure 10.1. This example, according to Knüfer, also shows the influence of an intermediate compression. The abscissa provides the nominal diameters from 0 to 1000 mm, the left ordinate the transportation costs in ¢/Nm^3 per 100 km, and the right ordinate the transport capacity in 10^3 Nm^3. The left curve, showing transportation costs with (top curve) and without (bottom curve) intermediate compression, indicates that the costs decrease to one third when the diameter of the pipe is decreased by 30%. The transport capacity increases in this segment by 33% because of the intermediate diameter compression. One can read from the graph for all chosen pressures, whether the increase of the transport via intermediate compression causes an increase or a decrease of the transport costs for 10^3 Nm^3 of hydrogen. The sharpness and the minimum at a chosen diameter of 500–600 mm is interesting. Optimization of such a large investment in respect to cost is clearly important. However, calculation of the

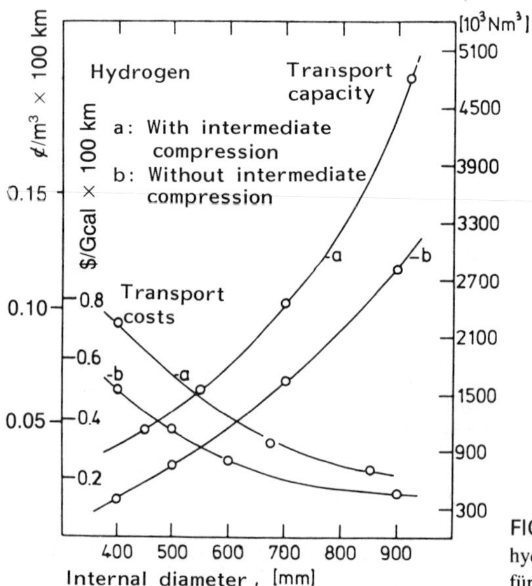

FIGURE 10.1. Transmission costs for hydrogen. From the Bundesministerium für Forschuns und Technologie.[4]

exact position of a minimum is not practically possible because location of the minimum changes, with the continual fluctuations of the capital cost (between 1974 and 1981 it was up 80%) and of the cost of electricity (which went up 300% during the same period).

Furthermore, it is not useful to optimize sharply the pipe diameter, because operational pressures in storage pipelines exposed to large daily and seasonal temperature changes cannot remain constant. Details are available in the publications given in the literature references.

We used the monograph by Herning[3] for our calculations. This monograph has the advantage of combining sufficient theoretical rigor with an ease of understanding for the practicing engineer. It minimizes approximations that are made for purely mathematical reasons and have no clear physical significance.

10.1.4.6. Calculation and Optimization of the Pipe Diameter

As outlined in the books quoted, the pressure drop along a pipeline of expanding and completely turbulent gas flow can be calculated from the continuity relation $w\alpha = w_1 \alpha_1$, and from the isothermal equation, $wP = w_1 P_1$. In these equations, all state-dependent quantities refer to the standard state of 0°C and 760 torr. Thus, one obtains:

$$P_1 \alpha_1 w_1^2 = P_0 \alpha_0 w_0^2$$

and the equation for the pressure drop:

$$\frac{P_1^2 - P_2^2}{2P_0} = \frac{1}{d} \frac{w_0 \alpha_0}{2g}$$

In fluid dynamics, SI units are still not being used. For example, the pressure (P) in high-pressure long-distance pipelines is expressed in atm instead of kp/m^2, and the length of the pipe is given in km. After substituting $P_0 = 1.03$ (=760 torr), after introduction of V_0 in Nm^3/hr, one obtains

$$\frac{P_1^2 - P_2^2}{L} = 13.9 \frac{\lambda v_0^2 \alpha_0}{(100d)^5}$$

The above equation assumes a transport temperature of 0°C, which is converted to the prevailing average transport temperature of 12°C by multiplying by the temperature ratio $\frac{285}{273}$.

Thus:

$$\frac{P_1^2 - P_2^2}{L} = 13.8 \frac{\lambda V_0^2 \alpha_0}{(100d)^5}$$

where P_1 and P_2 are the absolute pressures of hydrogen at the beginning and the end of the pipeline segment, respectively, L is the length of the hydrogen pipeline in kilometers, d is the pipe diameter in meters, V_0 is the flow rate in Nm^3/hr, α_0 is the standard density of hydrogen in kp/Nm^3, and λ is the coefficient of friction for the pipe.

The coefficent of friction for the pipe is a complicated function of the Reynolds number Re, and of the relative roughness $k/d = 0.014$.

One can use a simplification that the influence of Re vanishes for rough pipes and high Reynolds numbers:

$$\text{Re} = w\alpha d/\eta g = \text{const } d \ (\eta = \text{dynamic viscosity})$$

Thus, the calculation that will be carried out in detail for pipeline segment I between Gibraltar and the compressor station A, with Reynolds number Re = 7,595,425, a number above the critical value, and $L = 725$ kilometers, is simple:

$$P_G^2 - P_A^2 = 725 \times 13.8 \frac{0.014\ (1.9166 \times 10^2 \times 0.09)}{(100 \times 0.976)^5} \times \frac{288°}{273°} = 5272$$

This equation and the diameter of the pipeline at segment I give the initial pressure:

$$P_A = \sqrt{101^2 - 5272} = \sqrt{4929} = 70.2 \text{ atm}$$

The geometric volume of pipeline segment I is:

$$0.976 \times \pi \times 725{,}000/4 = 542{,}082 \text{ m}^3$$

which leads to the required amount to be stored:

$$0.7666 \times 10^6 \times 12 = 9{,}199{,}200 \text{ m}^3$$

After 12 hr of sunshine, the pressure in pipeline segment I will become:

$$P_{mI} = (P_G + P_A)/2 = (101 + 70.2)/2 = 85.6 \text{ atm}$$

The required storage pressure then becomes

$$P_{sI} = \text{amount to be stored/geometric volume} = 9{,}199{,}200/542{,}082 = 16.97 \text{ atm}$$

This means that the pressure will be

$$\text{Evenings} = P_{mI} = 85.60 \text{ atm}$$
$$\text{Mornings} = P_{mI} = 68.63 \text{ atm}$$

The hourly volume transported during the night is

At the beginning of the pipeline: 0.0000 Nm^3/hr
At the end of the pipeline: 0.766×10^6 Nm^3/hr

Thus, we calculate an average V_{mI} of

$$0.3833 \times 10^6 \text{ Nm}^3/\text{h}$$

The pressure after nighttime operation, with the lowest initial pressure of 68.3 atm, is calculated in accordance with the above example for pipeline segment 1:

$$P_{mI}{}^2 - P_A{}^2 = 218.4.; \qquad P_A = \sqrt{68.63^2 - 218.4} = 67.02 \text{ atm}$$

10.1.4.7. Section IIa of the Storage Pipe

According to the same scheme, we can calculate the numerical value of the operational cost for this section of the pipeline, which again consists of tubes of 1000 mm diameter with a length of 725 km, for which the pressure for station A is set at $p = 64$ atm. The amount of hydrogen flow passing through is then $V_{mII} = 1.15 \times 10^6 + 0.3833 \times 10^6/2$ Nm^3/hr $= 1.3416 \times 10^6$ Nm^3/hr. From this, one can calculate the fall in pressure in section II of the pipeline as $p_A^2 - p_B^2 = 2392.86$. The pressure at the end of the tube is $p_B = \sqrt{64^2 - 2393} = 42.8$ atm. The geometric volume of section II is $0.991^2\pi \times 725{,}000/4 = 558{,}830$ m^3, and from this the necessary amounts to be stored are $0.3833 \times 10^6 \times 12 = 4{,}599{,}600$ Nm^3. As was shown already for section I, there will be, when the day's production of hydrogen ends, a hydrogen pressure in section II that will have to be leveled off to the value of $p_{mII} = (p_A + p_B)/2 = (65 + 42.8)/2 = 53.9$ atm. The pressure that fits this situation amounts to p_{sII} = amount to be stored/geometric volume $= 4{,}599{,}600/558{,}830 = 8.23$ atm. We can calculate once more—as above for pipeline in section I—the daily variations of pressure for the pipeline in section II between pressure stations A and B. The mean hydrogen pressure will amount, in the evenings, to $p_{mII} = 53.9$ atm, and in the mornings to $p_{mII} = 45.67$ atm. In this section, the hourly amount to be transferred will be at night at the beginning of A: 0.7666×10^6 Nm^3/hr; and at the end of B: 1.15×10^6 Nm^3/hr. In the nightly transmission, the volume will therefore be $V_{mII} = 0.9583 \times 10^6$ Nm^3/hr.

10.1.4.8. Calculation of Pressure Decrease in Night Operation at the Lowest Pressure of 45.67 atm

In addition, according to the pattern above, we include figures for the decreased pressure equalization, again with a pipe roughness of $\lambda = 0.014$ and with a high R_e-value (3,740,229), and find that $p_{mII}{}^2 - p_B{}^2 = 1200.57$, from which it follows that the lowest first pressure at the end of pipe B, $p_B = \sqrt{45.67^2 - 1220.57} = \sqrt{865.18} = 29.4$ atm. This pressure would be too small (see the following calculation for section III) to operate the terminal Karlsruhe at 40 atm. Therefore, a larger-diameter is chosen here, a pipe 1220 × 12 mm of steel X50.

10.1.4.9. Calculations for Pipeline Section IIIA

This pipeline section should have a length of only 700 km. The flow rate in this section is 1.15×10^6 Nm³/hr at 25 atm. Under these conditions, $p_B{}^2 - p_K{}^2 = 665$; then $p_K = \sqrt{26^2 - 665} = \sqrt{11} = 3.3$ atm. Since this pressure is still too low despite the greater pipe diameter (internal diameter 1220 mm), we try a still wider pipe, internal diameter 1400 mm, and thus a pipe 1420 × 14.2 mm. The $p_B{}^2 - p_K{}^2 = 310.67$ and $p_K = \sqrt{26^2 - 310.67} = 19.11$ atm.

This inconsistency causes us to consider a variation of the dimensions of the pipeline in section II, or between pressure stations A and B, and choose IIA and IIB to denote the two possibilities involved. For IIB, we choose a pipe of internal diameter 1120 mm, instead of 1000 mm, with measurements of 1120 × 14.2 mm. $p_A{}^2$ stays the same at 64 atm, and length = 725 km. Then $V_{mII} = 1.13416$ Nm³/hr, and $p_A{}^2 - p_B{}^2 = 1475.83$; $p_B = \sqrt{65^2 - 1475.83} = 52.4$ atm. The geometric volume of pipeline section IIB then becomes $1.0916^2 \times \pi \times 725{,}000/4 = 678{,}092$ m³. The storage requires a quantity of $0.3833 \times 10^6 \times 12 = 4{,}599{,}600$ Nm³. As we have already seen, the pressure of pipeline section IIB will equalize after sundown to: $p_{mII} = (p_A + p_B)/2 = (65 \times 52.4)/2 = 58.7$ atm. In this case, the adjustable storage pressure works out to be:

$$S_{II} = \text{stored amount/geometric volume} = (4{,}599{,}600)/(678{,}092) = 6.78 \text{ atm}$$

Hence, the mean pressure works out to be higher in the evening, $p_{mII} = 58.7$ atm; in the morning, $p_{mII} = 58.7 - 6.78 = 51.92$ atm. The mean transmission is, as calculated above for section II, 0.95833×10^6 Nm³/hr. The pressure decrease calculated for night operation with the lowest initial pressure is ascertained again (as before) by $p_{mII}{}^2 - p_B{}^2 = 752.52$ and from that $p_B = \sqrt{51.92^2 - 752.52} = \sqrt{1943} = 44.1$ atm.

10.1.4.10 Optimization of Pipeline Section IIIB

This improvement through the optimization of pipeline section IIA encourages the variation of the pipe diameter of IIB, for which the internal diameter

1000 mm can be taken, and an operating pressure at point B of $p_B = 40$ atm can be applied, whereupon the transmitted amount is 1.15×10^6 Nm^3/hr. This results in $p_B - p_K = 1655.85$, and $p_K = \sqrt{41^2 - 1655.85} = 5.015$ atm. This pressure is obviously too low; therefore, we try again with pipe of internal diameter 1100 mm, 1120 × 11 mm pipes of X50. Then, $p_B{}^2 - p_K{}^2 = 1017$, and $p_K = \sqrt{41^2 - 1017} = 25.7$ atm.

10.1.5. Discussion of the Results of Calculations

The numerical calculations presented above are very detailed and therefore somewhat tedious and difficult to understand. The purpose of these calculations is to show convincingly the basis of the claim that energy transport as hydrogen gas at high pressure in pipelines could be practical, a claim that is not always believed by the representatives of alternative technologies. So that this conclusion might come over strongly, we shall give here a very brief synopsis of what we have said above.

Table 10.1 presents a survey summarizing all the variables calculated in which the optimal variables are designated by "X's" in the right-hand column.

It can be seen that the input pressure p, according to the optimal values of the nominal width and pressure (atm) with approximately 101 atm starting pressure in pipe section I, with 65 atm in pipe section IIB, and with 41 atm in pipe section IIIB, fits well with the desired total process. In the same sequence, the p_{min} values at the ends of these sections are 67, 44.08 and 25.7 atm. The optimal pipe diameters are in the same sequence 1000, 1100, and 1400 mm internal diameter, and they weight 363×10^3 tons in section I, 284×10^3 tons in section II, and 212×10^3 tons in section III, totaling 860×10^3 tons, an acceptable value.

The material in Figure 10.2 is remarkable, because it shows the interaction between stored and transported H_2 volumes during day and night shifts at one glance and quantitatively, a surprising aspect of the solar–hydrogen economy which has never been so clearly demonstrated.

A supplement to the pictorial presentation of Fig. 10.2 is given in Table 10.2, which presents the costs for the three tube sections, calculated at a 1973 level. It includes a 10% provision for emergencies, but excludes the costs for the two pressure-maintaining stations. The total cost amounts to $\$9.2 \times 10^8$. At the time of writing (1981), prices for steel construction and personnel costs are up by 80%, but this clearly has no bearing on the *comparative* costs, which will also have increased by the same amount in respect to electrical-grid transmission of energy. Table 10.3 shows the investment costs, salaries, and capital costs, assuming a 25-year lifetime. This includes the running costs with the calculated mortgage rates for the payment of bank loans. Table 10.4 pertains to the calculations for the principal compressor station at Huelva, divided into the

TABLE 10.1
Evaluation of the Calculations

Section	Internal diameter (mm)	Normal pressure (atm)	Entry pressure (atm)	End pressure (atm)		Pipe dimensions (mm)	Type of steel	Length (km)	Weight (tons)	Choice
				Max.	Min.					
I	1000	100	101	85.6	67	1016 × 20	X70	725	363,101	X
IIA	1000	64	65	53.9	29.4	1016 × 12.5	X70	725	226,511	—
IIB	1100	63	65	58.7	44.08	1120 × 14.2	X70	725	283,993	X
IIIA	1200	25	26	—	3.3	1220 × 12.5	X52	700	—	—
	1400	25	26	—	19.11	1420 × 14.2	X52	700	347,833	—
IIIB	1000	40	41	—	5.01	1016 × 10	X52	700	—	—
	1100	40	41	—	25.7	1120 × 11	X52	700	212,107	X

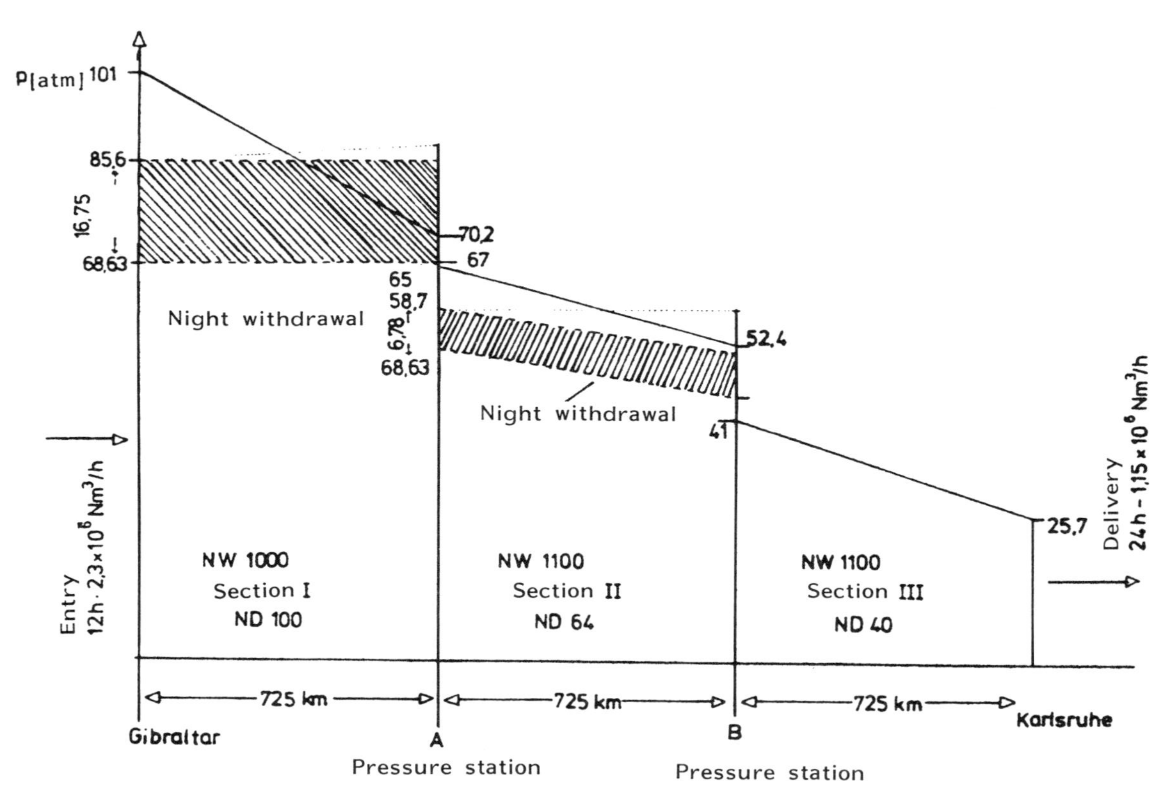

FIGURE 10.2. Quantitative diagram of the three-part pressure storage pipes including calculation for the fall of pressure in night use.

TABLE 10.2
Cost of Pipeline

	Length (km)	Price (dollars/m)	Total cost (dollars)
Section I			
Pipe material (363,101 tons)	725	—	190,628,000
Pipe-laying and excavation			
Easy course	290	160	46,400,000
Normal course	285	200	57,000,000
Difficult course	150	250	37,500,000
Section II			
Pipe material (283,993 tons)	725	—	149,096,500
Pipe-laying and excavation			
Easy course	270	100	27,000,000
Normal course	360	150	54,000,000
Difficult course	95	200	19,000,000
Section III			
Pipe material (121,107 tons)	700	—	106,053,500
Pipe-laying and excavation			
Easy course	300	100	30,000,000
Normal course	350	125	43,250,000
Difficult course	50	165	8,250,000
Steering cable with accessories	725	12.5	9,062,000
Damages	725	20	14,500,000
Corrosion protection, examination	725	5	3,625,000
River crossings	—	—	25,000,000
Engineering works	725	20	14,500,000
			834,865,500
10% unforeseen costs			83,536,500
	2150	427.40	918,402,000

cost of salaries, capital needs, and running costs. In the total, the mortgage payments, together with interest, have been taken into account, but the costs of solar technical developments before the beginning of the pipeline are neglected. The total comes to \$1.62/Nm^3, which is about 0.9¢/kWh. The figures are again for 1973.

The complete calculation will take into account the costs of the solar installation, but since there have been no real forerunners of such an installation, it is difficult to say what these costs will be. It must be taken into account that because the primary energy is solar, there are no fuel costs, but the investment costs will be higher than those of conventional plants. After the solar plant at

TABLE 10.3
Hydrogen Transmission in Storage Pipes—Calculations by Messer-Griesheim GmbH (1973)

Investment costs	$\$925 \times 10^6$
2150 km of welded steel pipe including all ancillary costs involved in manufacture and administration	
Salaries	$\$1.05 \times 10^6$
Capital costs (assuming 25 years as lifetime)	$\$65 \times 10^6$/yr
4% depreciation	
1.5% maintenance costs	
1.5% taxes and insurance	
Total: 7% of $\$925 \times 10^6$	
$\dfrac{\text{Salaries and servicing of capital}}{\text{Yearly hydrogen throughput}} = \dfrac{\$66 \times 10^6}{10^{10}\ \text{Nm}^3/\text{yr}}$	0.66¢/Nm^3
Financing	
Calculated interest, 4% of $\$925 \times 10^6$	0.37¢/Nm^3
Yield, 8% of $\$925 \times 10^6$	0.74¢/Nm^3
Total transmission costs in worst case:	1.77¢/Nm^3

TABLE 10.4
Compressor Station—Calculations by Messer-Griesheim GmbH

1. Compressor load: 2.75 million Nm^3/hr	
Initial pressure: atmospheric pressure; final pressure: 100 atm	
Preliminary compression with axial and radial compressors (13- and 7-step machines) with intermediate cooling at 1–25 atm	
Performance required: 2.75 million Nm^3/hr × 0.20 kWh/Nm^3 =	550 MW
Closed water cooling system, energy required	25 MW
Total energy required	575 MW
2. Investments in compressor station	
Compressor, including operation-ready installation (35 machines)	\$150 million
Cooling, pipes to machines, including installation	\$75 million
Motors and drive turbines and electrical switching	\$80 million
Buildings, foundations, and auxiliary buildings, including land and acquisition costs	\$75 million
Provision of cooling water, cooling plant, and water circulation system	\$55 million
Unforeseen costs	\$15 million
Total investments	\$450 million
3. Salaries	
4 Management personnel @ \$60,000/yr	\$240,000
20 Machinists (5 men × 4 shifts) @ \$20,000/yr	\$400,000
10 Mechanics @ \$20,000/yr	\$200,000
10 Operators and technicians @ 35,000/yr	\$350,000
Total (rounded off)	\$1,200,000

(*continued*)

TABLE 10.4 (*Continued*)

4. Capital requirement	
Compressor station with an assumed life-time of 10–20 years, including buildings and foundations with an arbitrary mean lifetime of 15 years	
7% Depreciation	
3% Repair and maintenance	
1.5% Taxes and insurance	
11.5% (without profit, without calculated interest on $450 million)	
Rounded off	$52 million
5. Energy costs	
Energy requirement, including provision of cooling water, according to section (1): 575 MW; electricity costs, assuming 1¢/kWh, 575 MW × 1¢/kWh × 1000 × 8700 hr/yr ÷ 10^{10} Nm^3/yr =	0.5¢/Nm^3
Ancillary operating materials (lubricating oil, chemicals, etc.): $5 million/yr ÷ 10^{10} Nm^3/yr =	0.05¢/Nm^3
Total energy and materials costs	0.55¢/Nm^3
6. Total costs of compression to 1–100 atm	
a. Capital service according to section (4): $52 million/yr ÷ 10^{10} Nm^3/yr =	0.52¢/Nm^3
b. Salaries according to section (3): $1.2 million/yr ÷ 10^{10} Nm^3/yr =	0.01¢/Nm^3
c. Energy costs according to section (5) =	0.55¢/Nm^3
Compression costs without calculated interest or profit	1.08¢/Nm^3
Plus 4% calculated interest costs on $450 million = $18 million	0.18¢/Nm^3
and 8% profit on $450 million = $36 million	0.36¢/Nm^3
Total compression costs including calculated interest costs and profit	$1.62¢/$Nm^3$

Almeria is in operation—with its three facilities (see Fig. 10.3)—it will be much easier, after a few years, to determine these cost more accurately.

In this connection, the recent observations of Kipker[5] are valuable; according to these observations, the flow characteristics of hydrogen gas are so much better than those of natural gas that in the 2000-km pipeline—apart from the principal compressor stations—there will be no need for any intervening compressors, whereas a natural gas pipeline would require at least one. In this connection, Kipker has also answered the question as to whether it is economical to collect the electrolytically produced oxygen, the volume of which is half that of the hydrogen produced. The physical characteristics of oxygen (viscosity and density in particular) are much less favorable than those of hydrogen, so much so, in fact, that on the 2000-km line, there would have to be six to eight compressor stations. Accordingly, it might be appropriate to refer here to the

FIGURE 10.3. Aerial photograph of the new European solar power station at Almeria (southern Spain). The experimental and demonstration plant consists of 3×0.5 MW_{el} parts (in the background is the tower with more than 100 convex mirrors regulated by microcomputers). The system was built by Martin-Marietta. The rays from the mirrors strike the black receiver at the top of the 50-m-high tower and carry the heat energy via a sodium cycle to a turbine. In the foreground are two different facilities on the farm principle that work with linear parabolic mirrors that are adjustable on only one coordinate.

diagram of a hydrogen economy in Figure 10.4, which dates from 1964[6] and proposes a separate, parallel pipeline for the O_2 transport.

Kipker[5] calculated the distance over which an oxygen pipeline could be developed and still offer a cost advantage in the production of hydrogen. He considered that for the transportation of, O_2 a steel pipeline of internal diameter 600 mm would be required and also the construction of several intermediate pressure stations. After subtraction of the transmission costs the anticipated profit due to the sale of oxygen will be about 2¢/Nm^3. Kipker bases his calculations on the assumption that a hydrogen consumer would pay about 7.5¢/Nm^3, so that the 1¢/Nm^3 for oxygen would allow a 15% decrease in the cost of hydrogen.

Persons who are knowledgeable in this field—though not many others—know that the experience with the transmission of oxygen through pipelines is

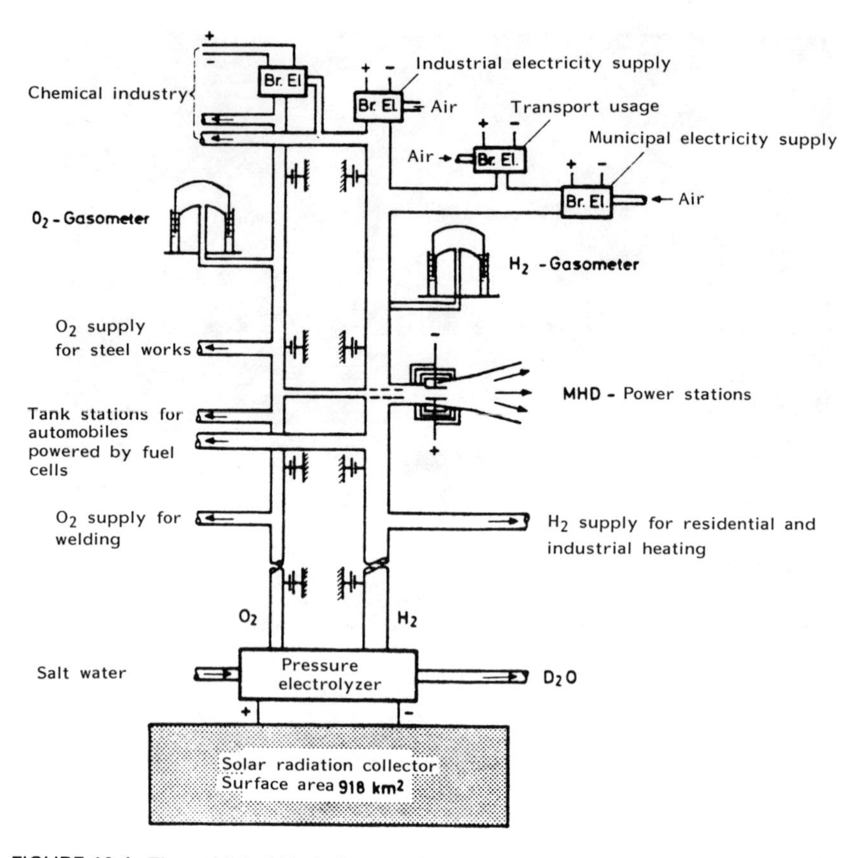

FIGURE 10.4. First published block diagram of a solar–hydrogen economy, authored by Justi[6] in 1964. It incorporates the essential characteristics of the concept.

much less happy than the corresponding experience with hydrogen because concentrated oxygen occasionally causes a fire in steel pipes, which are never wholly clean. On this basis, it would be desirable, using experience from the shipping of liquid natural gas, to load liquid oxygen into refrigerator ships in harbors near Huelva, Almeria (T_m of methane is 112K; of oxygen, 90K), and ship it cheaply to Wilhelmshafen.

These are worthwhile ideas for making the economics of hydrogen more favorable. However, let us not forget the suggestion of Justi[1] in 1974 that hydrogen should be introduced into the pipelines at 100 atm and transported at this pressure but, before use, should be run through a turbine and brought to a lower pressure by being made to do work (Figure 10.5). Thermodynamics show that the energy use in compression is proportional, not to the pressure *difference*,

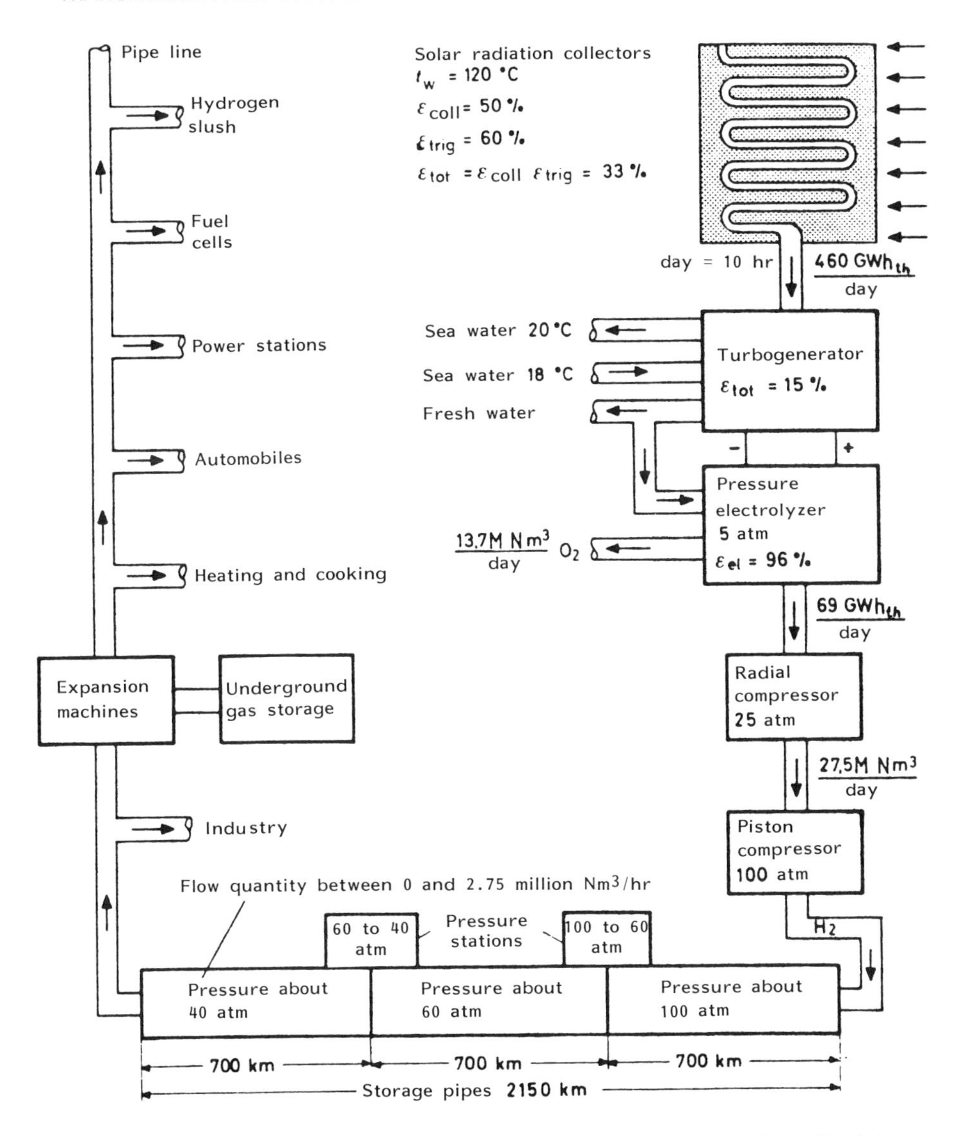

FIGURE 10.5. Block diagram of a solar–hydrogen economy from a study made by Messer-Griesheim GmbH in 1974.

but to the logarithm of the pressure *ratio;* thus, it is proportional to log 100/1 = 2.00. Similarly, log 44.0/1 = 1.60, so that ideally, in the case of fully reversible expansion, 80% (1.60/2.00) of the solar compression energy in Spain would be recoverable through the expansion work in the Federal Republic of Germany, preferably as electrical energy. Allowing 60% efficiency for the expansion and taking into account a nonideal equation of state for hydrogen gas, the energy in Spain recoverable in Germany would be about 48%, a notable cost saving.

10.2. THERMODYNAMIC OPTIMIZATION OF HYDROGEN TRANSPORT AND PIPELINE STORAGE

It is well known in thermodynamics that the work of condensation for an ideal gas is a minimum if it is carried out isothermally. For this reason, it is better to carry out the compression in a large number of successive small steps, rather than in one large step. During the stepwise expansion, cooling water is used to maintain a constant temperature. This technique is used in the liquefaction of air by the Linde process, in which the initial pressure is 200 atm and four steps are used. It has been found in this work that the theoretical expectation is indeed confirmed; i.e., the mechanical work required for the isothermal condensation is proportional, not to the pressure *difference*, $\Delta p_{2,1}$, but to the logarithm of the pressure ratios, p_2/p_1. Mathematically, the isothermal condensation work per unit (i.e., per mole) is given by $A = RT \log(p_2/p_1) = RT_1 (\log p_2 - \log p_1)$ (R = gas constant). Compression from 1 to 10 atm requires the same amount of work as that from 10 to 100 atm. Correspondingly, the compression work required to go from 1 to 40 atm is given by $A = RT_1 (\log 40 - \log 1) = RT_1 (1.6021 - 0) = RT_1 \times 1.6021$. Thus compression through a 60 atm pressure range requires work of only $A = RT_1 \times 0.379$ whereas compression through the 40 atm range requires $A = RT_1 \times 1.6021$ (five times more work than through the 60 atm range). Hence decompressing the pipe from 60 atm uses only 20% of the compression energy. The remaining 80% are wasted during discharge at the end of the line.

The preceding discussion suggests that by applying well-known and proven technology, 50–80% of the compressor energy can be regained by means of turbogenerators. The cooling of the hydrogen associated with this process can be utilized for such modern refrigeration needs as residential air conditioning or certain manufacturing processes.

10.3. QUANTITATIVE CALCULATION OF THE RECOVERY OF ENERGY

The cooling of hydrogen under expansion can be very roughly calculated by applying the Second Law of Thermodynamics: $T_1 - T_2 = T_1 [p_1/p_2{}^{1-(1/\kappa)} - 1]$,

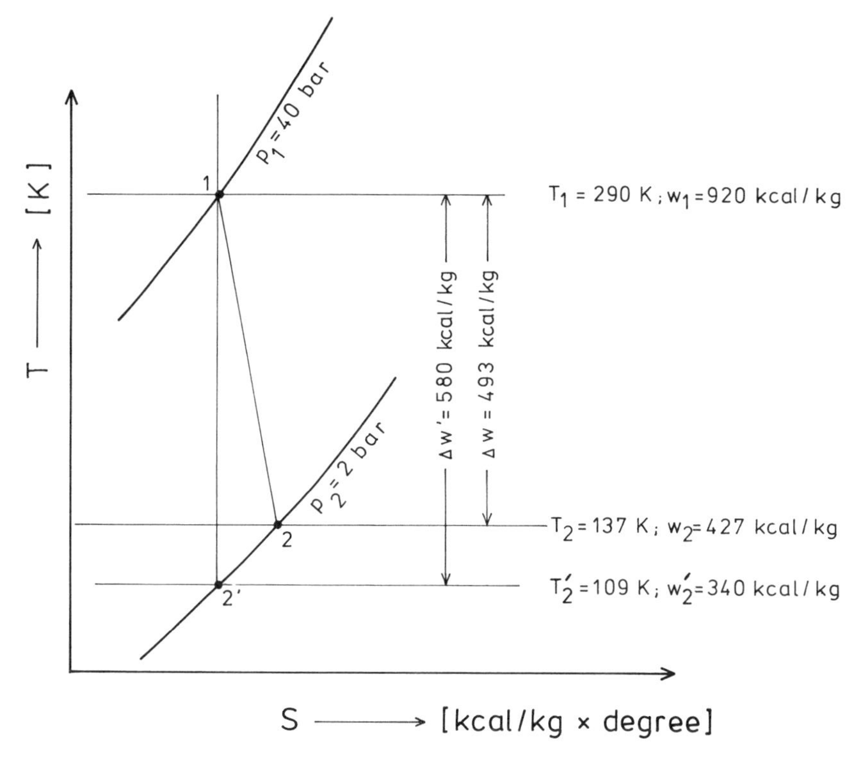

FIGURE 10.6. Part of a T,S diagram (hydrogen expansion).

where T_1 is the initial temperature, e.g., 290K or 13°C, and T_2 is the intended low temperature, p_1 and p_2 are the pressures before and after the expansion of the hydrogen, and κ is C_p/C_v, the ratio of the two specific heats at constant pressure p and volume V. In the case of hydrogen at 290K, $\kappa = C_p/C_v = 6.90/4.92 = 1.4024$, i.e, $1/\kappa = 0.7131$ and $1 - 1/\kappa = 0.2869$; because of the strong dependence of κ on temperature, however, this equation is not numerically reliable.

It is better to utilize the modern thermodynamic approach and get the necessary numerical values from available calculated diagrams, as, for example, the T,S diagrams that give relationships at constant pressure and volume and also the enthalpy, h (also called the "heat content," w) in kcal/kg.

The T,S diagram in Figure 10.6 combines a T range of 290–100K with an entropy range of 5–13 kcal/kg (abscissa). The expansion process is to take place at a temperature of 290K and p = 40 atm and will do very well for our calculation. First of all, we read from the diagram that at the point T = 290K and

40 atm, the enthalpy value $w_1 = 920$ kcal/kg. If we now try to make an ideal calculation, i.e., to calculate the work of expansion adiabatically (i.e., vertically) to $p' = 2$ atm, we find that the final value $w_2' = 340$ kcal/kg. Then, on expansion, there is an enthalpy gain of $w' = 920 - 340 = 580$ kcal/kg. Along with this would be an ideal temperature fall to $T_2' = 109$K, i.e., $-164°$ C. However, according to the Second Law, there will certainly be a number of losses involved, and these can be calculated by evaluating the partial efficiencies, as shown in Figure 10.6. In this part of the T,S diagram, the adiabatic expansions from T_1 to T_2', and correspondingly from p_1 to p_2', are denoted by vertical lines that represent the enthalpy difference, $\Delta w'$. This ideal enthalpy has to be multiplied by an efficiency, η_i, such losses being unavoidable in any heat engine, and this has the effect of reducing the value by a factor of 0.85. Correspondingly, the ideal final point $2'$ of the expansion is changed along the isobar p_2 = constant until it intersects with the isotherm T_2, so that the recoverable entropy decrease becomes $w = 0.85 \times 580 = 439$ kcal/kg. The final temperature that can be reached then changes from 109 to 137K.

In consideration of a total volume of hydrogen $V = 10^{10}$ Nm^3/yr at a hydrogen density of 0.09 kg/Nm^3—corresponding to a flow of 28.5 kg/sec—one obtains a performance rating of 590 MW for the turbine; at 95% mechanical energy efficiency, the turbine performance is reduced to 560 MW, in good agreement with the data of a recent project study by Messer-Griesheim. This performance is available for use so that the yearly work carried out would be 560 MW $\times$ 8760 hours $= 4.91 \times 10^6$ MWh.

The cooling potential can be calculated from the aforedescribed relationships as follows:

$$590 \text{ MW} = 590 \times 10^3 \text{ kW} = 590 \times 10^3 \text{ kJ/sec} = 140 \times 10^3 \text{ kcal/sec}$$

The yearly cooling work is then $140 \times 10^3 \times 8760 \times 3600 = 4.42 \times 10^{12}$ kcal. Finally, it may be said that if the low limit of the expansion temperature of 173K is not a desirable one, it would be possible to give up working adiabatically (or, more accurately, polytropically) and, with some entry of heat from the surroundings, to carry out expansion isothermally. Under these conditions, the efficiency of the expansion turbine would be lowered, to 56% at 2 atm and to 63% of the final pressure at 1 atm.

REFERENCES

1. E. Justi, *Akademie der Wissenschaft und der Literatur Mainz, 1949–1974*, pp. 41–53, Franz Steiner Verlag, Wiesbaden (1979).
2. H. Knüfer: *Fuel, Heat Power* **25:**12 (1973).

3 F. Herning: *Transmission of Solids in Pipelines*, Deutsche Ingenieur-Verlag, Düsseldorf (1954).
4. BMFT: *The Practical Scope of New Energy Schemes*, Part III, *Hydrogen*, Bonn (1975).
5. R. Kipker: *Chemie-Ingenieur-Technik* **48:**138 (1976).
6. E. Justi: *Conduction Mechanisms and Energy Conversion in Solid Bodies*, Verlag Vandenhoeck and Ruprecht, Göttingen (1965).

CHAPTER 11

The Storage of Hydrogen

The main problem in the use of solar energy is the daily and yearly variations, which are not in phase with the variations of energy demand. Therefore, for widespread application of solar energy, it is necessary to store energy collected during "on" times in the form of hydrogen.

11.1. THERMAL ENERGY STORAGE

A considerable amount of heat can be stored by exploiting the latent heat during phase transitions. When a liquid is cooled below its freezing point, the latent heat of melting is released. Per unit weight, this is usually a considerable amount of energy. Various eutectic[1] mixtures are known for this purpose. Other phase transitions such as crystallization of Glauber salt, $Na_2SO_{12} \cdot 10\,H_2O$, have also been discussed.[2] This salt stores, per unit volume, about 1.7 times the energy that can be stored in water.

For this kind of energy storage, it is desirable to have a low-cost material, a high specific latent heat, stability, and a low degree of electrochemical or chemical corrosion of the (usually) low-priced container material in the presence of the heat-storage liquid. Most of the materials that have been examined so far are eutectics of simple inorganic salts, the phase transitions of which lie between 150 and 850°C. Figure 11.1 shows the heat-storage possibilities of these systems,[3] which are several times higher than those of electrochemical systems, e.g., the well-known lead-acid storage battery, which stores about 30 Wh/kg. This com-

This chapter authored by Prof. H. H. Ewe, Hamburg, and Dr. H.-J. Selbach, Braunschweig.

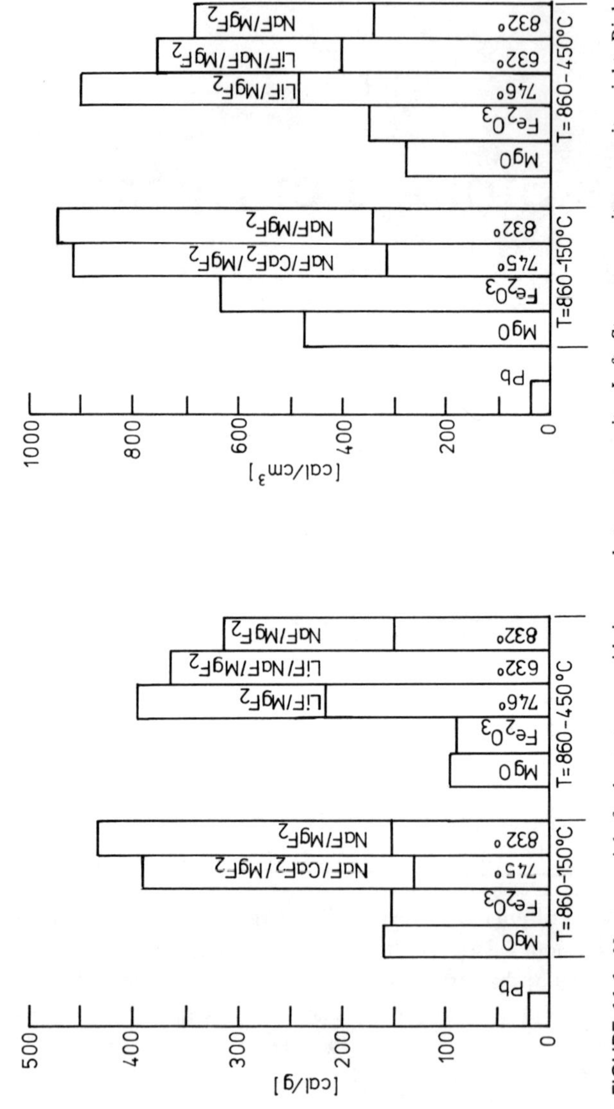

FIGURE 11.1. New materials for heat storage with increased storage capacity. Left: Storage capacity per unit weight. Right: Storage capacity per unit volume. From Philips.[3]

parison is not entirely appropriate because in the latter case, it is convenient electricity that is being stored.

The difficulties of this thermal energy storage are the necessary isolation of the system, the need for durability of the storage material, and the corrosion of the container walls that will occur if a relatively cheap material such as steel is used. The cost of these thermal storage units with salt melt is about 50¢/kWh. A further disadvantage of this process is that any attempt to transfer energy stored in this system involves the movement of large masses and is therefore uneconomical.

11.2. ELECTROCHEMICAL ENERGY STORAGE

The principle of electrochemical energy storage is well known in the form of lead-acid and nickel-cadmium storage batteries. The advantages are also very clear: There is no possibility of pollution either from noise or from exhaust gases; storage of high-grade electrical energy can be obtained with about 60% efficiency. On the other hand, lead-acid and nickel-cadmium storage batteries can store only relatively small amounts of energy, namely, 20–40 Wh/kg. Their lifetime and number of available discharge cycles are limited. If larger amounts of energy are to be stored, the battery needed would have to have the following minimal performance parameters: 50 W/kg, 200 Wh/kg, lifetime 4–6 years, 1000 complete charge–discharge cycles. The first two of these conditions are met by the new batteries that are tabulated in Table 11.1. As for the maximum useful life of the battery and the maximum number of cycles, no accurate data are yet available.[4–6]

TABLE 11.1
Parameters of New Storage Media[a]

Storage system	Output (w/kg)	Energy Density (Wh/kg)
Zinc-air	100	100
Ni-H_2 (compressed)	300	80
Ni-Fe	132	50
Zn-O_2	300	160
Al-air	150	240
Na-S	100	100
Na-air	200	400
Li-Cl_2 (liquid)	300	500
H_2-air	100	2000
Pb, acid	200	20–30

[a]From Technology of Efficient Energy Utilization.[5]

The cost of new storage batteries that have not yet been in practical use to any extent can be judged only very roughly. The sodium-sulfur cell has doubtless been subject to the most extensive development efforts of any battery, and its cost has already been estimated to be about $1/kWh (in comparison, the common lead-acid battery costs about $1–5/kWh). The sodium-sulfur battery works at about 350 °C, and this temperature is necessary so that sufficiently rapid reactions take place in the solid electrolyte.[8] The principal difficulties are presented by this solid β-Al_2O_3 electrolyte. This material tends to crack after a lengthy period of use and is sensitive to temperature variation. It is also a disadvantage to have to work at a relatively high temperature. Other storage batteries under development, for which the operational conditions are somewhat milder, are the iron-air battery and the cobalt-nickel battery; they seem at present to be more promising.[9,10]

The decisive disadvantage of all electrochemical storage batteries is their considerable weight. Further, some of the metals that are needed are not readily available in large amounts or are too expensive for big storage units.

Storage batteries may be the preferred means of storage for cars, emergency power sources, and other small-scale applications. They are limited to conditions in which the advantages of the storage of electrical energy out-weigh the disadvantages of the large weights involved.

11.3. SUPERCONDUCTING MAGNETS

Energy can be stored in the magnetic field created by a current passing through a coil. This type of energy storage is not practical if ordinary copper wire coils are used because the ohmic resistance of the copper transforms the energy of the magnetic field into low-temperature heat. If, however, superconducting coils are used, the ohmic resistance of which is zero, then the storage of energy over a long time span is possible. It is necessary, however, to cool the windings to a very low temperature (<20K). A plant for 100–10,000 MWh requires Nb-Ti coils with a radius of 50 m and a height of 50 m. This technology is not available in practice as yet, and indeed is not expected to be available until after the year 2000.*

11.4. ENERGY STORAGE IN FLYWHEELS

It is possible to store energy in a mechanical form as the rotational energy of flywheels, in which the greater are the mass, the radius, and the rotational velocity, the greater is the stored energy. These three quantities and their inter-

* Translator's note: Discoveries of the late 1980s make superconductance, available at − 100°C with prospects of much higher temperatures. Nevertheless, the materials involve rare earth oxides and may be too expensive.

relationship are limited by the large surface forces and the finite resistance of the material to disintegration. The usual flywheel is constructed of steel, and its storage capacity is only in the region of 6 Wh/kg. However, more recent developments utilizing highly anisotropic materials with extremely high resistance to fracture in a given direction (e.g., glass, graphite, or various phases of boron) have yielded higher storage capacities, up to about 60 Wh/kg.[11]

Because of their rather demanding construction, flywheels are restricted in their usefulness to special applications.[12]

11.5. STORAGE OF HYDROGEN

As already discussed earlier, great amounts of energy can be stored chemically in the form of hydrogen. The possibility of using hydrogen as a storage medium is one of the greatest advantages of the solar–hydrogen economy. The requirement is for low-cost storage with a very high safety margin and economical and practical weight–volume relationships. Hydrogen has a high energy density per unit weight. On the other hand, its specific heat value per unit volume is low due to its low density. Hydrogen would have to be compressed to 3500 bar to attain the same energy density as heating oil. Storage of gaseous hydrogen is therefore predominantly limited by volume considerations.*

11.5.1. Storage of Gaseous Hydrogen

11.5.1.1. Pressure Cylinders

At present, gaseous hydrogen is stored almost exclusively in steel cylinders at a pressure of 150–300 bar.[13]†

Precise data and the range of applications for hydrogen in cylinders are presented in Chapter 9.

11.5.1.2. Above-Ground Storage Tanks

Large amounts of hydrogen gas can be stored in stationary high-pressure tanks either above or below ground. In fact, above-ground tanks for the storage of natural gas are a familiar sight. Low-pressure tanks operate up to pressures of about 0.63 bar, and have a volume capacity of up to 600,000 Nm^3 (normal

* Translator's note: As is pointed out by A. C. Tseung, hydrogen storage in large rubberized containers at sea depth where the pressure is 100 atm may be a practical method because of the large area available.

† Translator's note: The possibilities of cylinders made of materials much lighter than steel (at present aluminum) must be taken into account. They have been applied to automotive transportation particularly by Zweig *et al.*[54]

cubic meters), which in the case of hydrogen gas means an energy content of 1.8×10^6 kWh. High-pressure containers work up to a pressure of about 14 bar and contain about 50,000 Nm^3. The cost of such storage amounts to about $\$12.5–15/Nm^3$.[14–16]

The capacity of above-ground pressure tanks cannot be greatly increased because of limitations imposed by the properties of the construction materials. Underground pressure tanks or underwater tanks can be used with much greater pressures.[17]

11.5.1.3. Caverns, Aquifers, and Natural Gas and Oil Fields

Depleted natural gas and oil fields offer a natural solution to the need for underground pressure tanks, as do aquifers and artificial caverns. In various parts of the world—particularly in the northern part of the Federal Republic of Germany—there are various salt domes, or gas caverns, formed by the dissolving of salt. Throughout the world, but particularly in the United States, about 700 such caverns are already in use for gas storage. A typical cavern lies at a depth of 600–1200 m and has a volume of 20,000–800,000 m^3; the operating pressure is about 60 bar, but could probably be increased to about 200 bar. The investment costs are entirely determined by those incurred for excavation; for large caverns, they are between \$0.5 and $1/Nm^3$.[18–20]

Storage in depleted oil and gas fields is certainly one possibility for the disposition of large amounts of gas. In the Federal Republic of Germany, however, there are few places in which geological formations would allow the setting up of such storage facilities in reasonably practical sites. There are two such storage domes at this time, one in Hamburg and one in Munich, and they have volumes, respectively, of 10^8 and 3×10^8 Nm^3. The advantage of these storage sites is their low investment cost, less than $\$0.25/Nm^3$.[19–22]

Similar to these oil and gas fields are water-containing porous layers (aquifers) covered with material impermeable to the gas being stored, and these can be used for the storage of gases by forcing out the water and introducing the gas. In the Federal Republic of Germany, there are already seven storage facilities of this type, with storage volumes between 3×10^7 and 3×10^8 Nm^3. The costs of such storage sites depend largely on the geological formation in which the aquifers are situated and differ greatly in various areas of the country. The storage costs for aquifer storage amount at present to about 1.65–8.2¢/Nm^3, including the investment cost of about $\$0.1–0.65/Nm^3$.[14,19,23–25] Information on storage capacities and the costs of various means of storage is presented in Table 11.2.

11.5.1.4. Storage in Pipeline Systems

In most cases, the facilities in which hydrogen is produced and those in which it is used are separated by significant distances and pipelines must be used

TABLE 11.2
Storage Capacities and Costs of Various Existing Means of Large-Scale Gas Storage

Gas storage container	Capacity (million Nm^3)	Investment costs $/$Nm^3$	Storage costs ¢/Nm^3
Aquifer storage	1–170	0.1–0.65	2.0–8.0
Dry low-pressure gas containers	0.1–0.5	—	2.0–8.0
High-pressure gas containers	Up to 0.33	13.0–18.0	—
Cavern storage	1–70	0.35–2	1–3

for the transmission and distribution of hydrogen. The working pressures of the pipeline can vary within a certain range, so that the amount of gas being stored varies and the pipeline acts as an energy-storage device.

The cost of this type of storage is in principle about the same as that of aboveground pressure tanks, but the simultaneous use of gas transport and gas storage gives rise to some small cost advantages.[26] Details of these pipeline systems are given in Chapter 9.

11.5.2. Storage of Low-Temperature Hydrogen

11.5.2.1. Liquid Hydrogen

The boiling point of hydrogen is 20.4K and the density in the liquid state 0.071 g/cm^3. The evaporation of 1 liter of liquid hydrogen yields 0.79 Nm^3 of hydrogen gas, and the necessary heat of evaporation is 129 Wh_{th}/kg.

The content of a 200-bar steel cylinder at 10 Nm^3 is therefore only 12.7 liters of liquid hydrogen. The liquefaction of hydrogen can be carried out by several processes with numerous variations. In large plants, the gas is cooled first with liquid nitrogen and in the second stage by the usual expansion devices. The actual liquefaction takes place through a throttle valve in which both the Joule–Thomson and the isothermal throttle effect take place (see Figures 11.2 and 11.3). The spins of the hydrogen nuclei in a hydrogen molecule are either parallel (ortho-H_2) or antiparallel (para-H_2), and the ratio of these two forms is temperature-dependent (see Figure 11.4). During the liquefaction of hydrogen, some of the ortho-hydrogen is converted slowly to para-hydrogen. This transformation is one of the barriers to any lengthy storage of liquid hydrogen because it usually takes place over several days and the transformation heat that is produced will give rise to an evaporation of about 70% of the hydrogen. In industrial hydrogen liquefaction, the equilibrium of the ortho- to para-hydrogen conversion is reached faster through the use of catalysts to reduce the later

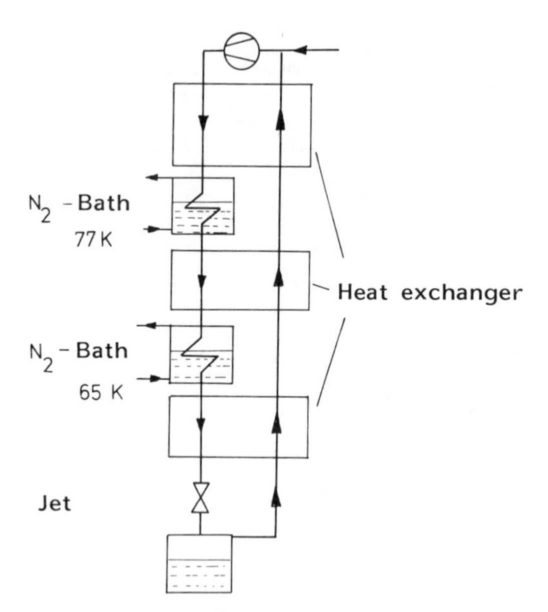

FIGURE 11.2. Schematic diagram of hydrogen liquefaction (Linde).

undesirable effects of evaporation to a low value. With respect to the ortho- to para-hydrogen conversion, an additional amount of energy of 20–30% is needed.[27–31] The amount of energy used in the liquefaction is shown in Figure 11.5 as a function of the initial pressure for a Linde process (i.e., a countercurrent process) and the Claude process (cooling by means of expansion at low temperatures).

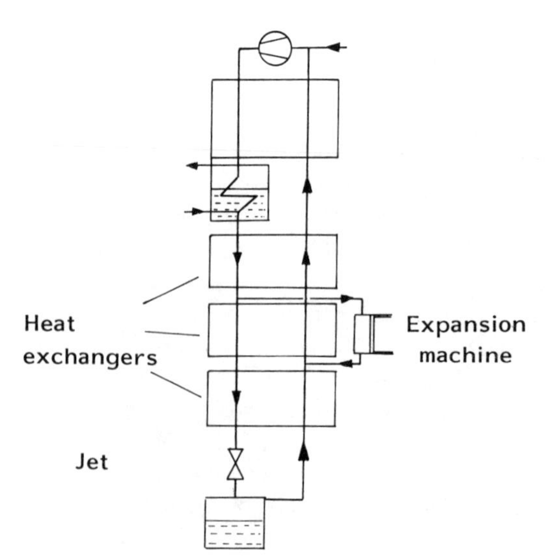

FIGURE 11.3. Schematic diagram of the countercurrent process with cooling through an expansion device (Claude).

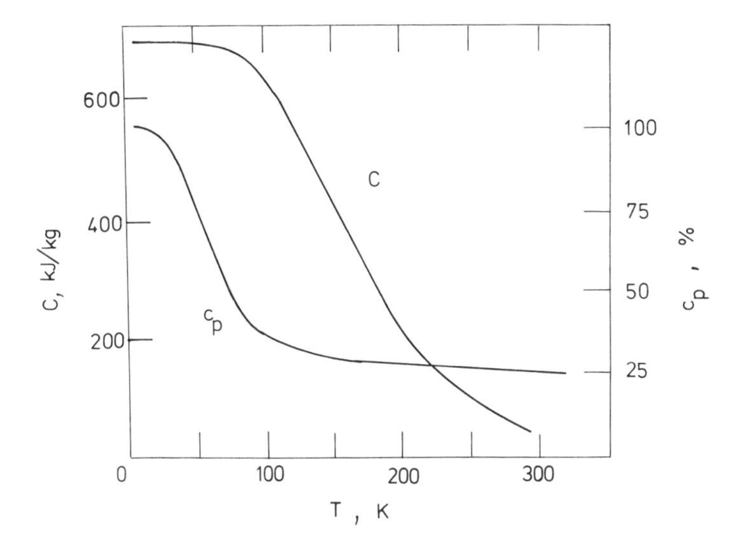

FIGURE 11.4. Concentration at equilibrium (C_p) and integral transformation heat (C) of para-H_2 in liquid hydrogen as a function of temperature. From Schraewer.[27]

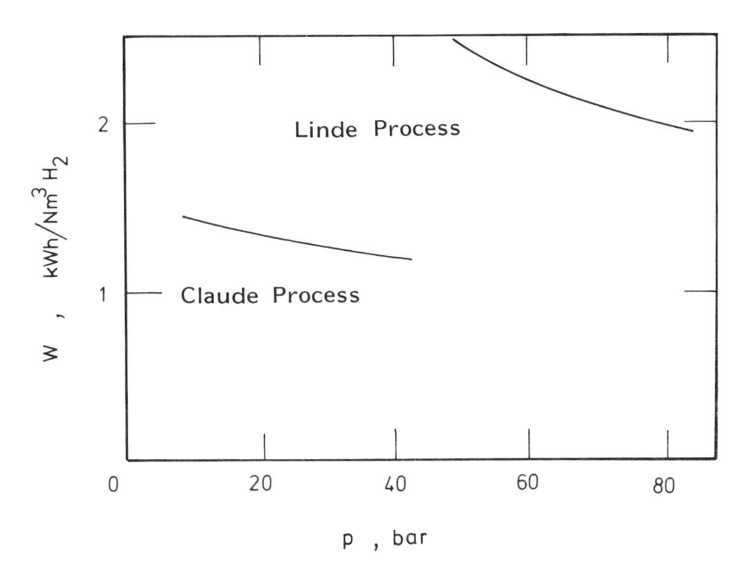

FIGURE 11.5. Energy expenditure (W) for the liquefaction of hydrogen by the Claude and Linde processes as a funtion of the initial pressure. From Schraewer.[27]

Liquid hydrogen must be kept in efficient cryostorage tanks. The transfer of heat from the outside to liquid hydrogen within the tank takes place in three ways: convection, conduction, and radiation. Convection is usually decreased by evacuating the space between the outer and inner parts of the vessel or by filling the space with some kind of insulation foam (see Figures 11.6, 11.7, and 11.8). Conduction is also considerably reduced by such means. Radiation losses, which are proportional to T^4, are reduced by utilizing reflecting material, mostly thin aluminum foil or silver mirrors. Since heat transfer by all three of these processes increases with increasing external surface area, cryostorage tanks are mostly cylindrical or spherical to minimize the surface area.

In the construction of large cryostorage tanks, in addition to the technical possibilities, the costs are an extremely important consideration. In the case of aboveground storage of liquid H_2, single-walled and double-walled tanks of various sizes are used (see Figures 11.6, 11.7, and 11.8). In the single-walled tanks, foam insulation is applied directly to the load-bearing walls. In space technology, it is mostly freon foamed insulation that is used. With external insulation, the tank is made of cold-rolled steel, but with internal insulation, ordinary construction steel can be used.[32] A 21-cm-thick insulating layer of polyurethane foam shows a heat-transfer coefficient of $k = 0.02$ W/m $\times$ K. The daily loss through evaporation is 1–2.5%.

Most liquid-hydrogen storage tanks in use today are double-walled tanks

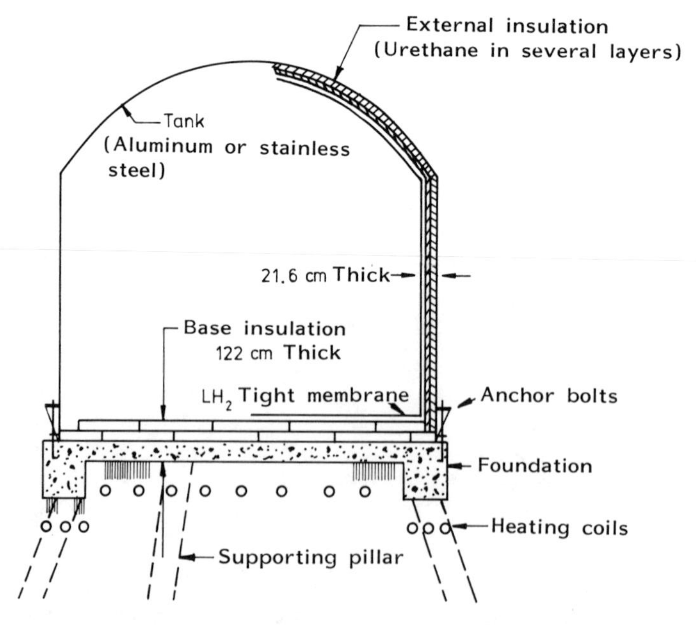

FIGURE 11.6. One walled, multilayer-insulated liquid hydrogen tank. From Hallet.[32]

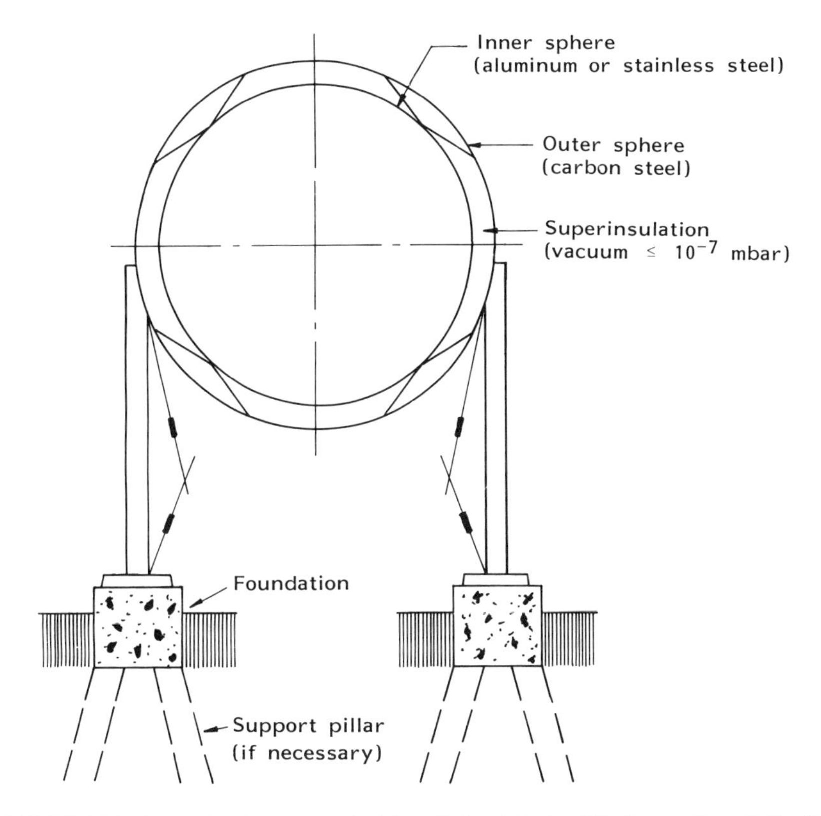

FIGURE 11.7. Schematic diagram of a double-walled tank for liquid hydrogen. From Hallet.[32]

with an evacuated Perlite insulation layer (see Figures 11.7 and 11.8). The tanks at Cape Kennedy have a capacity of 2.4×10^5 kg liquid H_2 and contain Perlite insulating layers. In Europe, there are some smaller cryogenic storage devices with a capacity of about 50 m^3 (at CERN), and these cost \$225 each and have a maximum excess pressure of 3 bar. The provision of one of these tanks with superinsulation would add about \$300,000 to the cost of a unit.[33] Table 11.3 shows data for such a double-walled tank.

For heat conductivity coefficients of $k = 0.0015$ W/m $\times$ K, the daily losses of hydrogen through conduction are small indeed and are near those of tanks with vacuum superinsulation. The investment costs of these tanks—because of the complex construction of the insulation and tanks—are about twice as high as those of simple single-walled tanks.

Liquid hydrogen—like liquid oxygen and liquid nitrogen—can be stored and transported in special tank cars or containers. At the place of consumption, the liquid hydrogen is stored in gasifiers. These gasifiers are small cryostorage

FIGURE 11.8. Double-walled tank for liquid hydrogen at Cape Kennedy.

TABLE 11.3
Capacity and Cost of Double-Walled Tanks with Evacuated Perlite Insulation[a]

Capacity (kg)	Insulation thickness (cm)	Heat flow (kW)	Evaporation loss ($\$10^6$/yr)[b]	Evaporation costs ($\$10^6$/yr)[b]	System costs ($\$10^6$/yr)
3×10^5	90	0.47	0.0034	0.9–1.1	0.9–1.1
9×10^5	90	0.94	0.023	1.9–2.2	2.5–3.0
18×10^5	90	1.50	0.018	2.9–3.4	4.7–5.6
36×10^5	90	2.40	0.015	4.7–5.6	9.0–10.8

[a] From Hallett.[32]
[b] At a hydrogen price of \$0.25–0.30/kg liquid H_2.

tanks with control devices that maintain a constant working pressure. Evaporation of the liquid hydrogen takes place as a result of the effect of the surrounding atmospheric heat.

11.5.2.2. Hydrogen Slush

Hydrogen slush is a mixture of liquid and frozen hydrogen that is in equilibrium with the gas at the triple point, 13.8K. The density of the icelike form is about 20% higher than that of the boiling liquid. To obtain the icelike form, one has to remove the heat content of the liquid at 20.3K until the triple point is reached and then remove the heat of melting; the "cold content" of the icelike form of hydrogen is some 25% higher than that of the saturated vapor at 20.3K. In producing the hydrogen slush, in comparison with the icelike form, one loses part of the increase in density and cold storage, but in mixing the icelike form with liquid hydrogen, one obtains a medium that is transportable in pipelines. The hydrogen slush can be produced from liquid hydrogen with the help of a vacuum pump and expansion through a jet.[34–36]

Compared with liquid hydrogen, the slushlike form of hydrogen has the advantage of a greater "cold content." The disadvantage of hydrogen slush lies in the present state of the technology for its production, which is not yet well advanced, so that it is impractical to calculate a realistic price. In any case, the production costs of hydrogen slush now and in the near future would be greater than the liquefaction costs of hydrogen because of the larger energy use involved. For this reason, the use of hydrogen slush is to be looked to only for special cases in which the higher density and greater cold content of hydrogen slush are really needed. Storage and transport would be carried out analagously to those of liquid hydrogen.

11.5.3. Physically or Chemically Bound Hydrogen

As has already been mentioned, one of the principal problems in storing hydrogen gas is its great volume and corresponding low energy density. The specific density of hydrogen can be increased considerably by dissolution or chemical binding of hydrogen in or to suitable materials, respectively. Of the various possible methods of this kind, only a few are economically and energetically sensible.

11.5.3.1. Metal Hydride Tanks

In the course of seeking a procedure for hydrogen storage, particularly for mobile applications, metal hydrides, and also those of various intermetallic compounds, have received a great deal of attention. In principle, any metal that combines with hydrogen and can be persuaded to give it up again easily is a suitable candidate for a hydrogen-storage medium. However, to obtain a useful and economically applicable means of hydrogen storage, several rather critical conditions are necessary: The storage material must be sufficiently inexpensive and have a reasonably large storage capacity and low weight. It must be able to undergo a large number of charge–discharge cycles with a small decrease in storage capacity. It should combine great reaction velocity with small amounts of heat produced by absorption and desorption, and good reversibility with satisfactory physical and chemical properties.

Hydride formation is an exothermal process, so that heat, i.e., energy, must be used up in the dissociation process. On energetic grounds, one needs a hydride with a small dissociation heat. Further, the temperature–pressure region in which the material gives off hydrogen is very important. Information concerning this process is given by concentration–pressure isotherms. Figures 11.9 and 11.10 show such isotherms for $LaNi_5$ and FeTi. $LaNi_5$ is at a maximum with six atoms-% of hydrogen, but the hydrogen content decreases with increasing temperature. Correspondingly, the equilibrium pressure increases rapidly so that at higher temperatures pressure containers become necessary.[37,38] FeTi shows several distinct regions. Up to the complete formation of FeTiH, there is a plateau, and then the isotherm rises sharply until $FeTiH_{1.6}$ is reached. The last part of the isotherm is of limited use for hydrogen storage because the necessary pressures increase so rapidly.[39–42]

Table 11.4 and Figure 11.11 present data relevant to the storage capacity of metals and alloys that form hydrides and are therefore suitable for storage of hydrogen. These data are to be regarded only as guidelines because the values are greatly altered by small amounts of impurities. For example, 0.92 wt.-% of

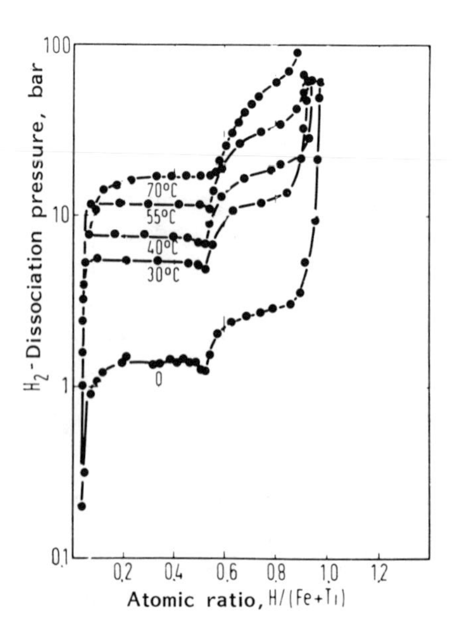

FIGURE 11.9. Isotherms of the hydrogen dissociation pressure as a function of the TiFeH ratio (the parameter is the temperature).

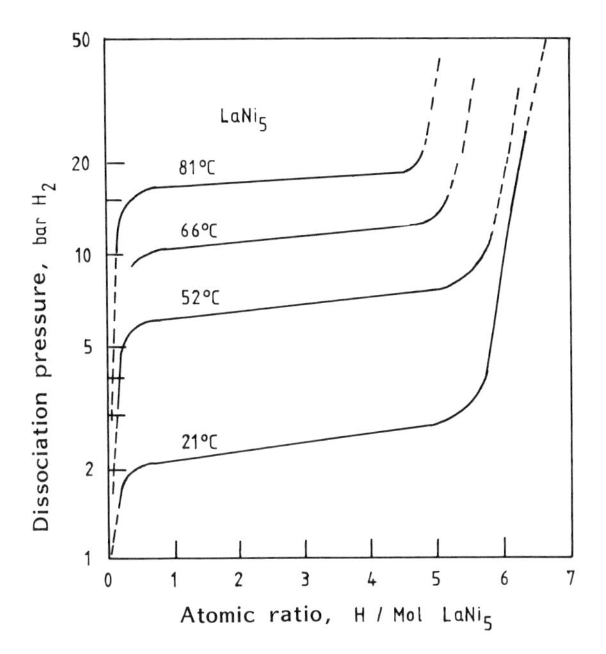

FIGURE 11.10. Isotherms of the hydrogen dissociation pressure of $LaNi_5$ as a function of the H_2 content with temperature as the parameter.

silicon in vanadium decreases the hydrogen-storing capacity of that metal by 10% and increases the dissociation pressure from about 4 to about 15 bar at 45°C. A similar sensitivity to impurities is shown by niobium.[37–46]

The most suitable alloys seem to be $LaNi_5$ and TiFe, but of these two, the first is restricted by its high cost to special situations. It is TiFe that is the object of efforts for technical development (alloys), and the storage capacity of this material for hydrogen amounts to 10 g H_2/kg TiFeH, which is equivalent to 65

TABLE 11.4
Data for Different Metal Hydrides

Physical characteristic	$LaNi_5H_{6.7}$	$FeTiH_{1.95}$	$MgNiH_{4.2}$	$Mg_4Cu_2H_6$
H_2 content (wt.-%)	1.5	1.8	3.8	2.6
Equilibrium temperature at 1 bar (°C)	10	19	250	240
p-T equation A	—	1760	3360	—
($\log p_{bar} = A/T + B$) B	—	6.92	6.39	—
Disintegration heat (kJ/mole H_2)	30.2	29.8	64.8	73.2

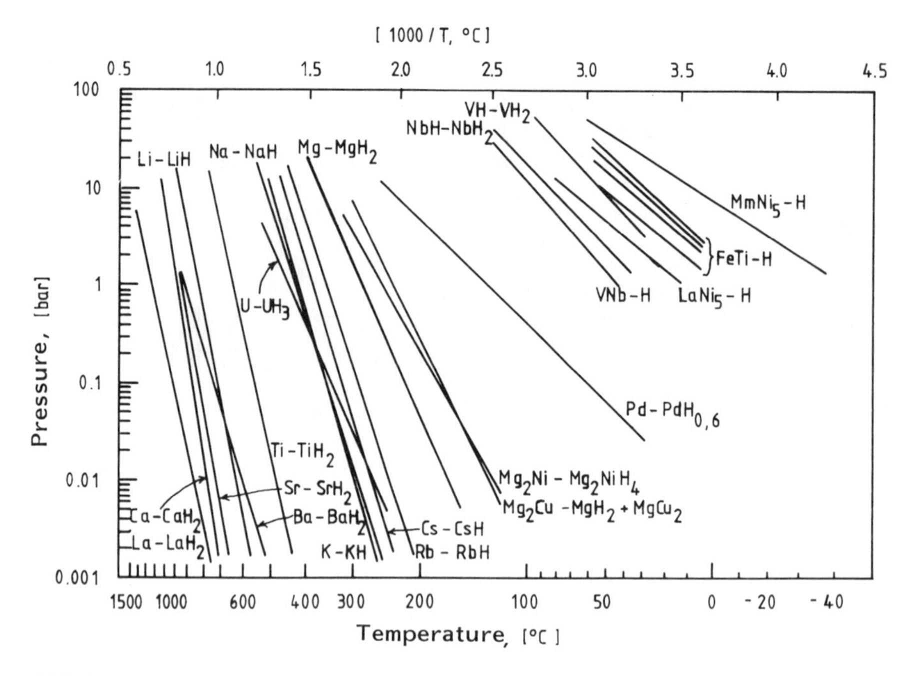

FIGURE 11.11. Temperature–pressure characteristics of different metal hydrides in double logarithmic scale.

g H_2/liter. The dissociation energy is 9.7% of the energy of the hydrogen stored. The low dissociation temperature means that this energy can be produced from the waste heat.

The storage capacities shown in Table 11.4 for TiFe alloys can certainly be reduced by various factors such as the particular tank available or the supporting matrix. However, research developments suggest that higher storage capacities could be obtained. A realistic approximate value for the storage capacity that takes into account the weight of the tank, the material for the support matrix, and other factors is in the range of 1–2 w.-% of hydrogen. It is not yet possible to say very much about the cost.[47]

11.5.3.2. Adsorption

Hydrogen can also be taken up by adsorption on the surface of highly dispersed adsorbants. Fe, Ni, and Pt adsorb about 1 H atom per metal surface atom. However, at best, only one atom in four is a surface atom, so that the storage capacity referred to for the entire storage material must be reduced to about 0.25 hydrogen atom per metal atom, and this gives rise to values that are unfavorable with respect to weight.[48] To make this approach to hydrogen storage a successful one, it will be necessary to have an extremely light and highly dispersed adsorbant. Thus, Justi [49,50] introduced high-surface-area charcoal. A coconut charcoal at 77K and 1 bar can take up 86 Ncm^3 hydrogen/cc charcoal, and this material has a weight of only 0.58 g/cm^3, i.e., about 1.5 w.-% of storage capacity. Justi's figure (Figure 11.12) shows that the adsorbed quantity greatly increases with lower temperatures and boiling points. Thus, if one goes from

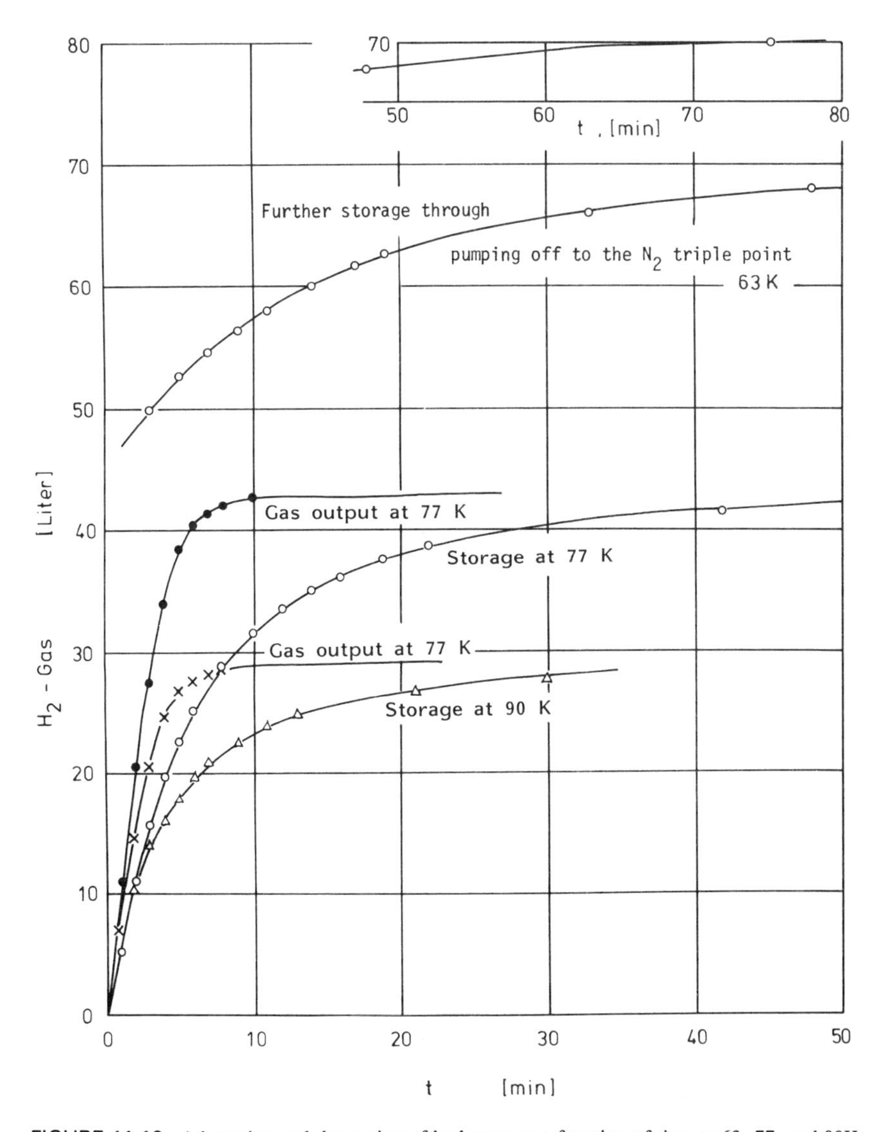

FIGURE 11.12. Adsorption and desorption of hydrogen as a function of time at 63, 77, and 90K. Granular adsorption of coal ATK III, bulk volume 500 cm^3, bulk weight 0.58 g/cm^3; before loading, heated at 470°C for 2.5 hr. From Justi.[49,50]

77 to 63K and utilizes the triple point of boiling nitrogen at a pressure less than 0.13 bar Hg, then the storage capacity increases by about two times. This method is under further investigation, and the introduction of modern commercial charcoal, as well as the use of 52K eutectically frozen liquid air, gives rise to the expectation that there may be some further progress. The small heat of adsorption (3.8–4.2 J/mole hydrogen)[51] is also advantageous.

The basic advantage of this method is that it makes it possible, by introducing an additional adsorptive phase, to avoid the problems of dealing with the very low boiling point of hydrogen (20K) and use cheaply producible liquid nitrogen, which boils at a temperature almost four times as high but is easily available everywhere (e.g., through the production of artificial fertilizers). The gain in cold temperature is paid for, of course, by a loss of density of the liquid hydrogen. However, the density of the adsorbed hydrogen at least reaches the order of magnitude of that of liquid hydrogen, as is shown from the results of Figure 11.12; namely, it is about one sixth of 0.071 g/cm^3. It is quite possible that further progress could bring this method to the breakthrough point and make possible the construction of lightweight cryostats that would be important for mobile applications.

11.5.3.3. Hydrogen-Containing Compounds

With liquid hydrogen-containing compounds, it is possible—in contrast to what is possible with hydrogen gas—to store hydrogen at a favorable fuel/container weight ratio. Good values can also be obtained with gaseous compounds that have low boiling points, because they can be maintained in liquid form with relatively low excess gas pressures.

Large quantities of hydrogen can be stored in the hydrides of nitrogen. These compounds have the great advantage that nitrogen is universally available and that in the decomposition of the nitrogen–hydrogen bond, the nitrogen can be released into the environment. Hydrazine is the most energy-rich of these compounds and also an easy one to decompose. It exists as a monohydrate having the formula $N_2H_4 \cdot H_20$. Hydrazine can be split by the use of Raney nickel in alkaline solution. There is a side reaction that produces some ammonia, which has to be removed by some counterreaction.[52] The hydrogen content of hydrazine is theoretically 8% of the weight. The decomposition is exothermic, so that no additional energy is used in the splitting of hydrazine.[53] However, the material is at present too expensive for broad application.

The cheapest nitrogen–hydrogen compound is, of course, ammonia. At normal pressure and temperature, ammonia is a gas, but it has to be compressed by only about 10 bar to become liquid at room temperature. There are several excellent schemes for splitting ammonia on a small scale. For example, at 600–800°C, ammonia is catalytically decomposed to nitrogen and hydrogen. The relative weights in such a process are as follows: An empty 40-liter ammonia flask weighs about 42 kg, and the contents when the flask is full about 21 kg. The 21 kg of ammonia contains about 41 Nm^3 hydrogen. About 20% of the hydrogen is used up in heating the reactor, so that one ends up with a storage capacity of 1.9 kg/Nm^3 hydrogen. One has to add to this the weight of the material on which the ammonia is dissociated, which depends to a great extent on the rate at which the gas is to be flowed through it, the mode of operation, and other factors, but is between 2 and 5 kg/Nm^3 per hr.

11.6. CONCLUSIONS

In contrast to many other processes, energy storage with hydrogen offers a great variety of possibilities. It must be remembered that with hydrogen, it is possible to store very large amounts of energy at low cost. Thus, for these large amounts, underground storage in aquifers, depleted natural gas or oil fields, or artificial caverns is suitable. There is a significant degree of "line storage" in a pipeline network set up for transporting gas over significant distances.

Storage of liquid hydrogen at low tempertures resembles electrochemical storage in storage batteries in being quite expensive. One has to look at a cost increase of 40–50%, and these methods of energy storage will probably be limited to mobile systems for which relatively small volumes are needed. In these applications, hydrogen-containing compounds such as hydrazine and ammonia may also be suitable, and the choice will be determined largely by price, weight, and safety factors.

REFERENCES

1. A. B. Meinel and M. P. Meinel: *Applied Solar Energy—An Introduction*, Addison–Wesley Publ. Co., Reading, Massachusetts (1976).
2. A. L. Robinson: *Science* **184:**785–787; 884–887 (1974).
3. Philips Research Press Release No. 724/1230/108 E (1974); 735/1001/122 E (1974).
4. J. O'M. Bockris, N. Bonciocat, and F. Gutman: *An Introduction to Electrochemical Science*, Taylor and Francis, London (1974).
5. Technology of Efficient Energy Utilization, Report of a NATO Science Committee Conference, Les Arcs, France, October 8–12, 1973.
6. F. v. Sturm: *Elektrochemische Stromerzeugung*, [*Electrochemical Current Production*], Verlag Chemie, Weinheim (1969).
7. K. G. Pankhurst: British Rail Research, personal communication, October (1974).
8. J. O'M. Bockris and A. K. N. Reddy: *Modern Electrochemistry*, Rosetta Edition, Plenum Press, New York (1974).
9. H. Cnobloch, *Neue Zürcher Zeitung*, Supplement, "Forschung und Technik," February 18, No. 40 (1976).
10. P. Brennecke, H. Ewe, E. Justi, and W. Rosenberger, *Journal of Power Sources* **16** (4): 271–284 (1985).
11. R. T.Dann: *Machine Design*, p. 130, May 17 (1973).

12. D. W. Rabenhorst, Intersociety Energy Conversion Engineering Conf. Proc., p. 38 (1971). D. W. Rabenhorst, presented at the 14th Annual Symposium, New Mexico Sections of the American Society of Mechanical Engineers and the American Society for Metals, University of New Mexico, Albuquerque, February 28 (1974).
13. R. Kipker: Neue Druckgasflasche für 300 bar Betriebsüberdruck [New high-pressure gas cylinders for 300 atm operation], *Gas Aktuell,* **9**:3 (June 1975).
14. Bundesministerium für Forschung und Technologie (publisher): *Auf dem Wege zu neuen Energiesystemen,* Teil III, *Wasserstoff und andere nicht fossile Energieträger* [*On the way to New Energy-Producing Systems*, Part III, *Hydrogen and Other Non-Fossil Energy Media*], Bonn (1975).
15. Anonymous: Europas gröβter Hochdruck-Gasbehälter [Europe's Largest High-Pressure Gas Container], Gasverwendung 7 (1973).
16. DIN 3397 and DIN 3396.
17. H. Justi: Verleich und Wirtschaftlichkeit der verschiedenen modernen Methoden der Gasspeicherung [Comparison and Economics of Various Modern Methods for Gas Storage], *Erdoel Kohle-Erdgas-Petrochem.* (Aug 1966).
18. B. Höfling *et al.:* Speichermöglichkeiten für Gase und Flüssigkeiten im Untergrund [Storage possibilities for the underground storage of gases and liquids], *Fortschr. Ber. VDI Z.* (June 1968).
19. *Jahrbuch für Bergbau, Energie, Mineralöl und Chemie* [Yearbook for Mining, Energy, Mineral/Oils, and Their Chemistry], Verlag Glückauf, Essen (1973).
20. J. H. Kelley *et al.:* Report of the NASA Hydrogen Energy Systems Technology Study, JPL 5040-1, Calif. Inst. Technology, Pasadena (Dec. 1975).
21. Anonymous, Neuer Erdgas-Untergrundspeicher in Betrieb, *Gaswaerme* Int. (Oct. 1973).
22. Bundesministerium für Forschung und Technologie. (publisher): *Einsatzmöglichkeiten von Energiesystemen,* III, *Wasserstoff* [*Opportunities for the Application of New Energy-Conversion Systems. III. Hydrogen*]. Bonn (1975).
23. A. B. Walters: Technical and environmental aspects of underground hydrogen storage, Proc. 1st World Hydrogen Energy Conf., Vol. II, 2B-65 (T. N. Veziroglu, ed.), Pergamon Press, New York (1976).
24. K. M. Coats: Some Technical and Economic Aspects of Underground Gas Storage, *J. Pet Technol.* (Dec. 1966).
25. D. P. Gregory and J. Wurm: Production and distribution of hydrogen as a universal fuel, Proc. 7th Intersoc. Energy Conv. Eng. Conf., 1329, San Diego (1972).
26. NASA-ASEE Summer Study Report: A Hydrogen Energy Carrier, Johnson Space Center, (1973).
27. R. Schraewer: Technology of hydrogen liquefaction, use of liquid hydrogen, its storage and transport or transmission, in: *The Hydrogen Energy Concept,* Ispra Courses H_2/75 No. 14 Sept. 29–Oct. 10, (1975).
28. J. C. Mullins, W. T. Ziegler, and B. S. Kirk: The Thermodynamic Properties of Para Hydrogen from 1 to 22 K, NBS-Techn. Rep. 1 (1961).
29. C. R. Baker and R. L. Shaner: A study of the efficiency of hydrogen liquefaction, Proc. 1st World Hydrogen Energy Conf., Vol. II, 2B-17 (T. N. Veziroglu, ed.), Pergamon Press, New York (1976).
30. C. R. Baker and L. C. Matsch: Production and distribution of liquid hydrogen, *Adv. Petr. Chem. Refin.* **10** (1965).
31. C. L. Newton: Hydrogen production, liquefaction and use, *Cryog. News* (Aug. 1967).
32. N. C. Hallet: Study Cost and System Analysis of Liquid Hydrogen Production, Final Report NASA CR 73 226.
33. J. R. Bartlit, F. J. Edesknty, and K. D. Williamson, Jr. Experience in Handling, Transport, and Storage of Liquid Hydrogen—The Recyclable Fuel, Proc. 7th Intersoc. Energy Conv. Eng. Conf., p. 1312, San Diego (1972).

34. R. Schraewer and W. Daus: Herstellung und Förderung von Wasserstoffmatsch, [Production and transport of hydrogen slush], Forschungsbericht NT 200 des BMWF (1974).
35. C. F. Sindt: A summary of the characterization study of slush hydrogen, *Cryogenics* (October 1970).
36. A. S. Raqual and D. E. Daney: Preparation and Characterization of Slush Hydrogen and Nitrogen Gels, NBS-Techn. Note 378 (1969).
37. J. H. N. van Vucht, F. A. Kuijpers, and H. C. A. Bruning, *Philips Res. Rep.* **25**(2)**:**133 (1970). T. B. Flanagan and S. Tanaka: Hydrogen storage by $LaNi_5$: Fundamentals and applications, Proc. Symp. Electrode Materials and Processes for Energy Conversion and Storage (J. D. McIntyre, S. Srinivasan, and F. G. Will, eds.), Electrochem. Soc. Proc., Vol. 77-6 (1977).
38. K. D. Beccu, H. Lutz, and O. de Pous, *Chem.-Ing.-Technik* **48**(2)**:**161 (1976).
39. G. Strickland, J. Milan and W.-S. Yu: The behavior of iron titanium hydride test beds: Long term effects, rate studies and modeling, *Int. J. Hydro-Energy,* p. 309 (1977).
40. D. M. Gruen, M. H. Mendelsohn, and I. Sheft: Absorption of hydrogen by the intermetallics $NdNi_5$ and $LaNi_4Cu$ and a correlation of cell volumes and desorption pressures, Proc. Symp. Electrode Materials and Processes for Energy Conversion and Storage (J. D. McIntyre, S. Srinivasan, and F. G. Will, eds.), Electrochem. Soc. Proc., Vol. 77-6 (1977). J. J. Reilly and R. H. Wiswall, Jr., *Inorg. Chem.* **13:**218 (1974).
41. J. J. Reilly and J. R. Johnson: Titanium alloy hydrides, their properties and applications, Proc. 1st World Hydrogen Energy Conf., Vol. II, 8B–3 (T. N. Veziroglu, ed.), Pergamon Press, New York (1976).
42. M. A. Pick and H. Wenzl: Physical metallurgy of TiFe-hydride and its behavior in a hydrogen storage container, *Int. J. Hydrogen Energy,* **1:**413 (1976).
43. H. H. van Mal: The activation of a lanthanium-nickel-five hydrogen absorbent in: *Hydrogen Energy,* Part A, 605 (T. N. Veziroglu, ed.), Pergamon Press, New York (1975), L. Schlapbach *et al., Int. J. Hydrogen Energy,* **4:**21–28 (1979).
44. L. C. Beavis *et al.:* The formation and properties of rare-earth and transition metal hydrides, in: *Hydrogen Energy* Part A, p. 659, (T. N. Veziroglu, ed.), Pergamon Press, New York (1975).
45. R. H. Wiswall Jr., and J. J. Reilly: Metal hydrides for energy storage, Proc. 7 Int. Soc. Energy Conv. Eng. Conf., p. 1342, San Diego (1972).
46. C. H. Waide, J. J. Reilly, and R. H. Wiswall: The application of metal hydrides to ground transport, in: *Hydrogen Energy,* Part B, 779 (T. N. Veziroglu, ed.), Pergamon Press, New York (1975).
47. H. Buchner and R. Povel: *Int. J. Hydrogen Energy* **7**(3)**:**250–266 (1982).
48. H. Ewe: *Electrochem. Acta* **17:**2267 (1972).
49. E. Justi: Method and apparatus for storing gaseous fuel for the operation of fuel cells, US.-Pat. 3,350,229 (1963).
50. E. Justi: DAS 1 265 802 (1968).
51. H. Ewe, P. Brennecke, E. Justi, and H. J. Selbach, Proceedings of the 3rd International Solar Forum, Hamburg, June 24–26, 1980, pp. 331–336, DGS-Sonnenenergie Verlags GmbH, Munich (1980).
52. H. Laig-Hörstebrock, H.-J. Schwartz, D. Sprengel, and A. Winsel: Development and Testing of a 3.5 kW Fuel Cell Stack, Ergebnisbericht z. Forschungsvorh. WI. B3-7221-EDU-202-68 des BMFT.
53. D. Sprengel: Inert Gas Operation of Gas-Diffusion Electrodes, 3rd Int. Symp. on Fuel Cells, Brussels, June 16–20, 1969, in: Société d' Etudes de Recherches et d'Applications pour l' Industrie and Société Commercial d' Applications Scientifiques (eds.), Proceedings, Presses Académiques Européennes, Brussels (1969).

CHAPTER 12

Safety Aspects of Using Hydrogen

For acceptance of the secondary energy carrier hydrogen, which in the future will be introduced on a large scale, knowledge of the safety aspects is of considerable importance. The safety and the dangers in dealing with hydrogen are determined by its physical and chemical properties. In this chapter, the characteristics of hydrogen that are relevant for safety measures are presented and compared with those of other energy carriers so that the risks in dealing with gaseous and liquid hydrogen can be estimated.

12.1. PHYSICAL DATA AND SAFETY-ENGINEERING QUANTITIES

In nearly every discussion concerning the safety aspects of the transition to hydrogen, the subject of the explosion of the dirigible "Hindenburg" at Lakehurst, New Jersey, in 1938 (Figure 12.1) arises. In the subsequent history of civil aviation, far more catastrophic aircraft accidents have occurred, but no ban on kerosene as a fuel has been imposed on the grounds of safety. It is not reasonable to base a decision about energy carriers on a single criterion, e.g., inflammability. Risks must be assessed in light of all the relevant characteristics that pertain to safety matters.

The use of hydrogen technology in the future will place specific requirements

This chapter authored by Dr. P. W. Brennecke and Dr. H. J. Selbach, Braunschweig.

FIGURE 12.1. Burning of the dirigible "Hindenburg" in Lakehurst, New Jersey, in 1938.

on safety techniques. To point out clearly the possible risks, a number of physical data have been collected in Table 12.1 and the safety-engineering quantities of gaseous hydrogen compared with those of methane (natural gas) and gasoline.[1-3]

From Table 12.1, it can be seen immediately that the wide limits on the explosion ranges of hydrogen–air mixtures (4–75 vol.-% hydrogen), the extremely low ignition energy of 0.02 mJ (in respect to 29.53 vol.-% hydrogen), and the high combustion and detonation velocities are disadvantages in the use of hydrogen as compared to methane and gasoline. The presence of water vapor, however, somewhat narrows the inflammability limits of gas mixtures of these kinds.

As to the properties of hydrogen that *reduce* the danger in potential accidents, the low density of about 0.08 kg/m^3 (under normal conditions) and the

TABLE 12.1
Physical Data and Safety-Engineering Quantities of Gaseous Hydrogen, Methane, and Gasoline

Properties or quantities	Hydrogen	Methane	Gasoline
Molecular weight	2.016	16.043	≈107.0
Density (kg/m^3)[a]	0.0837	0.65	≈4.40
Lower heating value (kJ/g)	119.93	50.02	44.5
Higher heating value (kJ/g)	141.86	55.53	48
Viscosity [10^{-4} g/(cm sec)][a]	0.875	1.10	0.52
Combustion range in air[b]	4–75	5.3–15	1–7.6
Combustion range in oxygen[b]	4–95	5–61	—
Detonation range in air[b]	18.3–59	6.3–13.5	1.1–3.3
Stoichiometric relation to air[b]	29.53	9.48	1.76
Minimum combustion energy in air (mJ)	0.02	0.29	0.24
Explosion energy[c]	2.02	7.03	44.22
Ignition temperature (°C)	585	540	228–471
Flame temperature in air (°C)	2045	1875	2197
Combustion velocity in air (m/sec)[a]	2.65–3.25	0.37–0.45	0.37–0.43
Detonation velocity in air (km/sec)[a]	1.48–2.15	1.39–1.64	≈1.4–1.7
Diffusion coefficient in air (cm^2/sec)[a]	0.61	0.16	0.05
Diffusion velocity in air (cm/sec)[a]	≲2.00	≲0.51	≲0.17
Joule–Thomson coefficient (deg/bar)	+0.04	−1.6	—

[a] At normal pressure and 20°C.
[b] In vol.-% H_2.
[c] In kg TNT/(m^3 H_2[a]).

high ignition temperature of 585°C may be cited. The low density allows hydrogen that has been released from a container to rise rapidly and thus dilute itself in the air. Thus, hydrogen is a safer fuel than gasoline in respect to its high ignition temperature. Furthermore, the small heating value per volume and the minimal radiation emittance of a hydrogen flame are obvious advantages for hydrogen in respect to safety engineering.

The effects of the diffusion velocity of hydrogen, which is considerably higher than that of methane and gasoline in air and is about 2 cm/sec, one can only judge from case to case. Thus, in enclosed spaces, it would be possible for inflammable hydrogen–air mixtures to form. In open spaces, a stream of escaping hydrogen rapidly dilutes itself to a concentration below that necessary for ignition.

In using liquid hydrogen, one must bear in mind the dangers that are associated with the low temperature. In particular, one must take into account the high expansion ratio of liquid hydrogen to gaseous hydrogen. Were there to be an accidental warming of liquid hydrogen (e.g., in a liquid hydrogen storage facility), very high pressures could be generated and could lead to breakage and heavy damage. If, as a result, there is an entry of air, oxygen would accumulate in the liquid hydrogen and significantly increase the danger of explosion.

12.2. PHYSICAL DANGERS

One of the important physical dangers in the use of hydrogen is the embrittlement of metallic structures (e.g., pipes) by hydrogen. Because of its small molecular size, hydrogen dissolves in some metals and may cause rapid deterioration of their material properties.[4] Apart from the dependence of the amount dissolved on the metal itself, there is the effect of pressure and temperature. When hydrogen diffuses into a metal, ductile materials become brittle and tear easily when subjected to a small load. If there is simultaneous mechanical load, the result is a comparatively rapid increase of microcracks in the structure. In carbon-containing steels, there is also the danger of methane formation and the decarburization of grain boundaries, which have a corresponding effect on the mechanical properties of the steel. At normal pressures and temperatures, the danger of hydrogen embrittlement is comparatively low; through an appropriate choice of material and creative engineering, the safety difficulties connected with these properties can be considerably diminished. The prospect that there will be a large-scale introduction of hydrogen demands a corresponding knowledge of the physical and metallurgical properties of metal–hydrogen systems.

12.3. CHEMICAL DANGERS

The most important dangers in the use of hydrogen arise from its strong chemical reactivity, an example of which is the well-known explosive reaction between hydrogen and oxygen. The energy needed to set off a hydrogen–oxygen explosion depends largely on the ratio of the two gases and the temperature. As can be seen in Figure 12.2, there is a large increase in the necessary energy as one goes away from the stoichiometric ratio (29.53 vol.-% hydrogen), this increase being about 10 times when hydrogen is increased to 58 vol.-%. With increasing temperature, the minimum energy needed to set off the explosion falls, so that above 500°C, the gas mixture ignites spontaneously; at the same time, the ignition range of the hydrogen–air mixture increases.[2]

Thus, it is desirable to study the combustion behavior of hydrogen more closely. Hydrogen burns in air with a pale blue, very hot flame (2045 °C). The invisibility of the hydrogen flame is often mentioned as one of the risk factors in the use of hydrogen, but the radiation intensity of the hydrogen flame is less than that of a methane flame. As a result, the heat caused by radiation in the vicinity of the hydrogen flame is less than that near a natural gas flame. The colorless flame could make it difficult to see small fires that might occur when

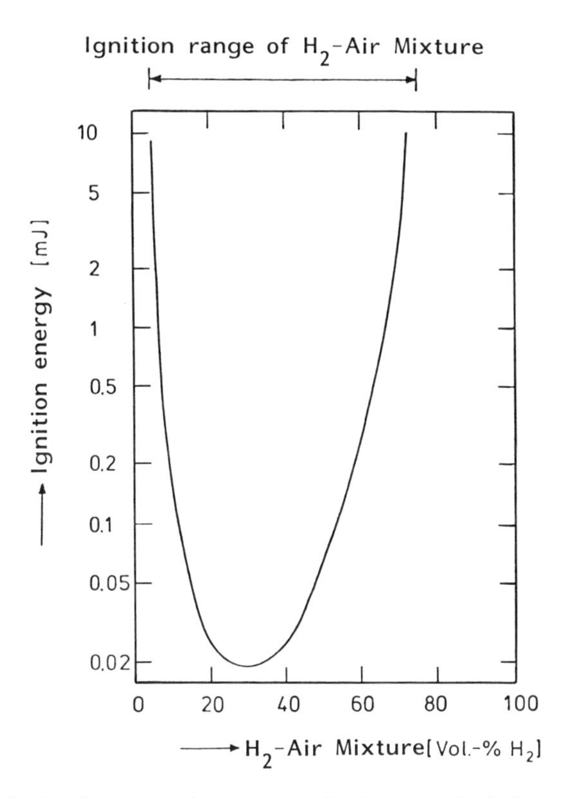

FIGURE 12.2. Combustion range and minimum combustion energy for hydrogen–air mixtures.

hydrogen leaks. The greatest safety risk is certainly the danger of explosion and detonation, but this can be largely avoided if appropriate construction, technical, and administrative precautions are taken.

12.4. SAFETY INSTRUCTIONS

The introduction of hydrogen presents some special technical points that will have to result eventually in some specific safety precautions and rules. Safety measures for hydrogen are not likely to be radically different from those associated with the use of conventional gases for combustion purposes (town gas, natural gas), but will have to incorporate knowledge of the specific properties of hydrogen.[5] This is certainly possible, because handling of gas mixtures containing large quantities of hydrogen has long been a familiar part of the commercial handling of gases. Taking into account that a comparatively large number of applications for hydrogen will be in the gaseous rather than in the liquid form, it will be necessary to first take precautions to prevent the formation of dangerous explosive hydrogen gas mixtures and to set up guidelines describing measures to deal with such situations.

12.5. EXPERIENCE IN SAFETY ASPECTS OF DEALING WITH HYDROGEN

Experience in handling and using hydrogen has already extended over decades, about 60 years with gaseous hydrogen and about 25 years with liquid hydrogen. The accidents that have occurred in this time do not show any greatly increased degree of damage when compared to those observed with other flammable gases or liquids.[6] If one takes into account that the present yearly production is about 10^{11} Nm^3 (normal cubic meters) hydrogen, then this fact becomes quite significant.

As an example of experience with safety measures in dealing with gaseous hydrogen, the pipeline in the network owned by the chemical company Hüls AG described in Chapter 9 may be considered. There are no particular safety aspects of this hydrogen pipeline system that differ from those of other pipeline networks in the chemical industry. In building this network, no particular safety measures were used except expansion joints, rupture-resistant joints, and a pressure-sensitive safety valve at the beginning and end of each section.[7] During the 30 years in which this 208-km-long pipeline network has existed, there have not been any explosions or detonations or any damage to any materials due to diffusing hydrogen. The operation has not been troubled by any safety-relevant situations or significant disturbances of any kind. This example shows that hydrogen pipeline systems are already safe.

12.6. CONCLUSIONS

With reference to the current application possibilities and areas in which hydrogen may be used, the safe handling of gaseous and liquid hydrogen can be assured by the construction and administrative measures taken for the conventional risks. Although large-scale technical application, according to the experience in the chemical and petrochemical industry, presents no particular safety problems with hydrogen, safety problems that may arise in its transport, distribution, and decentralized use will have to be fully identified and detailed safety analyses carried out. Of concern here are first of all the hydrogen embrittlement problem, cases in which there are breaks in lines where hydrogen is unintentionally released into the surroundings.

In summary, it can be stated that according to present-day knowledge, dealing with the coming hydrogen technology will present no significant safety risks that would increase the difficulty of introducing hydrogen as a secondary energy carrier.

REFERENCES

1. E. Justi: *Spezifische Wärme, Enthalpie, Entropie und Dissoziation technischer Gase* [*Specific Heat, Enthalpy, Entropy, and the Dissociation of Technically Important Gases*], Verlag J. Springer, Berlin (1938).
2. BMFT (publisher): *Auf dem Wege zu neuen Energiesystemen* [*On the Way to New Energy-Conversion Systems*], Part III, pp. 362–392, Bonn (1975).
3. J. Hord: *Int. J. Hydrogen Energy* **3**(2)**:**157–176 (1978).
4. M. Fischer and H. Eichert: *DFVLR-Nachrichten*, No. 34, pp. 33–37 (1981).
5. W. Tanner, in: Kommission der Europäischen Gemeinschaften [Commission of the European Community] (publisher), Seminar on Hydrogen as an Energy Vector: Its Production, Use and Transportation, EUR 6085 DE/EN/FR/IT, pp. 544–566, Brussels (1978).
6. R. Reider and F. J. Edeskuty: *Int. J. Hydrogen Energy* **4**(1)**:**41–45 (1979).
7. C. Isting and B. Thier: *Lichtbogen* **24**(176)**:**16–20 (1975).

CHAPTER 13

The Conversion of Hydrogen into Electricity by Means of Fuel Cells

13.1. INTRODUCTION

In a solar-hydrogen economy, it is likely that hydrogen will be arriving from distant sites which receive plentiful sunlight. The hydrogen will arrive either in pipelines as a gas or in tankers as a liquid, and will then be converted at the final site into mechanical or electrical energy.

13.2. HIGHLY REVERSIBLE PRODUCTION OF ELECTRICAL ENERGY FROM HYDROGEN BY MEANS OF HYDROGEN–OXYGEN FUEL CELLS

In electrochemical fuel cells, the chemical energy of hydrogen is converted directly into electrical energy without the intermediate step of the production of heat energy. Fuel cells are therefore not subject to the Carnot factor. Figure 13.1 shows schematically the principle of an alkaline hydrogen–oxygen fuel cell. The cell divides the overall reaction of the combination of hydrogen and oxygen to form water into the anodic oxidation of hydrogen and the cathodic reduction of oxygen. The two gases, hydrogen and oxygen, remain separated, and electron transfer between the two molecules (i.e., the well-known explosion reaction of

This chapter authored by Prof. H. H. Ewe, Hamburg.

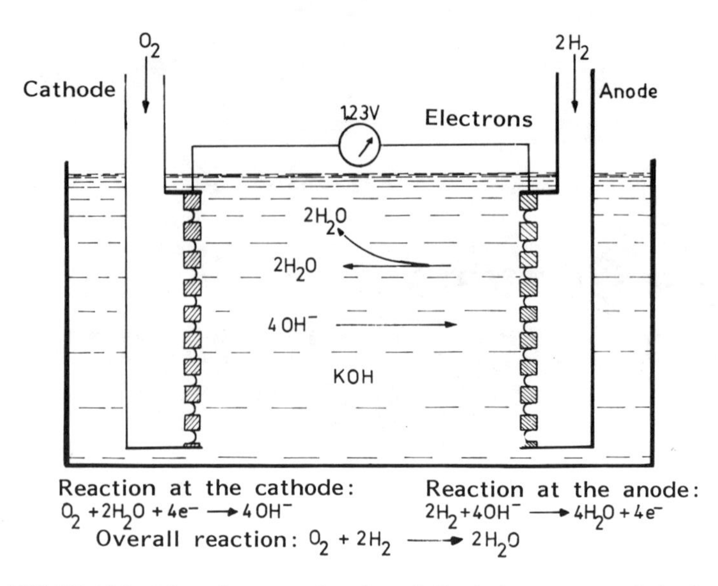

FIGURE 13.1. Schematic construction of an alkaline hydrogen–oxygen fuel cell.

hydrogen and oxygen with the accompanying large evolution of heat) does not take place. The gases exchange their electrons only via the metallic electrodes.

Because of the different electrochemical affinities of the gases for the electrodes with which they come in contact and the differing interactions of the gases with the electrolyte, the electrodes (which also act as catalysts) set up different potentials during the reactions. The potential difference arising at the electrodes causes an electric current to flow through the external load. This current is carried in the internal circuit by ions in the electrolyte and in the external circuit by electrons. The reaction partners, hydrogen and oxygen, are continuously added to the electrodes, and the water produced as a product of the reactions is extracted from the electrolyte by evaporation. The maximum possible potential, E_{max}, of the fuel cell arises from the free energy of the gaseous reactants, $\Delta G = \Delta H - T\Delta S$ to $E_{max} = \Delta G/nF$ and at room temperature reaches a value of 1.23 V. The figure for the efficiency describes the practicably realizable fraction of chemical energy that can be converted to electrical energy. The maximum value of this is given by the ratio of the free energy ΔG (at constant P and T) to the heat-content change. Thus, $\varepsilon = \Delta G/\Delta H$ has a maximum value of 0.83, but when the load on the fuel cell is increased, it decreases to between 0.5 and 0.7. At any rate, it always remains clearly above the value of 0.4, which is the maximum value for a heat engine.[1]

13.3. SCHEMATIC CONSTRUCTION OF A HYDROGEN–OXYGEN FUEL CELL

For the various possible fuels, there exist various possible fuel cells. For hydrogen technology, the main interest is in fuel cells that work according to the principle of gas diffusion electrodes. Direct conversion of the chemical energy of the hydrogen–oxygen reaction into electrical energy is obtained by interaction of the gases with the electrolyte and the electrodes. Because of certain geometric factors, the conversion of the gases takes place only in the vicinity of the three-phase boundary where the electrode catalyst, the gas, and the electrolyte come together. At the anode, a reaction occurs between a hydrogen molecule and two OH^- ions to form two water molecules and two electrons. At the cathode, two water molecules and four electrons produce four OH^- ions. These reactions occur as a result of a series of individual electrode reactions. With respect to the conversion of hydrogen gas, Figure 13.2 shows one possible reaction path: Hydrogen gas dissolves in the electrolyte and diffuses to the catalyst surface, where the reaction occurs. The hydrogen molecules dissociate into hydrogen atoms, and thereafter ionize to protons, whereby the electrons are transferred to the electrode material and travel around the circuit (passing through the external load) to the other electrode. Analogous reaction sequences take place at the oxygen cathode.[1–10]

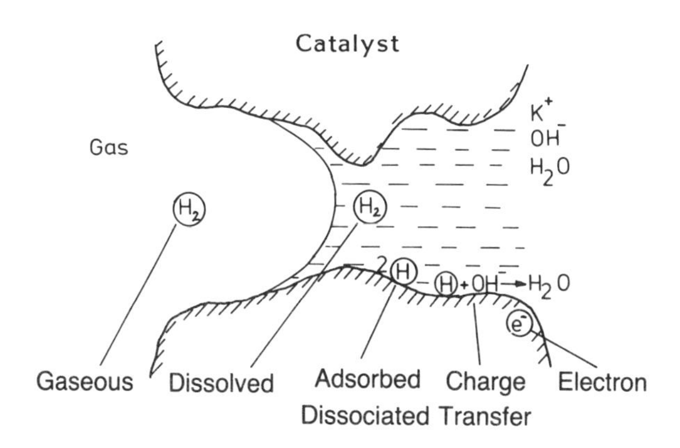

Overall reaction: $H_2 + 2OH^- \rightarrow 2H_2O + 2e^-$

FIGURE 13.2. Schematic diagram of the anodic oxidation of hydrogen taking place according to the reaction scheme shown.

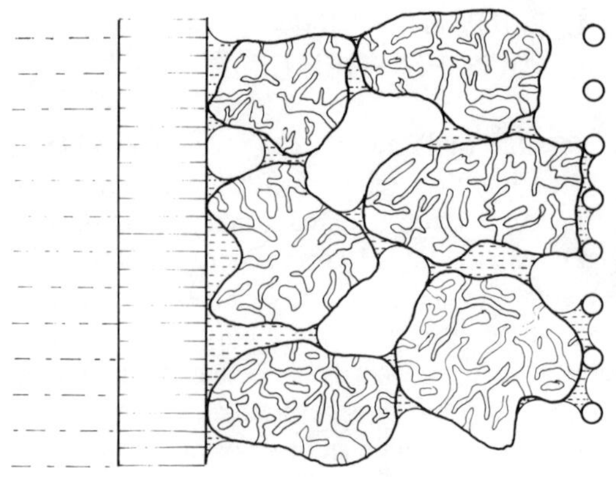

FIGURE 13.3. Principle of a two-layered supported gas diffusion electrode.

At each of these three-phase boundaries, there arise only very small currents, in the order of magnitude of some 10^{-6} A. In porous electrodes, there are about 10^5 pores/cm^2 of the external geometric surface, so that one can obtain relatively high current density up to about 1000 mA/cm^2. These porous electrodes contain one system of fine pores filled with electrolyte and a second system of coarser pores, which the excess gas pressure keeps free of electrolyte and therefore filled with gas. The two neighboring pore systems, which are shown schematically in Figure 13.3, provide a large contact zone between the gas and the electrolyte.

To avoid gas losses through the coarse pores into the electrolyte, there exists on top of the working layers a second layer of very fine pores. This second layer consists of finely sintered powder, asbestos, or some other fine-pored corrosion-stable hydrophilic material, which prevents the escape of gases into the electrolyte. This layer must therefore have no coarse pores and must be prepared very uniformly.[1–11]

13.4. ALKALINE LOW-TEMPERATURE FUEL CELLS WITH RANEY CATALYSTS

Figure 13.4 shows schematically a fuel cell with supportive electrodes, used in the investigations of Von Sturm and co-workers,[7,12] in which expanded metal is bounded by two fine-pore layers consisting of asbestos paper that are pressed onto the expanded metal by a catalyst in powder form and a current-conducting network. By using the filter press construction technique, one obtains a thin layering of the individual cells.

Besides the coarse- and fine-pore systems, electrodes for high current densities need highly active catalysts.[1,7,13,14]

In fuel cells intended for space flight, and for most of the other American developments, platinum catalysts are generally used.[14] Instead of using costly noble metals, Justi and Winsel[1] and co-workers use the cheaper Raney catalysts.[1] For the hydrogen anode, for example, Raney nickel can be used as a catalyst obtained through the leaching out of aluminum from nickel-aluminum alloy. This nickel catalyst is fine-structured and proves to have an inner surface of 30–150 cm^2/g. By special treatment, including the addition of small amounts of another alloy material or a certain amount of an oxidation–reduction couple, the catalytic activity and the stability of this catalyst can be increased and improved.[15–19] With respect to the oxygen cathode, Raney silver or silver, finely divided by means of chemical precipitation of hydroxides, is used.[1,20] As an example of the finely divided structure and large porosity of these catalysts, Figure 13.5 shows an electron-microscopic photograph of a Raney catalyst, formed from Al80–Ag20, by leaching out the aluminum component.[21]

Such a fuel element should set up a potential of 1.23 V if the gases are introduced at a pressure of approximately 1 bar. In fact, the actual potential that is observed, due to side reactions probably at the O_2 electrode, is about 1.1 V at a current density of zero. When a finite current passes through the cell, there

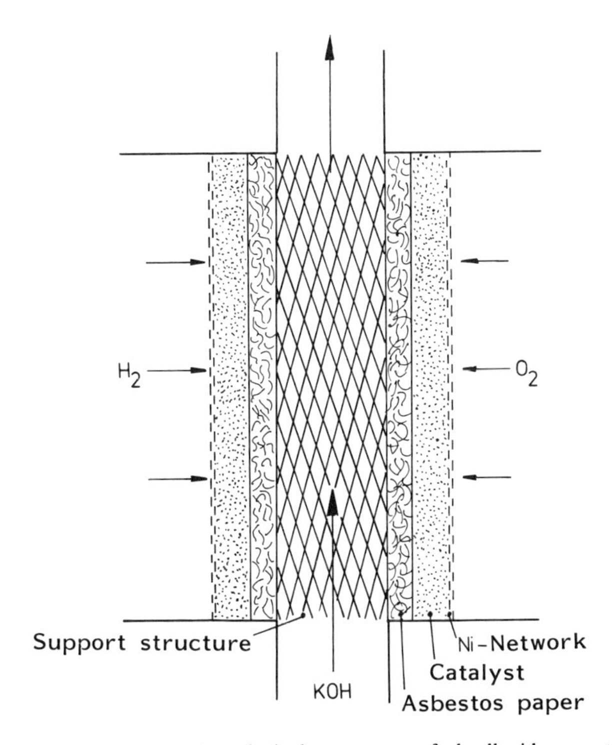

FIGURE 13.4. Schematic construction of a hydrogen–oxygen fuel cell with supported electrodes.

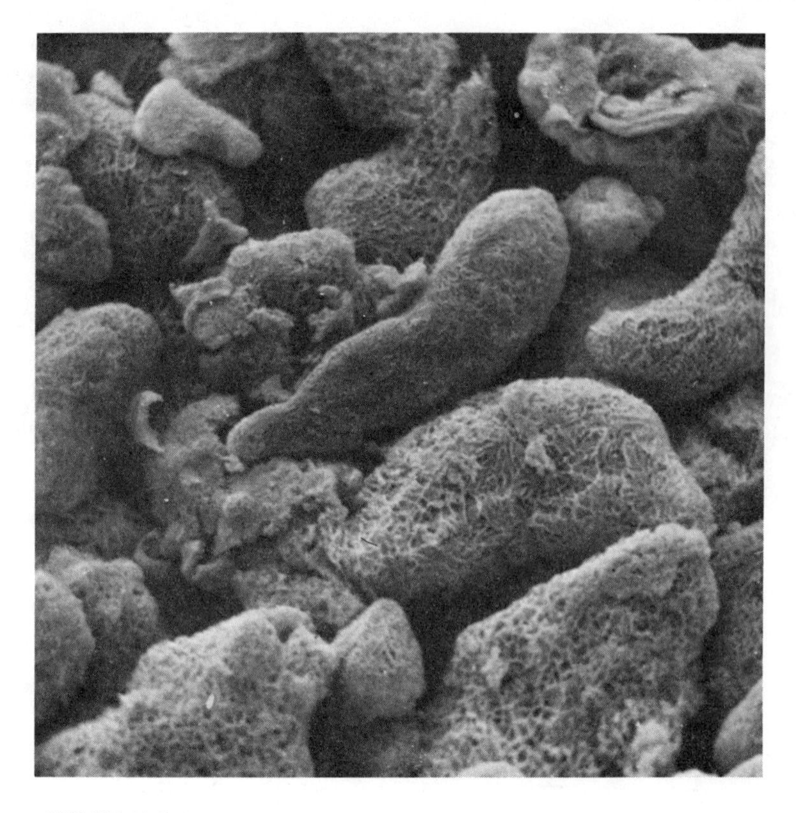

FIGURE 13.5. Electron-microscopic photograph of Raney silver grains. ×2300.

are further losses due to difficulties in electron transfer, polarization, concentration changes, transport difficulties, the presence of inert or impure gases, and other factors, and each and all of these cause the actual available potential of the fuel cell to decline as the current density increases. Nearly all these losses in potential can be reduced by raising the temperature of the fuel cell, but the upper limits for this increase are set by the increase in stress on materials and corrosion that the increase causes. Most fuel cells with aqueous electrolytes have a working temperature of between 70 and 90°C.[1,12,22,23]

Besides the actual fuel element with its modules of individual electrodes, the fuel cell itself contains auxiliary apparatus for temperature regulation, for the removal of the aqueous product and of impurities in the gases, and for handling the flow of electrolyte and that of the gases. Figure 13.6 shows a complete and fully automatic hydrogen oxygen fuel cell produced by Siemens AG. This element has been built up in modules from supported electrodes using asbestos paper cover layers, which are formed into modules using polysulfone. The actual fuel cell element is in the compartments at the right. On the left, there is a unit for removing the water produced during the operation of the cell

FIGURE 13.6. 2.4-Kilowatt H_2–O_2 fuel-cell plant of Siemens AG.

and also the reaction heat, and additionally, a pump for the electrolyte, the electrolyte container, and other regulatory devices. Table 13.1 presents technical data for this apparatus. The data in this table are conservative when compared with performance figures from some of the most recently improved cells. Fuel cells are operating in the laboratory with a current density of 400–700 mA/cm^2 at 90°C and can be overloaded to 1000 mA/cm^2 for short times (1 hr); thus, the output per unit weight can be greatly improved.[17]*

13.5. MEDIUM-TEMPERATURE FUEL CELLS WITH PHOSPHORIC ACID ELECTROLYTE

Beginning with fuel cells developed for space travel, the cost of the precious metal catalysts that are needed in these elements has been reduced in the United

*Translator's note: As of 1987, fuel cells with power unit weight of more than 1kW per kg are clearly in sight.

TABLE 13.1
Fully Automatic H_2–O_2 Fuel-Cell Plant of Siemens AG (1976)

Nominal performance	2.4 kW
Overload factor	1.3
Nominal voltage	24 V
Number of cells	33
Fuel	H_2, O_2, technical grade. 1.8 bar
Current density	310 mA/cm^2
Operating temperature	80°C
Electrolyte	6 N KOH
Battery weight	23 kg
Total weight	53 kg

States through systematic reduction of the platinum content of the electrodes. Of the numerous lines of investigation, only the project of United Technologies Corporation, Power System Division, South Windsor, Connecticut, for power plants with a peak output of up to 26 MW will be presented.[24–27]

This company has put about 800 people to work on the development of fuel-cell elements with a matrix of phosphoric acid electrolyte (≈98%). Since the electrolyte remains unchanged, there is no need to have an electrolyte-circulating system. The water content of the phosphoric acid depends on the temperature, so that with stable temperature, the water content is also stable.[26] The removal of both water and heat occurs from the gas phase, particularly in the air-run cathodes. These fuel cells work at 140–190°C. The electrodes contain platinum catalyst with rhodium and tungsten additions. The catalysts are bonded

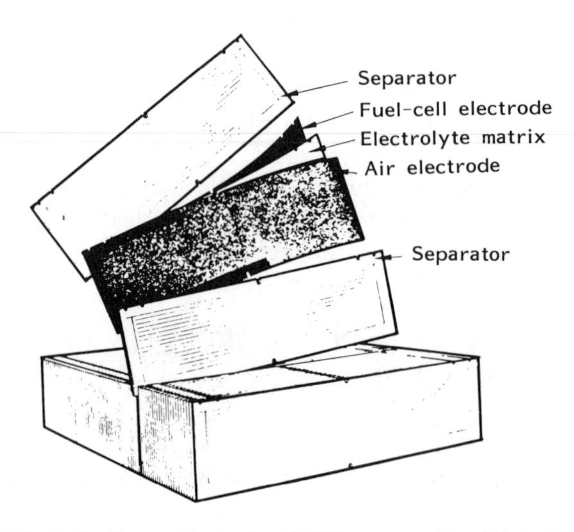

FIGURE 13.7. Fuel-cell assembly for the 26-MW power station of United Technologies.

with PTFE and introduced into the electrolyte matrix. Figure 13.7 is a schematic of the cell construction with electrodes that are $\frac{1}{4}$–$\frac{1}{3}$ m^2 in size. The stationary performance of such a cell at present is about 0.1–0.25 W/cm^2, but an overload of several times this amount can be withstood for several hours, the current densities reaching 1000 mA/cm^2.[27] In recent years, the following goals have been pursued by United Technologies: The cells should work with hydrogen produced by re-forming hydrocarbons and should become cheaper and more efficient. Costly working materials should be replaced with cheaper ones, e.g., high-quality steel with carbon-layered iron. The amount of costly catalyst per unit area should be reduced. Thus, in cells in the laboratory, electrodes have worked with less than 0.5 mg $platinum/cm^2$, and these have been placed in a thin layer on carbon and bonded with PTFE.[24–30]

The development of these fuel cells with phosphoric acid electrolytes is being carried out in two large programs. In the TARGET project (Team to Advance Research for Gas Energy Transformation), fuel cells of 10–250 kW are being built for single-family homes that use only gas as the energy medium. In 1971–1973, 12.5-kW batteries were tried in a wide range of test programs under practical conditions. In these tests, 65 units put in about 205,000 working hours and produced about 1,000,000 kWh. Typical test conditions involved the variation of load, meteorological conditions, and environmental protection demands. A typical apparatus is shown in Figures 13.8 and 13.9. On the basis of

FIGURE 13.8. View of a 12.5-kW TARGET fuel cell (PC 11) for use with hydrocarbons (particularly methane) and air. Left: fuel cell and re-former; right: inverter.

FIGURE 13.9. PC 11 fuel cell with housing open. Left: fuel cell; right: re-former.

this experience with project PC 11, project PC 16 was developed. This apparatus had an equal performance, but needed only about one quarter the volume and weight of PC 11 (see Figure 13.10). The apparatus was fueled by natural gas that was converted beforehand to hydrogen and carbon dioxide.[31,32] Table 13.2 contains the most important data from the TARGET fuel-cell program. The data were taken from the works cited and should be regarded as only indicative and provisional; better results are available.

At the end of 1971, United Technologies began to work with ten American electrical companies on a second program for development of 26-MW fuel cells to provide electricity for small towns. Fuel-cell power plants with a capacity of 26 MW were developed; each is 5.5 m high and occupies an area of less than 0.4 hectare. The power plant is made up so that the individual modules can be transported by truck and are therefore readily mobile for use in rapidly growing parts of towns or in temporary settlements such as those set up during the building

FIGURE 13.10. Enlargement of fuel cells PC 11 and PC 16.

of towns. The 26 MW plant is made up of four re-former modules, eight fuel-cell modules, and eight inverters. Two modes of operation are foreseen: either the 26 MW plant with partial-load operation and correspondingly favorable efficiency or 26-MW equipment for use in fairly short peak performances (up to 3 A/cm^2) with half the number of cells, and correspondingly diminished

TABLE 13.2
Data for the PC 11 and PC 16 Fuel Cells (TARGET Program)[a]

Performance	12.5 kW
Electrolyte	98% Phosphoric acid in porous matrix
Operating temperature	120–150°C
Surface area of an electrode	250–930 cm^2
Current density	≈100 mA/cm^2
Voltage	120 V
Fuel	Natural gas or light oils are re-formed at ≈800°C into H_2, CO_2, and CO; at lower temperatures, CO reacts with H_2O to give CO_2 + H_2
Oxidant	Air
Catalyst	Pt: Anode 1–2 mg/cm^2 Cathode 5–10 mg/cm^2
Operating lifetime	Up to 40,000 hr

[a] See Figures 13.8–13.10.

FIGURE 13.11. Model of the 26-MW fuel-cell power station of United Technologies.

investment costs, and higher heat generation and fuel costs. The efficiency of conversion should be in the range of 37–38% (see Figure 13.11).[33]

The development of these cells was carried out in a stepwise manner. The first phase was from 1972 to 1976 (cost about \$56,000,000) and led to 1-MW units, which gave good results. The last phase of this program was the 4.8 MW

TABLE 13.3
Projected Data for the 26 MW Fuel Cell Generator of United Technologies

Performance	26 MW
Area required	17 m × 27 m
Electrolyte	98% Phosphoric acid
Voltage	120 V
Current density	Up to 3000 mA/cm² (intermittent)
Catalyst	Pt, less than 1 mg/cm²
Operating lifetime	
Fuel cell	5 years
Re-former	10 years
Alternator	20 years
Fuels	Natural gas, propane, liquid hydrocarbons with low sulfur content + air
Installation costs	\$140/kW for normal load \$110/kW for peak load

unit, which was planned to be ready at the end of 1978 (cost about $52,000,000) and actually gave practical power in Tokyo in 1984. The further plan is to develop the 26 MW unit.

13.6. CONCLUSIONS

The efficiency reached with the hydrogen–oxygen fuel cells described herein is sufficiently high so that fuel cells now (1987) stand on the brink of a massive technical development. Significant practical improvements can be expected to occur in the near future. The principal problems with current fuel cells concern reproducibility, reliability, price, and long-term stability. The knowledge gained with some cells suggests that the technological difficulties can be overcome. The beginning of mass production of fuel cells, followed by rapidly growing experience, should lead to a reduction of the existing technological problems. The same thoughts apply to the investment costs, which as yet cannot be accurately estimated and will depend on the number of cells produced.

A further principal problem with fuel cells is providing them with suitable fuel gases—a problem for which hydrogen is undoubtedly the best solution. At present, provision with *pure* hydrogen is difficult because the hydrogen is supplied in steel cylinders (too heavy) or from the re-forming of gases that can be more readily handled, such as ammonia, methane, and benzene. In the course of development of a hydrogen economy, hydrogen itself would become more readily available, which would allow easier introduction of fuel cells.*

REFERENCES

1. E. W. Justi and A. W. Winsel: *Kalte Verbrennung—Fuel Cells* [*Cold Combustion*], Franz Steiner Verlag, Wiesbaden (1962).
2. G. J. Young (ed.): *Fuel Cells*, Reinhold Publishing Corp., New York (1960).
3. W. Mitchell Jr. (ed.): *Fuel Cells*, Academic Press, New York and London (1963).
4. W. Vielstich: *Brennstoffelemente—Fuel Cells*, Verlag Chemie, Weinheim/Bergstrasse (1965).

---

* Translator's note: Two other types of fuel cells exist, both of them working at much higher temperatures than those mentioned here. The first involves molten carbonate electrolytes ($T \approx 680°C$), and the overall reaction is to convert CO to CO_2. The second ($T \approx 1000°C$) involves solid oxide electrolytes, in which O^{2-} is the mobile ion. The reaction is to form steam from hydrogen and oxygen. In these cells, the high temperature reduces the need for good catalysts. Further, the re-forming reaction in which H_2 is produced from (readily available) natural gas occurs within the cell and needs no external re-former.

The cells both offer the promise of higher efficiencies than do the lower-temperature cells but involve higher corrosion and material costs.

The minimum cost among fuel cell types (S/kW) is reached by the alkaline fuel cell. It is likely that this cell would be used for obtaining electricity from H_2, were this readily available at a cost equal to that from natural gas.

5. H. A. Liebhafsky and E. J. Cairns: *Fuel Cells and Fuel Batteries*, John Wiley and Sons, New York (1968).
6. J. O'M. Bockris and S. Srinivasan: *Fuel Cells; Their Electrochemistry*, McGraw–Hill Book Co., New York (1969).
7. F. v. Sturm: *Elektrochemische Stromerzeugung [The Electrochemistry of Electricity Production]*, Weinheim/Bergstrasse (1969).
8. C. Berger (ed.): *Handbook of Fuel Cell Technology*, Prentice-Hall, Englewood Cliffs, New Jersey (1968).
9. V. S. Bagotskij, and D. L. Motov: *Dokl. Akad. Nauk SSSR* **71:**501 (1950).
10. E. Yeager and A. Kozawa: *Kinetic Factors in Fuel Cell Systems: The Oxygen Electrode*, AGARD, Oxford (1964).
11. F. T. Bacon, in: G. J. Young (ed.): *Fuel Cells*, p. 51–93, Reinhold Publishing Corporation, New York (1960).
12. F. v. Sturm, H. Nischik, and E. Weidlich: *Ingenieur Dig.* **5:**52–57 (1966).
13. G. Sandstede (ed.): *From Electrocatalysis to Fuel Cells*, Battelle Seattle Research Center, University of Washington Press, Seattle (1972).
14. K. V. Kordesch, in: O'M. Bockris and B. E. Conway (eds.), *Modern Aspects of Electrochemistry*, Vol. 10, pp. 339–443, Plenum Press, New York (1975).
15. A. Pzhenichnikov, R. Burshtein, D. Lainer, N. Kagan, G. Melikova, and F. Sabirov, Pat. UdSSR 147616 (1961).
16. H. Ewe, E. Justi, and A. Schmitt: *Electrochem. Acta* **19:**799–808 (1974).
17. K. Mund, G. Richter, and F. v. Sturm: *J. Electrochem. Soc.* **124:**(1):1–6 (1977).
18. M. Jung and H. v. Döhren: *Metalloberfläche* **25**(2)**:**42 (1971).
19. G. Richter, 3rd Int. Symp. on Fuel Cells, Brussels, June 16–20, 1969, in: Société d'Etudes de Recherches et d'Applications pour l'Industrie and Société Commerciale d'Applications Scientifiques (eds.), Proceedings, pp. 194–202, Presses Academiques Europèenes, Brussels (1969).
20. K. Höhne: *Siemens Forsch Entwickl. Ber.* **3**(1):31–35 (1974).
21. H. R. Schröder: Thesis, TU Braunschweig (1974).
22. *Fuel Cells: A Bibliography*, TID-3359, U.S. Energy Research & Dev. Adm. (1977).
23. C. J. Crowe: Fuel Cells—A Survey, NASA SP-5115, Washington, D.C. (1973).
24. W. J. Lueckel, L. G. Eklund, and S. H. Law, IEEE Winter Meeting, New York, Paper No. T 72 235-5 (1972).
25. C. Martin: *Aviat. Week Space Technol.*, p. 56 (1973).
26. H. Binder and G. Sandstede: *Chem.-Ing.-Techn.* **47**(2):51–56 (1975).
27. H. R. Kunz and J. McIntyre: *Electrochem. Soc. Proc.* **77**(6)**:**607 (1977).
28. W. Vogel, J. Lundquist, P. Ross, and P. Stonehart: *Electrochim. Acta* **20:**79–93 (1975).
29. O. J. Adlhart, Proc. 24th Power Sources Symp. May 1970, p. 182.
30. S. G. Abens, B. S. Baker: Fuel Cell Stacks, Report ERC-73 96-S of Energy Research Corporation, Connecticut 1976.
31. R. Villers: *Energy and the Fuel Cell*, United Technologies Corporation, Power Systems Division, South Windsor, Connecticut (1976).
32. Pratt & Whitney Aircraft, P & W Fuel Cell History, Corporate Report (1973).
33. W. J. Lueckel and J. R. Casserly: *Fuel Cells for Utility Service*, United Technologies Corporation, Power Systems Division, South Windsor, Connecticut (April 1976).

CHAPTER 14

The Catalytic Combustion of Hydrogen

14.1. INTRODUCTION

A considerable part of present energy consumption occurs in the production of useful or process heat. Over 70% of energy use in the Federal Republic of Germany occurs in the production of heat that is then used by households, small-scale consumers, and industry.[1] If the energy supply were converted increasingly to gaseous fuel and consumers were over time supplied with a fuel mix in which the hydrogen content increased, the market for heat would represent a considerable potential for the introduction of a hydrogen technology. However, substitution of hydrogen for oil, natural gas, and coal would necessitate changing the types of burners to those suitable for the technical needs of hydrogen combustion—an added capital cost that would have to be faced.

The combustion of hydrogen can take place either as direct combustion (flame combustion) at high temperatures or as catalytic combustion (flameless combustion) at relatively low temperatures. It follows that there are two different possibilities for the construction of hydrogen burners. After a brief treatment of the direct combustion of hydrogen, this chapter will concentrate on the fundamental properties of catalytic hydrogen burners and will describe the construction and technical properties of such burners that have already been built as prototypes.

This chapter authored by Dr. P. W. Brennecke, Braunschweig.

14.2. DIRECT COMBUSTION OF HYDROGEN

The conception of new heating installations, and the modification of conventional gas burners so that they can work with gaseous hydrogen, have to be seen in the context of knowledge of the combustion behavior of hydrogen, i.e., the interrelationship between the combustion properties of hydrogen and the actual combustion process in the burner.[2] During combustion with air, the burning properties of H_2 are determined by the following physical properties:

Ignition limits of hydrogen–air mixture	4–75 vol.-% hydrogen
Minimal ignition energy in air	0.02 mJ
Ignition temperature	585°C
Flame temperature in air	2045°C
Combustion velocity in air	2.65–3.25 m/sec

A decisive difference between the combustion of hydrogen and the combustion of conventional fuel (e.g., natural gas) is given by their very different ignition or combustion velocities. For example, the combustion velocity of hydrogen in air is 2.65–3.25 m/sec, but that of methane in air is only 0.40 m/sec.[3]

Thus, there are difficulties in controlling a hydrogen flame, pertaining to its steadiness. Conventional gas burners mix gas and air in definite proportions. The air–gas mixture is set so that the velocity with which it reaches the mouth of the burner (before combustion with oxygen takes place) is greater than that of a flame which might travel in the opposite direction. Thereby, a flashback (the rapid combustion of the material occurring at an area for which the reaction was not intended) is avoided. But, if the air–gas mixture velocity is much higher than the velocity of the flame, it will expel the flame prematurely, and therefore extinguish it. In this manner, the guidelines for the construction of a conventional burner are set.

The combustion velocity is a characteristic of the gas concerned and influences the technical construction of the burner. For a stable burner, the velocity of the gas stream to the mouth and the combustion velocity of the gas mixture must be closely matched. If the combustion velocity is too great, there is danger of a flashback; if it is too little, the flame will extinguish itself. In the case of hydrogen, the combustion velocity in the air of 2.65–3.25 m/sec means that the danger of a flashback is greater than that of the flame's extinguishing itself, i.e., being blown out.

The investigations that have been done hitherto on the modification of conventional gas burners to work with hydrogen have shown that fairly simple changes can achieve the appropriate modification, so that a burner for natural gas can be made to run on hydrogen.[4,5] Primarily, change must be made in the nozzle. With a burner for hydrogen, there need be no mixing between the gas jets and the actual combustion chamber. Thus, because of the high combustion velocity of the hydrogen–air mixture, the flame always reaches the gas-jet openings. By decreasing the length of the burner pipes and the diameter of the burner pipe at the level of the gas jets, the velocity of the burning air in the mixer head can be considerably increased, and this usually gives rise to a better combustion value with respect to excess of air in the exhaust gas.[4] Further modifications concern, for example, automatic safety barriers and special materials for sealing purposes.[5]

In general, there are no real problems with the direct combustion of hydrogen in air, and the present state of technical knowledge for optimizing the burners is sufficient.[6] The main task in future investigations is that of defining and revising the safety standards.

14.3. CATALYTIC COMBUSTION OF HYDROGEN

In addition to the direct combustion of hydrogen discussed above, the possibility exists for burning hydrogen catalytically (without a flame) and for utilization of hydrogen in heating and in the production of process heat. Methane, the minimum ignition energy of which in air is approximately 0.3 mJ, can be burned catalytically at temperatures of 300–450°C.[4] However, the minimal ignition energy of hydrogen in air is only 0.02 mJ, and catalytic combustion occurs at lower temperatures, even in the region of 150–200°C. The catalysts in use at present are predominantly high-cost noble metals such as platinum and palladium, which are distributed on porous carrier plates. Catalytic combustion occurs flamelessly within the pores, and the entire plate is heated.

To maintain the process of catalytic combustion, kindling of the hydrogen in air must be prevented. Hence, the three following combustion conditions are relevant.

1. If the velocity of the hydrogen–air mixture is greater than the combustion velocity of hydrogen, the flame will not be maintained; it will be blown out. With the high combustion velocities for hydrogen–air mixtures (2.65–3.25 m/sec), it seems improbable that the necessary flow velocity can be maintained over the whole catalytic surface to achieve complete combustion.
2. An alternative possibility arises when the hydrogen–air mixture is kept at a composition outside the ignition limits. Catalytic combustion of a hydrogen–air mixture with a hydrogen content of less than 4 vol.-% is in principle possible. Combustion of a mixture with a hydrogen content of more than 75 vol.-% would lead to incomplete combustion and is therefore not practical. Combustion of a mixture with less than 4 vol.-% hydrogen would require very large catalytic surfaces and support material and lead to technical and economic disadvantages.

3. The only technically suitable possibility for catalytic combustion at present seems to consist in keeping the catalyst surface at a temperature lower than the self-ignition temperature. One must take into account that hydrogen has a *positive* Joule-Thomson coefficient.[7] During expansion of the pressurized effluent hydrogen, a resulting temperature increase in the gas of +0.04°C/bar takes place.

The above mentioned combustion conditions lead to the conclusion that the surface of a catalytic burner must be chosen so that the rate of heat generation does not exceed the dissipation rate of the heat. Air or water can simply be used to dissipate the produced heat. The construction of the burner must ensure that the entire hydrogen–air mixture undergoes complete combustion.

14.4. PROPERTIES OF CATALYTIC HYDROGEN BURNERS

Catalytic burners are really premixing burners in which the reaction temperature and reaction rate are determined by means of the catalyst. The hydrogen–air mixture is introduced into a porous support that contains the catalyst material (e.g., platinum or palladium), and combustion takes place at the activation temperature of the catalyst. The essential difference in *catalytic* combustion—in comparison with direct combustion—is that the temperature in the reaction chamber can be varied. The actual combustion reaction takes place on the surface of the solid bodies present, and these are in a much better position to dissipate the heat produced. With variable heat transfer to the combustion chamber, heat dissipation can be adjusted so that the temperature can be influenced. On this basis, catalytic hydrogen burners can be regarded as particularly safe for use in space heating.

The most important property of catalytic burners is the production of heat at comparatively low temperatures. These low operating temperatures have the consequence that in the combustion of a hydrogen–air mixture, *there is a negligible production of poisonous nitric oxides*. Because of the large ignition limits of hydrogen–air mixtures (4–75 vol.-% hydrogen), it is possible that catalytic burners can be constructed in such a way that the emission of pollutants would be minimized. In the investigations that have so far been made, it has been found that an air fraction of about 35% in a hydrogen–air mixture will give rise to the smallest amount of polluting material. Thus, very large quantities of a hydrogen–air mixture could be catalytically converted without any ecologically negative effects. The only combustion product would be water vapor.

A further advantage of catalytic combustion is that—depending on the intended application—the heat produced by means of the burners can be varied by using different active materials as catalysts and different surfaces areas. This gives rise to advantages of construction that can be applied to different objectives. Whatever the application, catalytic converters can be technically adjusted and operated.

Catalytic burners are particularly advantageous if the combustion temperature is less than 100°C, i.e., when the catalytic surface or other cooled surfaces of the burner (particularly auxiliary surfaces) are kept at temperatures which are below the condensation temperatures of the various combustion products (water vapor and nitrogen). In this case, the maximum possible energy of the hydrogen, i.e., its higher heating value, is used, and there arises the possibility of heating without any necessity for a chimney and with practically 100% use of the energy. Thus, among the positive side effects of such a heating system would be:

1. Because of the low combustion temperature, the operating lifetime of the catalyst supports would be greatly extended.
2. There would be increased moisture in the heated room and a corresponding possibility of recovering pure water.

14.5. STATE OF DEVELOPMENT OF CATALYTIC HYDROGEN BURNERS

Commercial catalytic hydrogen burners are not yet available. Research and development work has been increased and extended in scope in recent years. Many of the recent developments are already in an advanced stage of experimentation, and prototypes are being built.

For the catalytic combustion of hydrogen, two types of burners exist, and these can be described in terms of the temperature ranges in which they work:

1. The low-temperature type consists of porous ceramic material as support, the inner surface of which is covered with catalysts. With this type, operating temperatures of 150–250°C with a low-temperature combustion rate of hydrogen can be attained.
2. The high-temperature type uses a high-temperature carrier material such as silicon carbide, into which the catalyst is introduced. Operating temperatures reached with this type are rather higher—e.g., 250–500°C—and a significant amount of the combustion energy is emitted in the form of heat radiation.

The construction of a catalytic hydrogen burner is shown in Figure 14.1. The hydrogen–air mixture enters the burner from the back, then proceeds through distributors and temperature shielding and finally into the porous burner plates, where the catalytic reaction occurs to form water vapor.

To give some idea of the present spectrum of the experiments and devel-

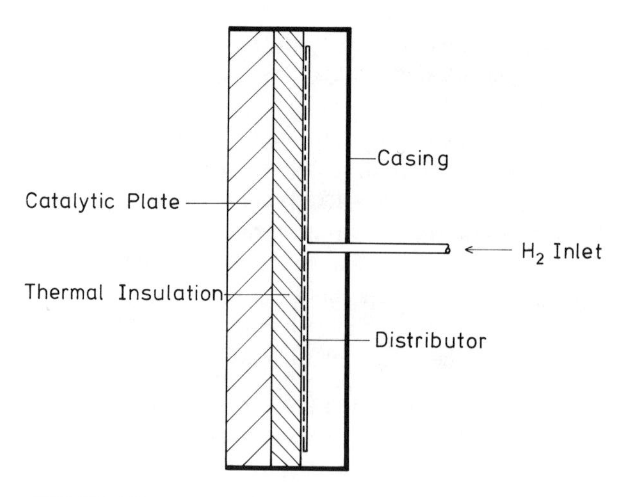

FIGURE 14.1. Schematic design of a catalytic hydrogen burner.

opment work that is going on, a brief account will be given of the most important research groups and the results of their work.

Justi and Selbach (Technical University of Braunschweig, the Federal Republic of Germany) have attempted (in collaboration with Winsel) to apply catalytic recombinators that are used in galvanic secondary batteries, such as the lead battery.[8] These are used for catalytic combustion of the hydrogen given off in the charging of these batteries.[8] This causes recombination of the explosive gas mixture that forms in these batteries. The recombination catalyst consists of finely divided palladium on a hydrophobic support. As protection against poisoning, the catalyst layer is coupled with a carbon surface. The foremost technical property of this arrangement is its self-limiting character: If there is too much of the explosive gas mixture, one does not get too high a rate of reaction with overheating and explosion; rather, the catalyst limits itself to a certain temperature because the reaction distributes itself over the surface appropriately. With the use of this recombination catalyst, a working model of a catalytic hydrogen heating installation was developed and researched. After the first successful runs, some further examinations of problems with catalytic hydrogen combustion that have not yet been cleared up are being carried out.

Mercea, Grecu, Fodor, and their collaborators (Institute of Isotopic and Molecular Technology, Cluj-Napoca, Rumania) are examining the technical possibilities of catalytic hydrogen conversion for space heating.[9,10] The investigations that they are carrying out are concentrated on meeting the demands that have to be satisfied from the viewpoint of the heating system and the special development and construction problems that arise in this application with catalytic hydrogen burners. The experiments are carried out in a 19.2-m^3 test room (surface area: 7.5 m^2). In the course of their investigations, Mercea and co-workers have been able to show that:

1. Catalytic conversion of hydrogen at nearly constant temperature (temperature variation: $\pm 2°C$) occurs; i.e., the combustion process takes place stationarily.
2. A constant temperature of the room can be achieved.
3. Only a very small amount of hydrogen reaches the room air (C_{H_2} = 0.00015), as shown by analysis; i.e., there is practically complete combustion of the hydrogen.

Experiments carried out hitherto lead to the result that by catalytic hydrogen combustion, appropriate room temperatures can be reached. During the course of experiments with these burners in closed rooms, it is necessary to have a ventilation system that introduces a constant amount of oxygen into the room. To keep the water vapor content of the room air from getting too high, it may be necessary to make provision for water condensation to occur on appropriate surfaces. Another important part of the work of Mercea and co-workers is that of developing supports for catalysts using platinum. The particular object of the research has been porous support plates fashioned of ceramic and fiberlike materials whereby with two different patterns of temperature distribution in the support plates, the efficiency of the hydrogen burner is determined as a function of the amount of water vapor that is introduced.

Haruta, Sano, and co-workers (Government Industrial Research Institute of Osaka, Japan) concentrate their work on the choice of suitable catalysts for flameless hydrogen combustion and also on the conception, building, and operation of prototype catalytic burners.[11–14] The goal of these developments is to develop cheap catalytic materials along with the expensive noble metals. The new materials have to show a high and maximally constant catalytic activity during operation. Experiments show that oxide compounds of the transition metals cobalt, nickel, manganese, and copper are suitable catalysts for hydrogen combustion at about 150°C. Further examinations of these catalysts are necessary. Further, Haruta and co-workers evaluated the properties of platinum, palladium, and a special cobalt-manganese-silver oxide compound, all of which are placed on nickel and ceramic supports, in regard to their possible use in hydrogen burners. The second main point of the investigations of this group is the technical development and examination of hydrogen burners. Thus far, two prototypes have been built, characterized by the pattern of hydrogen distribution within the burner. As the experiments show, specially designed distributors give rise to a constant temperature distribution in the catalyst plates, thereby producing improved delivery of specified amounts of heat. Prototypes that have been fitted out in this way deliver heat in the range of 0–1.74 W/cm^2 of catalytic surface at a very high efficiency of combustion ($\approx$96%).

The problems that would be associated with the introduction of hydrogen as a heating fluid and the areas of further research and development work that will have to be carried out are as follows[6]:

1. Development of catalytic burners for use with a wide range of mixtures of methane and hydrogen. These would have the objective of reducing the need to change burner design with an increase in hydrogen content.
2. Development of cheap, long-lived catalytic materials for catalytic combustion of hydrogen at low temperatures.
3. Research on concepts of space heating with catalytic hydrogen burners.
4. Definition and examination of safety standards for heating systems with hydrogen, including the use of hydrogen in industrial heating arrangements.

14.6. SAFETY ASPECTS OF CATALYTIC COMBUSTION

In operating a catalytic hydrogen converter, there is a certain element of danger if the combustion of hydrogen is not complete, because it could eventually build up to an explosive mixture that could give rise to an unwanted ignition of hydrogen in the presence of a spark, instead of its catalytic combustion. Both dangers have to be avoided by constructing burners appropriately.

The catalytic surface must be sufficiently large, and there must be no dead spaces in the burner. Apart from this, the surfaces themselves must be cooled with either air or water. The surfaces must be large enough so that one-time ignition cannot maintain itself. Thus, a short-lived ignition of the gas stream would not be critical, so long as the material being used can stand somewhat higher temperatures than those reached during catalytic combustion.

A further danger is the poisoning of the catalytic surface. In this way, the catalytic activity is lost, partially or perhaps completely, so that unburned hydrogen can build up to an explosive mixture. There is no doubt that before any burner can be applied on a larger scale, experiments of longer duration will have to be carried out to examine the long-term burner efficiency and to reveal any potential limitations. The use of hydrogen burners would become more problematical if the hydrogen available were not pure and would have to be used in the presence of another impurity.

14.7. PROSPECT

Transition from the present gas heating systems to those based on hydrogen, i.e., catalytic hydrogen combustion, will not be simple. Before really concrete statements can be made about what still has to be done, the results of the research and development work that is now going on will have to be thoroughly evaluated.

REFERENCES

1. Bundesministerium für Wirtschaft (publisher): Energieprogramm der Bundesregierung [Energy Program of the Federal Government]—Dritte Fortschreibung vom 4.11.1981, Bonn (1981).
2. U. Bonne: *Int. J. Hydrogen Energy* **8**(4):295–299 (1983).
3. B. Lewis and G. von Elbe: *Combustion, Flame and Explosion of Gases,* Academic Press, New York (1961).
4. Bundesministerium für Forschung und Technologie (publisher): *Einsatzmöglichkeiten neuer Energiesysteme—Teil III: Wasserstoff [Opportunities for New Energy-Conversion Systems—Part III: Hydrogen*], Bonn (1975).
5. J. Pangborn, M. Scott, and J. Sharer: *Int. J. Hydrogen Energy* **2**(4):431–445 (1977).
6. Deutsche Forschungs- und Versuchsanstalt für Luft- und Raumfahrt (publisher): Wasserstoff als Sekundärenergieträger—Vorschlag für ein Forschungs- und Entwicklungsprogramm [Hydrogen as a Second Energy Carrier—Suggestions for a Research and Development Program], DFVLR-Mitteilung 81–10, Stuttgart (1981).
7. E. Justi: *Spezifische Wärme, Enthalpie, Entropie und Dissoziation technischer Gase [Specific Heat, Enthalpy, Entropy, and the Dissociation of Technically Important Gases*], Verlag J. Springer, Berlin (1938).
8. E. Justi and H.-J. Selbach, in: *Akademie der Wissenschaften und der Literatur Mainz, Jahrbuch 1981,* p. 101, Franz Steiner Verlag GmbH, Wiesbaden (1982).
9. J. Mercea, E. Grecu, and T. Fodor: *Int. J. Hydrogen Energy* **6**(4):389–395 (1981).
10. J. Mercea, E. Grecu, T. Fodor, and S. Kreibik: *Int. J. Hydrogen Energy* **7**(6):483–487 (1982).
11. M. Haruta and H. Sano: *Int. J. Hydrogen Energy* **6**(6):601–608 (1981).
12. M. Haruta, Y. Souma, and H. Sano: *Int. J. Hydrogen Energy* **7**(9):729–736 (1982).
13. M. Haruta and H. Sano: *Int. J. Hydrogen Energy* **7**(9):737–740 (1982).
14. M. Haruto and H. Sano: *Int. J. Hydrogen Energy* **7**(10):801–807 (1982).

CHAPTER 15

Industrial Applications of Hydrogen

Hydrogen is already in use as an important raw material in chemical industry and particularly in the petrochemical industry. Its various applications have been known and investigated for a considerable period of time.[1–3] Hydrogen for industrial use is produced almost exclusively from the fossil fuels: natural gas, oil, and coal, which will therefore be depleted sooner and become more costly as time goes on. It is as true of the fossil fuels used as raw materials in the chemical industry and the petrochemical industry as it is of those used in other applications: they cannot be replaced. In the long term, therefore, a distinction must be made between the use of fossil fuels as *sources* of heat energy in combustion and their use as *raw materials* to produce hydrogen and other chemicals. It may well be that industrial needs for very large quantities of hydrogen will give rise to the beginning of a comprehensive new hydrogen technology.

In looking at the future uses of hydrogen in this chapter, the various *present* industrial applications of hydrogen will be treated individually. Apart from the well-known processes—e.g., ammonia synthesis—there are processes for the conversion of solid hydrocarbons into liquid intermediate products and energy media, as, for example, in the conversion of coal to oil. In metallurgy, there is the prospect of major use in the direct reduction of iron ore to iron. The latter applications would lead to medium- and long-term needs for very large quantities of hydrogen.

This chapter authored by Dr. P. W. Brennecke, Braunschweig.

15.1. AMMONIA SYNTHESIS

About 60% of hydrogen used in the chemical industry at present is used in the synthesis of ammonia. About 20% of ammonia production goes into the production of synthetics (e.g., as an intermediate in the production of thermoplastics) and other nitrogen-containing intermediate products. Most of the ammonia produced is converted to artificial fertilizer. The steadily rising requirement for fertilizer in the Third World is leading to an increase in the needed capacity of plants for ammonia synthesis and thereby to a rising demand for hydrogen in the near future for this need alone.[4]

The synthesis of ammonia is carried out by the direct combination of nitrogen and hydrogen at high temperature and high pressure according to the equation

$$N_2 + 3H_2 \rightleftarrows 2NH_3 + 92.5 \text{ kJ}$$

On a large scale, the synthesis of ammonia is carried out by the Haber–Bosch process. In this process, 1 volume of nitrogen is mixed with 3 volumes of hydrogen at temperatures of about 500°C and pressures of about 200 bar and brought into contact in a furnace containing a catalyst (iron, aluminum oxide, and some potassium oxide), giving rise to the evolution of heat and the formation of ammonia.[5] The ammonia produced is liquefied in part by washing under pressure, but mostly by cooling the gas mixture. Modern systems for this high-temperature process can deal with 6000 m^3 of gas mixture per hour, need little servicing, and operate continuously for years.

The economic viability of ammonia synthesis depends primarily on the cost of the hydrogen used. Because of the relatively low price of oil in the past and because of the present structure of refineries, the needed synthesis gas has come predominantly from natural gas, benzene, or heavy fuel oil. Actual hydrogen production has been carried out predominantly by a steam re-forming process or by partial oxidation of heavy oils.[6] Texaco and Shell have each developed separate processes for the production of synthetic gas from heavy fuel oil by partial oxidation using oxygen itself. The hydrogen produced by both processes is quite expensive because of the oxygen needed and the high plant costs. The gasification of solid fossil fuels (e.g., coal) for the synthetic production of gas is not cost-competitive if natural gas and oil prices remain low, so that coal gasification has been carried out only in countries where coal is very cheap. The processes used for this purpose, which bear names such as Koppers–Totzek, Lurgi, and Winkler, differ mainly in the limits set on the types of coal used and the consequent amount of oxygen that has to be added.

Since the price of fossil fuels has increased greatly in recent years, as mentioned in the introduction, and will continue to increase in the longer-term future as they become increasingly scarce, it would be better to conserve them as raw materials for chemical manufacture. There will likely come a time when hydrogen produced electrolytically (using non-fossil electricity) is cheaper than fossil-based hydrogen. At this point, it is appropriate to mention Czuppon *et al.*,[4] who have considered the relationship between energy needs in modern ammonia synthesis and energy use in the United States and discussed the relationship between these two problems.

15.2. SYNFUEL PRODUCTION

There will be a need for greatly increased hydrogen production in the near future, because hydrogen is the principal raw material needed for the production of carbon-containing synfuels. This must be seen in the context of the need for both fuels and raw materials for chemical manufacture. The building of plants for specific processes must be considered, as must the associated research and development work. The production of synthetic liquid fuels from coal must be considered[7] in relation to the increasing practice of converting higher-boiling oil fractions into lower-boiling ones, for this process requires hydrogen. For the middle term, the hydrogen needed for this process can be obtained from the partial oxidation reaction of heavy oil fractions by means of reaction with steam. However, the long-term goal should be to avoid using up part of the fossil fuels to produce the hydrogen needed in their liquefaction (for this removes part of the low-grade fuel that could be converted). It would be better to obtain the required hydrogen directly from water alone.

Thus, when one considers the application of hydrogen in the petrochemical industry, one has to distinguish between a short-term and a long-term strategy. The short-term strategy will be based on the present-day energy supply from oil. While we are still in the oil era, hydrogen is needed in (1) the desulfurization of fuel oil and (2) the conversion of heavy fuel oil (sulfur fractions) to lighter components. In a longer-term strategy, the component of oil in the energy mixture will have been reduced and replaced by coal, e.g., in the production of a fuel for vehicles. In this case, large amounts of hydrogen would be needed for (1) the gasification of coal and (2) adding hydrogen to coal to make oil substitutes.

The desulfurization of fuel oil involves the use of hydrogen as part of the refining of the fossil fuels. According to present technology, sulfur can be removed from heavy oil either directly by distilling at atmospheric pressure with later removal of sulfur residues indirectly or by atmospheric distillation followed by vacuum distillation and desulfurization–distillation. The direct process removes 70–80% of the sulfur, depending on the input of hydrogen. In the indirect process, only the vacuum distillate undergoes desulfurization, and thus only about 35–40% of the sulfur is removed.

The removal of sulfur from heavy oil is also carried out by means of catalytic processes involving hydrogen. Such processes are being used in Japan, the Middle East, and the United States on a very large scale. Desulfurization plants may be called "hydrorefining" or "hydrotreating" plants. In their construction, they resemble conventional cracking plants. On the other hand, in hydrorefining processes, the fact that there is a catalyst present with high selectivity for the desulfurization reaction

$$H_2 + S \rightarrow H_2S$$

plus the changed process conditions (pressure and temperature) tend to suppress the cracking reaction. As the need for both desulfurization and cracking increases, more of the product consists of saturated hydrocarbons. Thus, the need for hydrogen should grow at a greater rate than the need for a greater degree of desulfurization, or the yield of desulfurized fuel oil would fall.

With certain oils from the Middle East (Kuwait) with a sulfur content of 5.5%, obtaining a 60% degree of desulfurization requires a hydrogen input of 95 Nm^3 H_2/ton of oil (1Nm^3: 1 standard cubic meter at 1 bar and 0°C). To increase the degree of desulfurization to 87%, the amount of hydrogen must be increased to 240 Nm^3 H_2/ton of oil.

The conversion of heavy fuel oil containing sulfur to lighter components can be carried out in hydrocracking plants, which have been in operation for many years. The heavy oil is converted into lighter fractions by using hydrogen and process heat. Hydrocracking occurs in two stages. In the first stage, nitrogen and sulfur are removed from the heavy oil. The actual requirement for hydrogen depends on the desired degree of conversion of the distillate into light hydrocarbon fractions, but it is in the range of 250–450 Nm^3 H_2/ton of oil.

The use of hydrocracking plants for the production of gasoline from heavy oil assumes a readily available supply of cheap hydrogen. The product produced by hydrocracking can be used as a component for mixing with fuel or as a raw material in ethylene and methane synthesis. With respect to the products obtained from the reaction of heating oil with hydrogen, the following empirical overall equation[8] is applicable and sets up a quantitative relationship for the requirement for hydrogen in the complex, empirical reaction:

$$\begin{aligned} &1 \text{ kg fuel oil (type S)} + 0.494 \text{ Nm}^3 \text{ H}_2 + 3.35 \text{ MJ} \rightarrow \\ &\qquad 0.0854 \text{ kg C}_1 \ldots \text{C}_3 \\ &\qquad + 0\ .0980 \text{ kg C}_4 \\ &\qquad + 0.2185 \text{ kg light gasoline} + 0.4880 \text{ kg heavy gasoline} \\ &\qquad + \text{about } 0.89 \text{ kg naphtha} \\ &\qquad + \text{about } 0.1040 \text{ kg residue} \\ &\qquad + \text{about } 0.0410 \text{ kg H}_2\text{S and NH}_3 \end{aligned}$$

The coal-gasification processes that are in use or in development at present can be divided into two classes, steam gasification and hydrogenation gasification.[9] Hydrogenation gasification uses hydrogen itself as the gasifying agent and produces coal gas with a high methane content in an exothermic reaction. In this reaction, considerable amounts of residual coke are obtained. These can be converted to gaseous products in the same process by subsequent steam gasification.

In the overall process of coal gasification, two basic types of reactions occur:

1. Heterogeneous reactions, in which the gasifying material, e.g., H_2, and also the product gases react with the solid material.
2. Homogeneous reactions in the gas phase, in which the primary gaseous reaction products react among themselves and with gasifying material (e.g., formation of methane).

The interaction of these two reactions determines the course of the gasification process, its heat of reaction, and the composition of the product gases that arise.

The technical process whereby coal gasification is carried out is largely controlled by the contact between the gas and the solid material, as well as by the streaming conditions, both forward and counterstreaming, between the solid material and the gasifying medium. The most important processes that are used in this technology are called the fixed bed, the fluidized bed, and the entrainment.

In the fixed bed reactor, pieces of coal move by gravity from the top to the bottom while oxygen and steam move upward. Reaction zones are established in this reactor. In the coolest zone, coal is pyrolized; in the middle zone, coal is gasified; and in the hottest zone, coal undergoes partial combustion. With fine-grain coal and higher flow rates of oxygen and steam, gasification occurs in the fluidized bed, in which the temperature is practically a constant throughout the bed and individual reaction zones can no longer be distinguished. The same may be said of the enstrained bed reactor, in which fine-grain coal is converted at high temperatures in the flame. Under these temperature conditions, the conversion velocity of the coal depends on the relative velocity at the gas–solid interface. Consequently, special measures are taken, e.g., two burners are opposed so that there is vigorous agitation of the reaction mixture in the reactor. In the entrained bed reactor, a balanced streaming of the coal and the gasifying medium is ensured. Particular care must be taken in removing the unconverted mineral components at high temperatures in the reactor, and because of the low melting point of the ash content, the slag is removed as a liquid. At lower gasification temperatures and with the use of certain coals, the ash can be removed in the solid form.

With respect to optimizing the configuration of the gasification reactor, the main point is the means by which the reaction is provided with heat. When the heat is produced within the gas generator, the reactor is called "autothermal" type; when the heat is produced externally, the reactor is of the "allothermal" type. Heat is produced in autothermal gas generators by the exothermic chemical reactions. Among these are reactions of carbon with oxygen, either in the form of air or as pure oxygen, and the reaction of carbon with hydrogen. These reacting gases are then mixed with the actual gasification medium, steam. When oxygen is added to the steam, additional CO or CO_2 is produced; when hydrogen is added, additional methane is obtained.

Industrial application of coal gasification is directed at this time to the production of ammonia for the manufacture of artificial fertilizer, as well as to so-called "indirect hydrogenation." In the latter, by means of syntheses that follow the gasification process, liquid automotive fuels can be produced. In terms of a hydrogen technology in development, the production of liquid energy media by hydrogenation after coal gasification would have particular significance.

At present, indirect hydrogenation on a large scale is used for the synthesis of methanol and also in the Fischer–Tropsch synthesis. Thus, in 1980, about 12×10^6 tons of methanol (CH_3OH) were made from synthetic gas ($H_2 : CO = 2.3 : 1$ in bed reactor). Methanol is currently used as starting material for numerous chemical processes and is expected to play an important part in the future as an alternative fuel, or as the starting material for the Mobil Oil Company's synthesis of oil. Instead of the old high-pressure methanol synthesis by the BASF or the UK Wesseling process (350 bar, 350°C, ZnO/Cr_2O_3 catalyst), modern low- and medium-pressure processes are available, and these are used in all new plants. The ICI process, which is used in about one third of the world production of methanol, works at 100 bar and 250°C with a copper-zinc-aluminum catalyst. The Lurgi process is a low-pressure one, with pressures as low as 40–50 bar and temperatures of about 260°C.

By combining the gasification process with methanol synthesis, and by optimizing pressure gasification techniques, it is possible to achieve a minimum raw material, and energy consumption of 43.9 GJ/ton of methanol (1.5 tons of coal with a heating value of 29.3 GJ/ton). The Koppers–Totzek process carried out at 1 atm can be combined with a methanol plant, and the consumption of coal is then 51.2 GJ/ton of methanol.

The Fischer–Tropsch process converts synthesis gas ($H_2 : CO = 1.7 : 1$ in a migrating bed reactor; $H_2 : CO\ 3 : 1$) in an entrained bed reactor into mostly straight-chain aliphatic hydrocarbons with a large variety of molecular weights, starting with gaseous products and going up to high-molecular-weight hydrocarbons. With this process, in 1980, about 2.4×10^6 tons of aliphatic hydrocarbons were produced. The first basic reaction is the actual Fischer–Tropsch conversion:

$$CO + 2H_2 \rightarrow (-CH_2-) + H_2O$$

The second basic reaction is the homogeneous water gas reaction, which takes place as a following reaction and with particular ease on an iron catalyst:

$$CO + H_2O \rightarrow H_2 + CO_2$$

Taking these successive reactions into account, one then obtains, using an iron catalyst for the Fischer–Tropsch reaction, the overall reaction:

$$2CO + H_2 \rightarrow (-CH_2-) + CO_2$$

In synthesizing automotive fuels such as gasoline and diesel oil by means of a Fischer–Tropsch synthesis, either of two methods can be used. Predominantly smaller molecules can be selectively synthesized and then joined to make larger molecules, or large hydrocarbon molecules can be synthesized and then separated by distillation or cracked into the desired fractions.

In addition to gasification, a second process for the liquefaction of coal must be considered: direct hydrogenation. In this refining process, coal is dissociated by high temperature and converted directly into liquid products by the simultaneous addition of hydrogen. The various hydrogenation processes that have been developed follow this basic procedure: The coal is first powdered so that 90% of it is less than 1 mm and 60% less than 60 μm in diameter. At this point, a catalyst in an amount weighing a small percentage of the weight of the mixture is added. The mixture is then pulverized or mashed with residual oil (about 40% middle-fraction oil and 60% heavy oil), and the coal increases in volume as a consequence of the addition of the oil. After passage through several preheaters, the coal–oil slurry is introduced into hydrogenation reactors arranged in series. According to the way in which hydrogen is added, there can be two different hydrogenation procedures. In the hydrogenation procedure of Bergius and Pier (started in 1913), molecular hydrogen is added to coal in the presence of catalyst. The coal is first ground and mixed with a suitable solvent. Hydrogenation proceeds in the so-called "sump phase" with the addition of hydrogen at high pressure and at temperatures between 400 and 500°C. The reaction is accelerated by catalysts; it is often influenced favorably by certain mineral components in the coal. Various oils are produced from coal by this method, and each has a different distillation condition. The composition of the oils obtained depends on the conditions of the process, particularly the temperature and pressure, the amount of hydrogen added, and the time in the reactor, all of which can be suitably varied.

In the second process, developed by Pott and Broche in the 1920s, coal is brought to a pressure of 100–150 bar and temperature of about 400°C and treated with a solvent that gives off hydrogen. The solvent chosen is, for example, a tetralin–cresol mixture, or a middle oil from the hydrogenation of coal and pitch. In these hydrogen donor solvents, the organic components of coal are largely dissolved; the undissolved material is removed by filtration, yielding a pitchlike extract low in sulfur and ash. In the extraction, the solvent gives hydrogen to the coal. The amount of hydrogen released from the solvent is rather small, but it saturates the valences that are freed from the coal during the polymerization stage.

After the hydrogenation products have been left at the top of the reactor, precipitation is caused by lowering the temperature by as much as 40°C. The resulting residual mud contains residual coal, ash, catalyst, asphalts, and other components that boil at more than 325°C. In newer coal hydrogenation plants, the residual mud is subjected to refinement under vacuum distillation; the residue, which then remains liquid at higher temperatures, is subjected to gasification to obtain hydrogen. Subsequent passage through a cold separator results in a reduction in temperature of the hydrogenation product and finally in so-called "stripping" which is lowering of pressure and separation of the condensed gases. Hydrogen is then extracted from the gases. The oil from the coal is then finally separated into products with different boiling points (light, medium, and heavy oils) in a distillation plant.

Hydrogenation by the Bergius–Pier method was used until 1945 in 12 plants, which had a total production of liquid products of 4×10^6 tons/year. The chief products were fuels, in particular aviation fuels (kerosene).

Hydrogenation by the Pott–Broche method was used until 1945 in only one plant. The material added was anthracite. The principal product was an extract that could be used as electrodes for aluminum production. The plant had a production of about 30,000 tons/year of extract.

The hydrogenation plants left in Germany after World War II were gradually abandoned. Work was taken up again in the United States and, after the oil crisis of 1973, in the Federal Republic of Germany as well. New and improved concepts of hydrogenation were applied. Since 1973, several modified hydrogenation processes that have come to be known collectively as the "new German technology" have been developed in the Federal Republic of Germany.

15.3. DIRECT REDUCTION OF IRON ORE

Apart from the chemical and petrochemical industry, the iron and steel industry offers a possible market for hydrogen, on technological, economic, and ecological reasons.[10] The structural change occurring worldwide in the ferrous metal industry makes it likely that there will be a dramatic increase in use of the electrical process combined with direct reduction processes for steel production. In the iron and steel industry, hydrogen should be applicable as a reduction medium in the production of raw iron and sponge iron. It would be possible to substitute hydrogen for the coke now used to reduce iron ore and thus change from production of raw iron in furnaces to a direct iron ore reduction process (production of sponge iron). The transition from classic coke metallurgy ("high temperature furnaces") to a sponge iron technology would be accompanied by a change in the steel-producing sector to electro-steel production, because "sponge iron" can be produced better in high-performance arc furnaces, which are undergoing rapid development. This change would require a large amount of hydrogen, and these would be expected to increase sharply in the coming years.

The fundamental difference between coke metallurgy and "sponge iron" technology, which caused the change in steel-making practice, is further explained in the following observations on the direct reduction of iron ore. Direct reduction takes place largely in the solid state, so that the product, sponge iron, is also in the solid state and indeed has an appearance somewhat similar to that of the iron ore from which it came. Further processing of the sponge iron, which converts it to steel, occurs principally in the arc furnace, in which the iron is melted and separated from the gangue. The oxygen potential of the iron—and here the new technology of direct reduction-steel production in an electric furnace differs from the classic coke metallurgy—is decreased gradually and without fluctuations from the ore to the finished product. In the blast furnace, iron oxide is reduced to iron. Further, compounds of other elements (sulfur, phosphorus, silicon, manganese) are reduced, requiring additional reductants. These elements and appreciable amounts of carbon dissolve in the liquid iron. Carbon and the other elements in liquid iron must then be removed by oxidation with oxygen, in preparation for steel production. This process uses oxygen, the element that was previously separated from the iron oxide to obtain iron. The direct path from iron to steel, i.e., direct reduction, turns out to be favorable from an energy-balance view.

The technology of the direct reduction of iron ore is at present in transition from the research to the production stage.[8] The plants that have been built thus far work largely by using natural gas, which is heated using fossil fuels and thereby converted to a reduction gas (i.e., a gas containing hydrogen). In future hydrogen technology, this reducing gas would be replaced entirely by hydrogen. Thus, in a hydrogen economy, the many reduction processes now in use would be replaced by the single direct reduction process.

The question of which direct gaseous reduction process is most suitable for the use of hydrogen can be reduced to the details of the technology concerned (shaft furnace process, retort process, fluidized bed process, and rotating furnace process). This simplification is valid because the various processes that work according to the same principle (e.g., the shaft furnace principle) differ only in the method by which the reducing gas is produced. One criterion for selecting the process is that the energy needed per unit of production be at a minimum. Shaft furnace processes have by far the higher specific production capacities per unit volume of vessel. The production is about 9 tons/day per m^3, about twice as high as that obtainable with the retort process and more than 6 times as high as that for the fluidized bed process of U.S. Steel or the rotary furnace process of Krupp. The higher specific reduction capacity is likely to lead to more favorable investment costs.

With respect to another important selection criterion, the specific energy demand, the shaft furnace processes are also the most favorable in terms of their requirements for hydrogen. There are two reasons for this. The first is the ability of the shaft furnace to function as a heat and substance exchanger, as shown by the successful application of high-temperature ovens. The second is the shaft furnace's nearly complete use-up of the reduction gases through return of the gases from the mouth of the furnace.

In the aforementioned direct reduction processes based on natural gas, the iron ore is reduced with a gas mixture that consists predominantly of hydrogen and carbon monoxide. The ratio of the reactants can be varied. In the reduction by H_2 and CO, the following chemical reactions occur, respectively:

$$Fe_2O_3 + 3H_2 \rightarrow 2Fe + 3H_2O - 816.4 \text{ kJ}$$

$$Fe_2O_3 + 3CO \rightarrow 2Fe + 3CO_2 + 288.9 \text{ kJ}$$

Following the stoichiometric relationships, iron ore reduction with hydrogen would require 610 Nm^3 H_2/ton of iron and with carbon monoxide as the reducing gas, 604 Nm^3 CO/ton of iron (1 Nm^3: 1 standard cubic meter at 1 bar and 0°C). Which of the reactions and what ratio of the two gases in the reduction gas will result in optimal performance depends on various aspects. One would have to look at details of the reaction mechanism and derive the various dependencies that would arise.[11] In the case of exothermic (CO) reactions, on one hand, and endothermic (H_2) reactions, on the other, one would strive to have a corresponding mixture of CO and H_2 so that the heat needs balanced out. From reaction kinetic considerations, however, it would be advantageous to use highly purified hydrogen. This latter viewpoint becomes all the more important when one realizes that along with the higher rate of reduction that occurs in the process using hydrogen, there is also more efficient use of the gas. A further advantage is that in the direct reduction with hydrogen, the product gases consist simply of water vapor. This advantage should by no means be neglected, in view of world pollution problems.

Iron ore reduction in the shaft furnace occurs under normal or lowered pressures, so that the piped hydrogen probably must only be expanded. The reduction temperature is between 850 and 900°C, and in this range, the temperature-dependent reaction velocity is sufficiently great. Furthermore there is no leaking or sintering of the ore, which must be avoided because of the gas flowing through the shaft. Hydrogen is heated (to about 900°C) before it is introduced into the furnace, to provide sufficient heat. Apart from the heating of the ore to the reduction temperature and the compensation of the reduction heat, there is water vapor to be removed from the ore and other losses to be dealt with. The gas that leaves the oven contains some hydrogen, and following condensation and the removal of the water vapor, the hydrogen is run back through the process again. The conditions chosen here for the reduction process lead to a relatively low total energy need of about 11.7 GJ/ton of sponge iron; this corresponds to a hydrogen demand of about 1070 Nm^3 H_2/ton of sponge iron. The efficiency of reduction in the shaft furnace is about 95% when the sponge iron is removed. The iron is cooled to a temperature of about 80°C before it contacts the atmosphere, because of the danger of reoxidation. This is done in a separate cooling gas process. The process yields heat amounting to 0.4 GJ/ton of sponge iron that could be re-used to preheat the reduction gas, since the cooling process uses cold H_2. Use of this process would reduce the overall energy need to about 1030 Nm^3 H_2/ton of sponge iron. Whether it is cost-effective to build the additional apparatus will have to be determined by further research and cost calculations.

It must finally be pointed out that it is possible to process sponge iron without problems and to do so conveniently in a plant reducing steel in electric furnaces. The danger of reoxidation is eliminated by preliminary cooling to 80°C. In the direct processing of hot sponge iron in the electric furnace, any transport should be carried out with a thorough elimination of air. To this end, the sponge can be made up into hot bricks as it comes out of the shaft oven, thereby greatly reducing the possibility of reoxidation.

In further processing in the arc furnace, direct reduction of sponge iron gives rise to the possibility of continuous addition. For continuous addition of sponge iron, there is a special technique available by which the speed at which the material is added can be kept accurately in balance with the added electrical energy. This technique has been successfully tried in high-power arc furnaces with trapping charges of up to 200 tons; charges consisting entirely of sponge iron were processed showing that scrap iron is not needed. As a consequence of continuous addition, the charging time is decreased and the current passed can be kept at full strength for practically the whole time. Because of the purity of the sponge iron, performance is increased in comparison with that of a scrap metal plant. On the other hand, the electrical energy used is increased, and this disadvantage must be balanced by the increase in performance.

15.4. OTHER POSSIBILITIES FOR THE INDUSTRIAL USE OF HYDROGEN

Apart from the use of hydrogen for ammonia synthesis, refining of oil, liquification of coal, and direct reduction of iron ore, hydrogen is needed in the chemical industry for producing oxyalcohols, cyclohexane, and benzene by the dealkylation of toluol, and in the hardening of fats. For some processes (e.g., fat hardening), only very pure hydrogen (i.e., electrolytic hydrogen) can be used.

The future requirements for hydrogen in these industries depend mainly on their growth rates. The increased requirements for hydrogen would be relatively small compared with the magnitude of the increases to be expected from potential developments in the petrochemical industry and in the direct reduction of iron ore.

15.5. PROSPECTS

Practically all the processes that have been suggested here for the use of hydrogen are known to be technically possible. For example, the direct reduction of iron ore with hydrogen is a technology that is ready for full development. For these reasons, there would be no basic difficulties in making a transition to the extensive application of hydrogen at an industrial level as part of the transition to a hydrogen economy.

REFERENCES

1. G. Kaske: *Chemie-Ingenieur-Technik* **48**(2):95–99 (1976).
2. J. Pottier, D. Marque, and C. Tellier, in: Commission of the European Communities (publisher): *Seminar on Hydrogen as an Energy Vector—Its Production, Use and Transportation,* pp. 531–543, EUR 6085 DE/EN/FR/IT, Brussels (1978).
3. C. J. Huang, K. Tang, J. H. Kelley, and B. J. Berger, in: J. W. White, S. M. Briles, and W. McGrew (eds.): *Hydrogen for Energy Distribution: Symposium Papers,* pp. 69–83, Institute of Gas Technology, Chicago (1979).
4. T. A. Czuppon, A. E. Cover, and I. J. Buividas, in: J. W. White, S. M. Briles, and W. McGrew (eds.): *Hydrogen for Energy Distribution: Symposium Papers,* pp. 111–135, Institute of Gas Technology, Chicago (1979).
5. A. F. Holleman and E. Wiberg: *Lehrbuch der anorganischen Chemie,* Verlag Walter de Gruyter, Berlin (1964).
6. Bundesministerium für Forschung und Technologie (publisher): *Auf dem Wege zu neuen Energiesystemen—Teil III: Wasserstoff und andere nichtfossile Energieträger* [*On the Way to New Energy Systems, Part III: Hydrogen and Other Non-Fossil Energy Media*], Bonn (1975).
7. Deutsche Forschungs- und Versuchsanstalt für Luft- und Raumfahrt (publisher): *Wasserstoff als Sekundärenergieträger—Vorschlag für ein Forschungs- und Entwicklungsprogramm* [*Hydrogen as a Secondary Energy Medium: Proposals for a Research and Development Program*], DFVLR-Mitteilung 81-10, Stuttgart (1981).
8. Bundesministerium für Forschung und Technologie (publisher): *Einsatzmöglichkeiten neuer Energiesysteme—Teil III: Wasserstoff* [*Opportunities for the Application of New Energy Systems, Part III: Hydrogen*], Bonn (1975).
9. H. Jüntgen and K. H. van Heek: *Kohlevergasung—Grundlagen und technische Anwendungen* [*Coal Gasification—Basics and Technical Applications*], Verlag Karl Thiemig, Munich, 1981.
10. J. F. Balter, G. Stocker, and Y. Gousty, in: Commission of the European Communities (publisher): *Seminar on Hydrogen as an Energy Vector—Its Production, Use and Transportation,* pp. 482–505, EUR 6085 DE/EN/FR/IT, Brussels (1978).
11. M. V. C. Sastri, R. P. Viswanath, and B. Viswanathan, *Int. J. Hydrogen Energy* **7** (12):951–955 (1982).

CHAPTER 16

Hydrogen as a Fuel in Automotive and Air Transportation

Oil and oil-related products such as gasoline and diesel oil are used to power cars and vehicles of all kinds because of their high energy density and ease of storage. Our standard of living depends in great measure on the automobile. It is further dependent on the mass transport of goods by water, rail, highway, and air; it also depends on the use of machines in agriculture and the resulting high rate of utilization of arable land. The fact that the exhaustion of the world's supply of oil is now in sight makes it necessary to develop new fuels or modes of transportation if we are to maintain our standard of living. Even without the imminent depletion of our oil supply, however, the increasing air pollution from automobiles, particularly carbon monoxide and nitrogen oxides in the atmosphere (as well as other polluting products of the combustion of gasoline), forces us to give thought to developing other fuels for transportation.

Cars and other mobile transport vehicles depend on engines for mechanical energy. The performance of an engine must be sufficiently powerful and easy to regulate. The engine must satisfy numerous criteria with respect to safety, economy, ease of handling, environmental concerns, and manufacturing constraints. Given these criteria, the most suitable propulsive source is the electric motor. Trains, in particular, are driven electrically. For the railways, the development of another fuel is unlikely. The best means of supplying the necessary electricity is through overhead lines.

The question whether cars can successfully be propelled by battery-powered

This chapter authored by Professor H. H. Ewe, Hamburg.

electric motors is more problematical. The only batteries commercially available at present are lead-acid and iron-nickel batteries; both are heavy, and the result is that the range of battery-powered cars is less than 200 km. Although this range is satisfactory for a number of applications, e.g., delivery vehicles and post office vans, the batteries take up so much of the vehicle's load capacity that the payload is quite small. It is true that new battery systems have been under development for some years now, and the new batteries (e.g., sodium-sulfur, zinc-air) have much higher energy and power densities and should therefore be quite suitable for utilization in electric cars. However, no decisive breakthrough can be said to have occurred as yet; i.e., no really cheap, light, and long-life batteries suitable for ordinary use have been developed.[1,2] A number of others (including molten salts) are in sight. Vehicles that do run on batteries at present are powered by lead-acid batteries and are employed in special conditions (e.g., forklift trucks) in which there can be no evolution of exhaust gases, e.g., in underground passageways and buildings.

The following discussion concerns the provision of power for electrical drive by a different scheme, namely, the combination of fuel cells and electric motors; in the case of internal combustion engines, the possibility of converting them from the use of gasoline and diesel oil as fuels to the use of clean-burning hydrogen is discussed.

16.1. ELECTRICAL PROPULSION OF VEHICLES BY MEANS OF FUEL CELLS AND ELECTRIC MOTORS

The principles of fuel cells were described in some detail in Chapter 13. When fuel cells are used to propel cars, the electricity generated in turn powers the electric motor. The size of the vehicle is determined primarily by the weight of the fuel cells, including the weight of the auxiliary apparatus for circulating the electrolyte, removing the water, cooling, and other functions. As for the fuel cells themselves, the decisive problem is what to use as catalysts. In alkaline electrolytes nickel and silver are used as catalysts; in acid electrolytes, platinum, tungsten carbide, and carbon are used. The catalysts determine the current density and thereby the size or weight and in particular the operating life of the cell.[3,4] The fuel cells available at present* have a specific weight of some 10 kg/kW and an operating life of more than 2000 hr (corresponding to a total driving distance of about 100,000 km).[5] In addition to this, however, one must take into account the weight of the motor (2–4 kg/kW), the drive train (2–4.5 kg/kW), and the auxiliary equipment associated with the fuel cells (≈5–10 kg/kW), so that one can calculate a specific weight of about 25–30 kg/kW. This weight is so high compared with that of a diesel engine or a conventional internal com-

* Translator's note: Very much lighter fuel cells are now (1987) in sight. Eventually, a power to weight ratio better than 1kW/kg seems very probable.

bustion engine (which, including oil, gear units, and cooling water) amounts to about 3 kg/kW that the internal combustion engine must be expected to be the usual power source for cars until there are substantial increases in the cost of gasoline.[6,7]*

Further difficulties with electric cars driven by fuel cells arise when one considers costs and the auxiliary apparatus. Since there has been experience only with the manufacture of individual units, it is not possible to calculate realistic costs. Cost *estimates* vary widely according to the number of units per year assumed to be produced. Using fuel cells of between 1 and 5 kW, the cost would be between $250 and 2500/kW.† As a counter to these high costs, the advantages of fuel cells can be cited: The efficiency of energy conversion to electricity, assuming conditions near those of the normal driving cycle, is 50–60%; the exhaust gas is pure water vapor without mixtures of nitrogen oxides.[8–10]

16.2. POSSIBLE USE OF INTERNAL COMBUSTION ENGINES POWERED BY HYDROGEN FOR AUTOMOTIVE TRANSPORTATION

An internal combustion engine powered by hydrogen would be based on the conventional gasoline-powered engine, although it would have to be modified because of the different characteristics of hydrogen. As a fuel, hydrogen differs in the following properties from gasoline and diesel oil (also see Table 16.1): First, since hydrogen is a gas, no gasification is necessary; hydrogen mixes readily with air. Hydrogen–air mixtures have heating values about 15% lower than those of gasoline–air mixtures, and therefore larger cylinders would be necessary. Hydrogen–air mixtures have wider combustion limits (with respect to temperature, ignition energy, and concentration ratios) and there is, consequently, a greater danger of backfiring and knocking. Faster combustion may involve more rapid development of pressure.

The principal problem of the hydrogen motor is backfiring into the exhaust pipe. This difficulty can be remedied by using a leaner mixture, which also reduces the emission of nitrogen oxides. The disadvantage is that greater displacement is needed to give the same performance. By recirculating the exhaust gases, it is possible to avoid backfiring and knocking. A further possibility is to introduce hydrogen at high pressure during the compression stroke (similar to fuel injection used in diesel motors).[11–15]

* Translator's note: The application of this view gives very different results, depending on whether one takes into account the ancillary costs of pollution or (as is usual) assumes this part of the cost of using fossil fuels is suppressed.

† Translator's note: The projected cost of fuel cells in commercial production is now (1987) between $100/kW (for alkaline cells working on pure H_2) to $2000/kW (for cells involving polymer electrolytes with expensive catalysts).

TABLE 16.1
Comparison of Hydrogen and Other Fuels in Terms of Several Properties That Are Important to Their Use in Internal Combustion Engines

Property	H_2	Comparison Fuel
Heat value	34.2 kWh/kg	12.9 kWh/kg (gasoline)
Minimal ignition temperature	574°C	538°C (butane)
Combustion range (volume % in air)	4.0–74.2%	1.9–8.6% (butane)
Combustion limits of a fuel–air mixture (volume % fuel in air)	0.14–10%	0.4–1.4% (gasoline)
Heat value of a stoichiometric fuel–air mixture	3190 kJ/m^3	3800 kJ/m^3 (gasoline)
Ignition energy	0.02 mJ	0.25 mJ (gasoline)
Maximal laminar combustion velocity	2.7 m/sec	0.4 m/sec (propane)

These and other processes have already been examined and successfully applied, as numerous communications from firms and research institutes attest. In particular, Daimler-Benz has been testing a hydrogen-powered city bus. The bus was built on a conventional chassis and is propelled by a modified internal combustion engine originally designed for use with gasoline. The engine is provided with a high-performance ignition device, and is equipped with an accessory water injector which decreases the combustion velocity, avoids backfiring and knocking, and reduces emission of nitrogen oxides. It yields 44 kW (60 hp).[16]

Figure 16.1 shows a 2.6-kW engine that was developed at Oklahoma State University. In this four-stroke one-cylinder engine, air alone is compressed and hydrogen gas is injected, thus achieving reasonably modulated combustion without irregular detonations and unusually high temperatures or pressure jumps.[17]

From these and other examples of hydrogen-powered vehicles, it can be seen that the internal combustion engine can be fueled with hydrogen and that the various changes necessary can be made without altering the present technology of manufacturing engines. Since engines work without a choke, higher efficiency will be obtained than would be expected using gasoline. Precise data on the improvements that could be achieved are not yet available because of the limited experience with hydrogen-powered engines.

Some advantages of hydrogen engines, however, are already clearly recognizable, particularly their lack of noxious emissions. Thus, hydrogen as an alternative fuel is non-poisonous compared with gasoline, some components of which (e.g., benzene), are carcinogenic. In the exhaust of a gasoline-fueled internal combustion engine, there are various harmful substances, including carbon monoxide, sulfur dioxide, unburned hydrocarbons, and lead compounds. In the exhaust of a hydrogen-powered car, there is none of the carbon monoxide

FIGURE 16.1. View of a Clinton-4 engine built to be powered by hydrogen. From Murray and Schoeppel.[17]

that now befouls our cities, some 50% of which is attributable to automotive transportation. There is also no sulfur dioxide, soot, or unburned hydrocarbons. The only pollutant present to any measurable degree in the exhaust from hydrogen-powered cars is nitrogen oxide. Figure 16.2 compares the nitrogen oxide emission of an internal combustion engine burning various fuels. When hydrogen is burned with a large excess of air, the emission of nitrogen oxide is very small. These extremely lean mixtures exceed the limits of operation of a *gasoline-fueled* engine, so that operation with such high amounts of air in the mixture is not possible in such systems, though it is with hydrogen. Conversely, as the stoichiometric mixture ratio is approached in a hydrogen engine, the nitrogen oxide

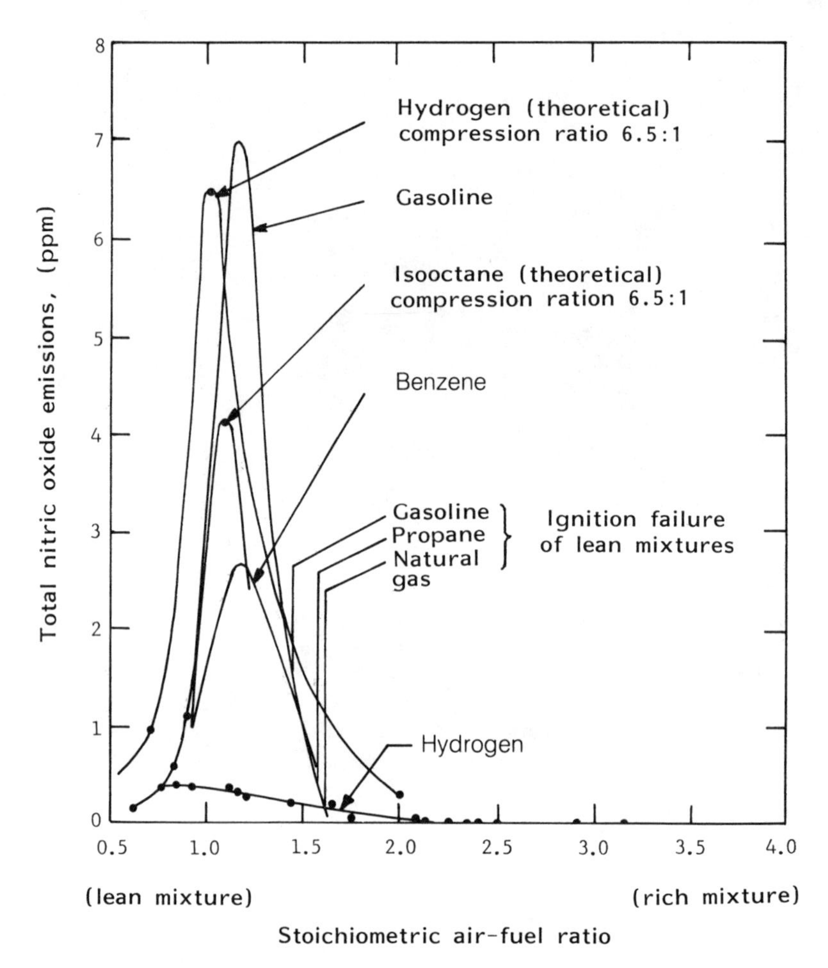

FIGURE 16.2. Nitrogen oxide emissions shown as a function of the airfuel ratio for various mixtures of fuels. From Pischinger and Streicher.[14]

emission increases drastically. Since backfiring occurs readily in the intake manifold, direct measurements are rather difficult to make. Because of this nitrogen oxide emission, operation of hydrogen cars with a mixture ratio of 1.7–2.0 is suggested.[18] In this range, the combustion ratio of hydrogen is favorable and similar to that of gasoline in an engine.

Another possibility for reducing nitrogen oxide is to recirculate the exhaust gases by returning part of them to the intake manifold so that the engine runs on a mixture of hydrogen, air, and exhaust gas containing a relatively large amount of water vapor.[19] Nitrogen oxide emission can also be reduced radically by layering the mixture, as Murray and Schoeppel[15] have shown (cf. Figure 16.2). The compression ratio of the hydrogen engine is 1 : 6. In this mode of operation, hydrogen is introduced directly into the combustion chamber near the upper dead center during the compression stroke so that no homogeneous mixture is created.

The construction of an internal combustion engine for use with hydrogen poses relatively few problems, compared with the conversion to electric cars. Existing production lines could be kept in service to a great extent because the necessary changes are largely confined to the ignition system and the fuel storage and delivery system. Thus, conversion of the internal combustion engine to hydrogen will be more practical, and certainly easier, than the transition to electric power with, for example, fuel cells. However, one would have to anticipate difficulties with the storage of hydrogen in automotive transportation. High-pressure steel cylinders are too heavy, but it seems likely that cylinders of lightweight metals will be available in the near future. High gas pressures will always present some measure of risk, however, because the cylinders will have to withstand accidents and because improper handling due to thoughtlessness or carelessness cannot be excluded. The use of liquid hydrogen in cryogenic tanks would be greatly advantageous with respect to pressure, weight and volume of the tank, and the amount of hydrogen that could be carried, but the skill that would be needed in refueling and the requisite training for the personnel concerned would be much greater than the handling of other types of fuels require.

Metal hydride storage would seem to be advantageous (no danger in accidents), although there is a weight disadvantage similar to that of high-pressure steel cylinders. On the other hand, the operating pressure for the hydrides is less than 50 bar, much lower than the pressures in the steel cylinders. Figures 16.3 and 16.4 show a schematic drawing and a photograph of a hydrogen-powered bus. The hydrogen is transported in three hydride storage devices. Two of these devices contain Ti-Cr-Mn alloys, which take up approximately 2 wt.-% hydrogen and release it for injection into the engine at quite low temperatures (under 20°C, the equilibrium pressure is 2–3 bar). For cold starts, initial heat can be provided by exhaust gases or heat exchangers. The third storage device contains hydride-forming materials such as Mg, Mg_2Ni, and Mg_2Cu, which require temperatures from 200 to 400°C (at pressures over 2 bar) in order to give off hydrogen. These temperatures are not available when the car is started, but can be obtained with the exhaust gases as a heat source after the engine has warmed up. These low- and high-temperature hydride storage devices have densities, including the tank housing, of 1.8–1.98 and 2.7–5.0 MJ/kg, respectively. If the total weight of the hydride tanks is some 340 kg, then about 5.4 kg hydrogen can be stored; this corresponds to about 21 liters of gasoline and is equivalent to a range of operation of about 150 km in the California driving cycle.[16] The weight and the relatively high cost of the hydride storage devices limit the range of hydrogen buses compared with those powered by diesel oil. Despite this considerably reduced range, the additional costs for a hydrogen-powered bus, almost entirely attributable to the hydride storage devices, are estimated to be $15,000.[6] It is clear that the use of hydrogen-powered vehicles will for now be limited to conditions

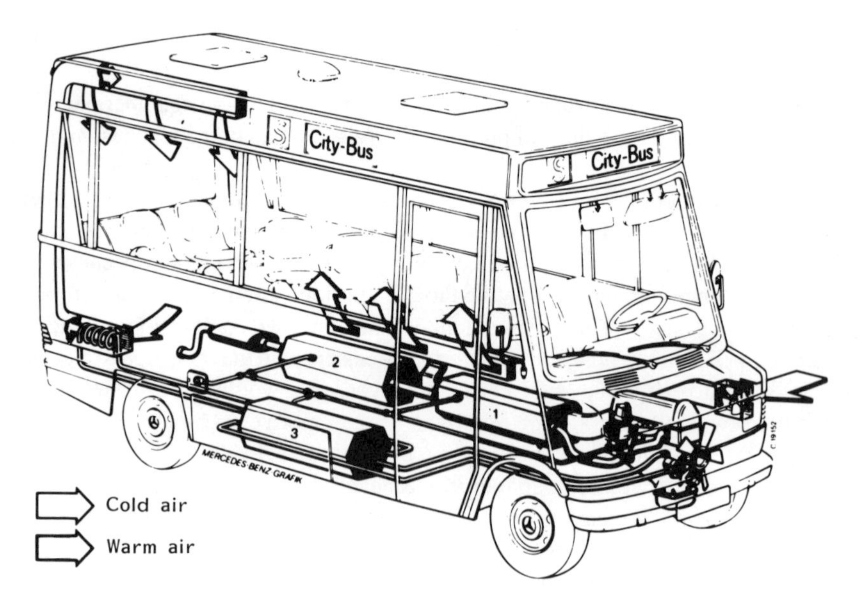

FIGURE 16.3. Schematic drawing of the hydrogen-powered Mercedes-Benz city bus. (1) High-temperature hydride storage device heated by exhaust gases (also usable for space heating). (2) Low-temperature hydride storage device heated by exhaust gases (used for further cooling of the exhaust gases and for water condensation). (3) Low-temperature hydride storage device with liquid heat exchanger (also usable for air conditioning).

FIGURE 16.4. Photograph of the hydrogen-powered Mercedes-Benz city bus.

under which harmful emissions (including the long-term difficulty with CO_2) from diesel or gasoline engines can no longer be accepted.

16.3. HYDROGEN AS AN AIRCRAFT FUEL

Experts agree that aircraft gas turbines can be powered by hydrogen with little difficulty. Thus, little attention need be given to the turbines themselves. Because of the high weight and volume required for storage of gaseous hydrogen, the only acceptable form of hydrogen as a fuel for aircraft is as the liquid. The characteristic properties of liquid hydrogen, in particular the temperature at which it must be maintained and its lower energy density per unit volume compared with aviation gasoline, would necessitate innovations in aircraft construction to provide cryo-insulated and above all considerably larger fuel tanks.[6,20–23]

The hydrogen-fueled aircraft that have been considered to date were all modifications of kerosene-fueled models. Even using these as test aircraft, striking advantages were found for liquid hydrogen as a fuel. Because of the lower weight of liquid hydrogen, the takeoff weight of these aircraft is markedly lower (260 tons for the Boeing 747 vs. 351 tons with kerosene), and the range is correspondingly greater (Table 16.2).

At present, development of special aircraft fueled by liquid hydrogen would require that lighter but larger cryogenic tanks be incorporated in the wings and in other compartments. These cryogenic tanks do take up considerably more space than kerosene tanks, but require similarly light insulation because an aircraft that has been fueled does not stand around for long, but is soon in the air, where the ambient temperature is low. For this reason, the relatively high evaporation rate that one obtains while the aircraft is on the ground is acceptable.

TABLE 16.2
Comparison of Some Data for Current Aircraft Fueled with Kerosene or Liquid Hydrogen[a]

	Boeing 747		Lockheed L1011	
Payload (tons)	55.8		25.4	
Range (km)	9265		6300	
Speed (Mach)	0.86		0.82	
	Kerosene	Liquid H_2	Kerosene	Liquid H_2
Take-off weight (tons)	351	260	195	144
Fuel (tons)	122	41	62	21
Wingspread (m)	59.4	59.4	47.2	43

[a] From Bundesministerium für Forschung und Technologie.[6]

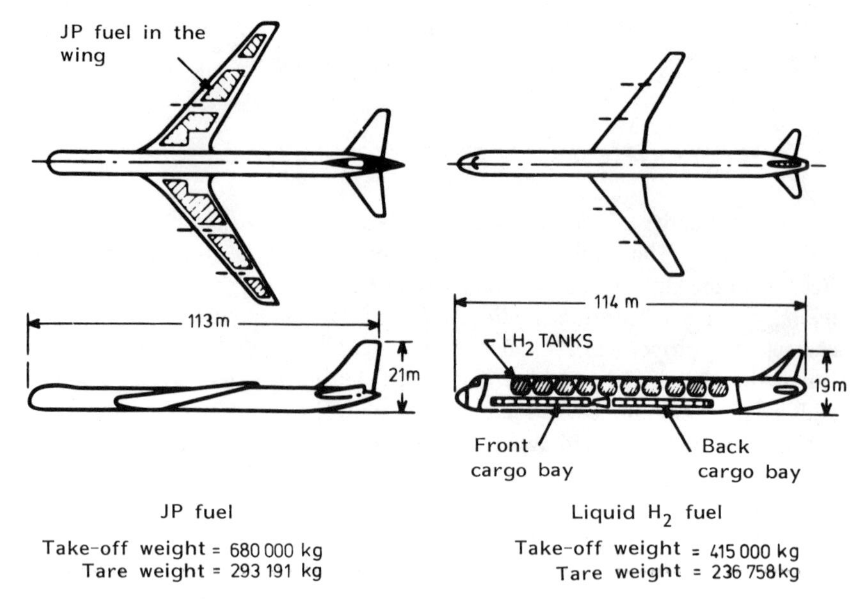

FIGURE 16.5. Comparison of the takeoff weight and empty weight of a subsonic air transport. Payload: 120,000 kg. Range: 9400 km. In changing from JP to liquid hydrogen, the takeoff weight is reduced from 680,000 kg to 415,500 kg and the empty weight from 293,191 to 236,758 kg. From Small *et al.*[22]

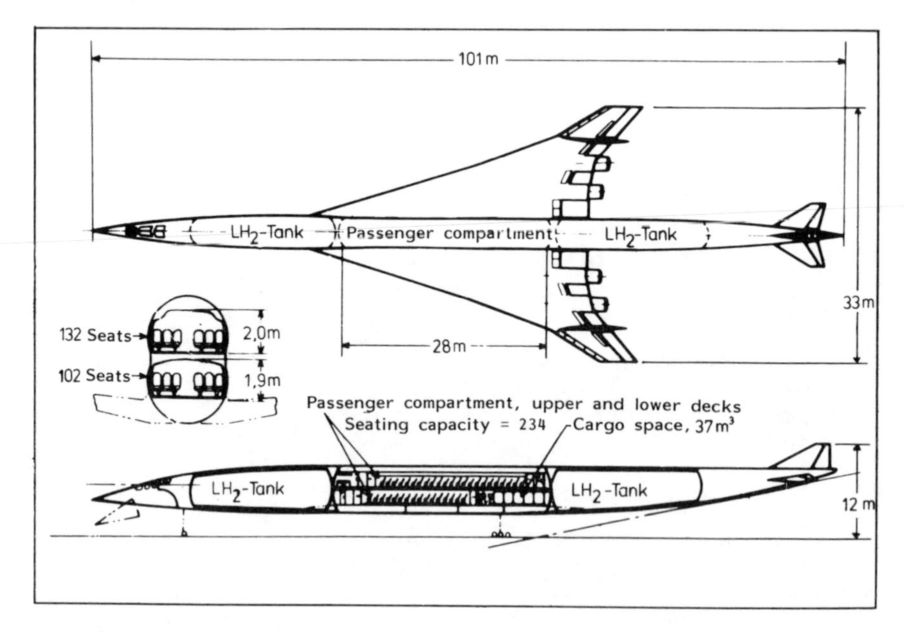

FIGURE 16.6. A supersonic aircraft powered by liquid hydrogen. The number of passenger seats is 234. (LH_2) Liquid hydrogen. From Brewer [23] and Escher.[24]

The necessary gasification of liquid hydrogen during flight can be obtained by utilizing exhaust heat from the turbines or heat from the surrounding air.

Newly designed aircraft incorporating such special features as large tanks that would allow them to be powered by liquid hydrogen have not yet risen from the drawing board. Typical suggestions are shown in Figures 16.5 and 16.6 (see also Small *et al.*,[22] Brewer,[23] and Escher[24]).

The advantages of liquid hydrogen for the fueling of aircraft are much more apparent when one considers supersonic aircraft. Figure 16.7 shows the distribution of weight for aircraft moving in the Mach 3 speed range. One of the advantages is that liquid hydrogen can be used to cool brine, which then circulates around the skin, a technique similar to that utilized in the Apollo spaceship. If such cooling were available, relatively heavy alloys of titanium and other metals could be replaced by aluminum and boron alloys, which are much lighter. The volume of aircraft would be greater because of the lower energy density per unit volume of liquid hydrogen compared with the fuel now used, JP5, but the range would be increased, as shown in Figure 16.8. Thus, the short range of present-day kerosene-fueled supersonic craft is a disadvantage that would be overcome by fueling them with hydrogen.

16.4. SAFETY ASPECTS

With hydrogen-fueled vehicles or aircraft, the most dangerous times are when they are being refueled and when they are parked. At these times, the

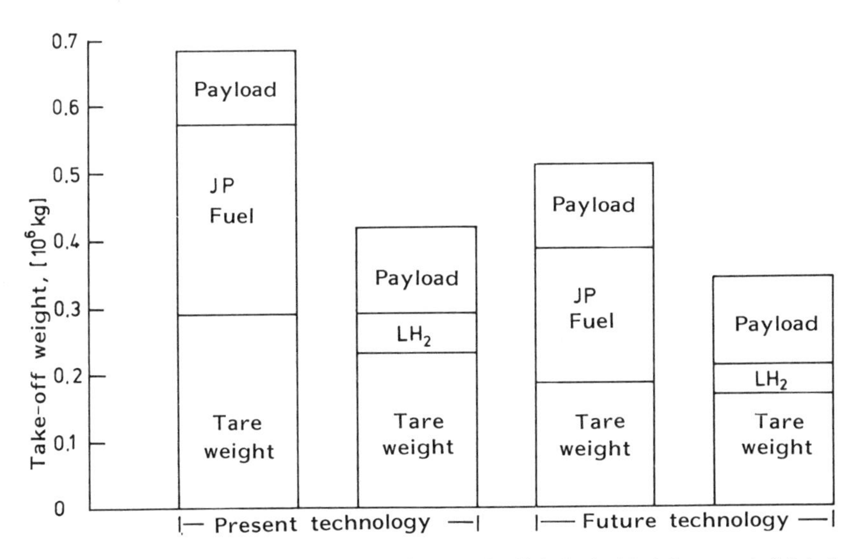

FIGURE 16.7. Comparison of the weight components for flight in the Mach 3 range. At left is the present state of technology: at right is future technology with hypersensitive aerodynamic steering and use of composite materials. (LH_2) Liquid hydrogen. From Small *et al.*[22]

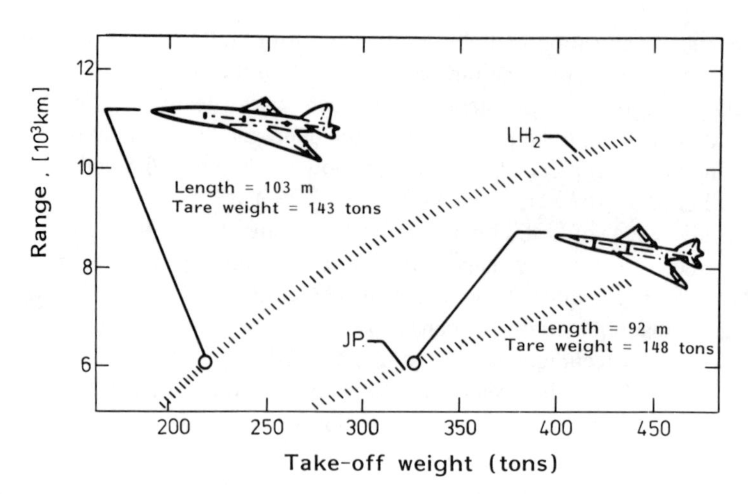

FIGURE 16.8 Comparison of the range as a function of the takeoff weight at Mach 3 with 300 passengers and titanium construction. (LH_2) Liquid hydrogen. From Small *et al.*[22]

accumulation of oxyhydrogen gas in or around the vehicle or aircraft must be avoided. In refueling, therefore, it must be remembered that small concentrations of hydrogen fuel in enclosed places can be explosive because of the low ignition energy and the broad range of explosive mixtures. On the other hand, there is the advantage that gaseous hydrogen dissipates rapidly upward. Even in highway accidents involving hydrogen-carrying tankers, the danger of explosion has proved to be small because of the strong construction of the hydrogen tank.

Space flight, with its enormous rockets powered by liquid hydrogen, has shown that these problems can be overcome.

16.5. COSTS

The propulsion of vehicles with liquid or gaseous hydrogen will be associated with increased costs compared with those associated with the use of gasoline because hydrogen tanks are much more expensive to manufacture than conventional fuel tanks. This is particularly true of hydride storage devices and very well insulated cryogenic tanks. Fuel consumption with steel cylinders and hydride tanks will also be increased because of the weight of the tanks that must be transported. Apart from all this, there is the cost of hydrogen itself, which depends greatly on how it is produced but which must always be compared with the ever-increasing price of crude oil, and with the cost of pollution (~$10/GJ) if the oil were synthetically produced from coal.

These higher costs, above all those of the tanks, can be contrasted with the great advantages that would accrue from powering the aircraft with liquid hy-

TABLE 16.3
Prices of Liquid Hydrogen[a]

Gasoline	1.21¢/$kW_{th}h$
Liquid H_2 from coal	1.10–1.40¢/$kW_{th}h$
Liquid H_2 from nuclear power plants	2.44–2.37¢/$kW_{th}h$

[a] From Jones[25] (as of 1974) (see also Johnson[26] and Michel[27]).

drogen. These advantages are, first, the reduction in weight of the fuel and the consequent great increase in range or in cargo or passenger payload. The cost of the liquid hydrogen can be obtained from the known cost of gaseous hydrogen and the cost of liquefaction. The latter depends on investment costs, costs of plant construction, and the actual energy use for the liquefaction process, which is about 9.6 kWh/kg liquid hydrogen with a heating value of 34.2 $kW_{th}h$. In 1974, Jones gave the prices shown in Table 16.3 for liquid hydrogen.[25] These prices are far lower than those in effect at present* because they are largely those obtainable in the United States and from very large plants, but they nevertheless show that the price of liquid hydrogen could become comparable with that of gasoline in the next few years. By way of comparison, note that the largest European hydrogen liquefaction plant is in France and that the cost is $1/liter for liquid hydrogen, i.e., 44.8¢/$kW_{th}h$.

16.6. PROBLEMS OF THE TRANSITION

The transition from gasoline, kerosene, and diesel fuel to hydrogen presents no problems with respect to the construction and production of the combustion engines. The problems that will arise will be those attributable to refueling, to the fabricating of hydrogen tanks, and to the fact that the two kinds of fuels cannot be mixed. Initially, hydrogen will therefore be used as a fuel only for certain earthbound fleets that operate in a relatively small area and are refueled in a central location.

For ease of handling, the hydride storage device is much better than the cryogenic tank. However, since it will be necessary to reduce its weight and cost considerably, some degree of technical development is still necessary. Since the internal combustion engine is far more advanced than the fuel cell-electric motor combination, the fact that the former is less efficient and noisier is unlikely to overcome the advantages of having such a huge infrastructure turning out cheap internal combustion engines.

It seems likely at present that the first use of hydrogen as a fuel will be in

* Around 5¢/$kW_{th}h$ in massive purchase.

aircraft, rather than in ground vehicles, and that the hydrogen will be in the liquid form because of the great advantage in takeoff weight it affords. For the preparation of liquid hydrogen, large liquefaction plants will have to be built. It seems likely that the early flights will be between two large airports with high traffic density, e.g., Frankfurt and New York. Large-scale application could be expected to follow as soon as the cost of running aircraft on hydrogen is less than that using kerosene. The technology is already there, as illustrated by the large space vehicles and also by aircraft modified to burn liquid hydrogen for demonstration purposes. These examples show convincingly that large amounts of hydrogen can be dealt with technically and applied without significant risk.

REFERENCES

1. F. Beck: *Chem.-Ing.-Tech.* **54**(9):809–817 (1982).
2. S. A. Weiner: The sodium-sulfur battery, a progress report in: J. B. Berkowitz, A. D. Little, and H. P. Silverman (eds.), Proceedings of the Symposium on Energy Storage, p. 141, the Electrochemical Society, Inc., Princeton, New Jersey (1976).
3. A. Winsel (ed.): *Ullmanns Encyklopädie der technischen Chemie,* Vol. 12, pp. 113–136, Verlag Chemie, Weinheim (1976).
4. F. v. Sturm: *Electrotechnische Zeitschrift* **99**(10)**:**615–624. (1978).
5. K. Strasser: *Elektrotechnische Zeitschrift* **101**(22)**:**1218–1221 (1980).
6. Bundesministerium für Forschung und Technologie (publisher): Programmstudie: *Auf dem Wege zu neuen Energiesystemen* [*On the Way to New Energy Systems*], 2, pp. 137–142, Bonn (1975).
7. H., G. Plust: Aktuelle Probleme bei elektrischen Antriebssystemen für Strassenfahrzeuge [Practical problems with electrically driven road vehicles], *Automobiltechnische Zeitschrift* **73:**9 (1971).
8. W. Vielstich and H. Schmidt: Brennstoffelemente [Fuel Cells], *Chem.-Ing.-Tech.* **42**(20):1266–1273.
9. F. v. Sturm: *Elektrochemische Stromerzeugung* [*Electrochemical Electricity Consumption*], Verlag Chemie, Weinheim (1969).
10. E. W. Justi and A. W. Winsel: *Kalte Verbrennung* [*Cold Combustion*]—*Fuel Cells*, Franz Steiner Verlag, Wiesbaden (1962).
11. R. A. Erren: Method of charging internal combustion engines, U.S. Patent No. 2.164.234, Application Sept. 17 (1938).
12. R. O. King *et al.:* The hydrogen engine: combustion knock and related flame velocity, *Trans. E.I.C.* **2:**143 (1958).
13. H. C. Watson, E. E. Milkins, and J. V. Deslandres: Efficiency and emissions of a hydrogen or methane fueled spark-ignition engine, Paper B-1-9, XV_{th} Congrès de la FISITA, Paris (1974).
14. F. Pischinger and K. Streicher: Charakteristische Verbrennungserscheinungen an Zweitakt-Gasmotoren [Characteristic combustion phenomena with two-stroke engines], Motortechnische Zeitschrift MTZ No. 5, p. 230 (1965).
15. R. G. Murray and R. J. Schoeppel: Emission and performance characteristics of an air-breathing hydrogen-fueled internal combustion engine, SAE Paper 719009 (1971).
16. Daimler-Benz AG, Pressemitteilung, November (1977); H. Buchner and R. Povel: *Int. J. Hydrogen Energy* **7**(3)**:**259–266 (1982).
17. R. G. Murray and R. J. Schoeppel: Paper presented at the Intersociety Energy Conversion Engineering Proceedings, Boston, August 3–6 (1971).

18. S. Furuhama and K. Yamane, Combustion characteristics of hydrogen fueled spark ignition engine, JARI Technical Memorandum No. 10, p. 14 (1972).
19. Y. G. Finegold *et al.:* The UCLA hydrogen car design, construction and performance, SAE Paper 730507 (1973).
20. P. F. Korycinski and D. B. Snow: Hydrogen for the subsonic transport, NASA Langley Research Center, Langley Field, Virginia (1979).
21. F. S. Kirkham and C. Driver: Liquid hydrogen fueled aircraft—prospects and design issues, NASA Langley Research Center, Langley Field, Virginia (1979).
22. W. J. Small, D. E. Fetterman, and T. F. Bonner, Paper presented at the Intersociety Conference on Transportation, Denver, Sept. 23–27 (1973).
23. G. D. Brewer: Liquid hydrogen fueled commercial aircraft, Symposium on Hydrogen for Energy Distribution, pp. 541–550, Lockhead, San Diego, California, July 24–28 (1978).
24. W. J. D. Escher: Prospects for liquid hydrogen fueled commercial aircraft, Report PR-37, Escher Technology Associates, St. Johns, Michigan (Sept. 1973).
25. L. Jones: Personal communication (July 1974).
26. J. E. Johnson: Paper presented by the Cryogenic Engineering Conference, Atlanta, August 10 (1973).
27. J. W. Michel: Hydrogen and exotic fuels, Oak Ridge National Laboratory, Oak Ridge, Tennessee, Report ORNL-TM-4461 (June 1973).

Index

Adsorption, and hydrogen storage, 259
AEG Telefunken, and the production of solar cells, 121
Air conditioning
  and climatization, 98
  solar (diagram), 99
Aircraft
  and a hydrogen economy, 34
  and economics of hydrogen, 320
  fueled by hydrogen, 317
  and safety with liquid hydrogen, 319
Air transportation, and hydrogen, 309
Albuquerque, NM, and solar farms, 148
Alkaline fuel cell, 274
Allis–Chalmers, cell for water electrolysis, 184
Amazon Basin, effect upon CO_2 concentration, 9
Ammonia synthesis, 298
Amorphous silicon, low cost production, 125
Angle, effect upon collection of solar light, 93
Ansaldo Co.,
  European work on power towers, 145
  heliostat, 146
Applications of hydrogen combustion, 295
Aquifers, for hydrogen storage, 248
Arc sponges, and sponge iron, 307
Area, for coal power plants, 4
Arrangement
  for dyes in monolayers, 169
  of monomolecular layers for light collection, 146
Ashermann, and the nonstoichiometric niobium nitride, 200
Atmosphere, and its CO_2 content, 11
Atomic power plant, on floating island, 13
ATP, its various parts, 159
Automotive transportation, and hydrogen, 309, 311
Authier, description of silicon solar cells, 129

Bacteria
  and energy farming, 163
  from plants, photosynthetic, 163
Barrier layer cells, and direct energy conversion, 80
Barrier layer photo cells, 116
Barstow, CA, and 10-MW_{el} solar capability, 149
Bassham and Calvin, their contributions to photoconversion, 159
Baum, and the power tower in Southern Russia, 140
Bergius and Pier, hydrogenation of coal by hydrogen, 303
Betz, efficiency factor in wind energy, 78
Biofouling, 111
Biophotolysis, 108
Bipolar electrolyzers (diagram), 181
Block diagram
  semitrailer for rail car transport for hydrogen, 212
  for solar–hydrogen economy, 236
Bloss and Pfisterer, wet process for cadmium sulfide production, 140

Bockris
  and block diagram of heat power plant for ocean gradients, 110
  suggestion to Westinghouse Company (1962), 36
Bogus, his examination of cadmium sulfide surface layers, 118
Breeder reactors, and the Ford Foundation, 62
Breeder technology, development, 63
Brewer, and hydrogen-powered supersonic aircraft, 318
Buch, evaluation of the siting of coal power plants, 3
Bus, hydrogen-powered, 312, 316

Cadmium sulfide
  cells, 133
    cross section, 137
    effect of various atmospheres, 134
  manufacture, 138
  service life, 135
Cadmium telluride, use in solar energy conversion, 119
Calculations for hydrogen pipelines, 219
Capital investment, and hydrogen pipelines, 220
Carnot factor, and energy conversion, 95
Catalytic combination of hydrogen, 289
Catalytic combustion
  aspects, 294
  of hydrogen. 294
Catalytic hydrogen burners, 290, 291
  schematic design, 292
Caverns, for hydrogen storage, 248
Cavity receiver construction, for solar collectors, 150
Chapin *et al.*, their invention of the silicon photo cell, 117
Chemistry, contribution to light collection, 167
Chlorophylls, and photosynthesis, 151
Chloroplast, and electron microscope (picture), 152
Claude, plant for OTEC (1930), 110
CO_2
  effect upon power plant choice, 44
  pollution, 8
  and predictions for the future, 10
  production, and the greenhouse effect, 45
  and seasonal change, 11
Coal
  combustion
    function of fuel cells, 17
    of hydrogen, 25
    and hydrogen fuel cell, 35
  power plants
    area for, 4
    evaluation of Buch, 3
  predicted production of world, 50
  supply, and the Federal Republic of Germany, 48
Collecting, and solar energy technology, 140
Collectors, flat-plate, 140
Combustion
  of hydrogen, 289
    applications, 295
    catalytic, 289
    range, 268
Compressor stations
  calculation by Messer–Greisheim, 233
  cost for hydrogen pipelines, 219
Concentrators, for solar light, 139
Conduction losses, in sending electricity over long distances, 200
Construction of new power plants, and shortages of cooling water, 6
Control centers, for hydrogen pipeline network in Germany, 208
Conversion efficiency, and solar cells, 128
Conversion of energy, net production, 69
Coolant water, and power output in power plants, 7
Cooling water, and nuclear plants, 19
Corrosion mechanisms, in heat convertors, 104
Cost
  fuel cells, 281
  hydrogen for aircraft, 320
  pipeline, 232
  silicon photovoltaics (prediction), 134
  for transmitting energy in various gases, 202
  transport of energy, 205
  transportation of energy (comparative), 15
Counter current forces, with cooling through expansion, 250
Cross section, of cadmium sulfide cells, 137
Cryostorage tank, 251
Current–voltage curves, among water electrolyzers, 192
Cylinders, and transportation of hydrogen, 21
Czochralski method, for silicon cells, 129

Dahlberg, description of photovoltaics, 65
Daimler–Benz, and the hydrogen city bus, 312
Davis and Kulcinski, and reactor concepts, 55
DeBeni and Marchetti
 and thermochemical cycles (1970), 38
 and thermochemical processes, 196
Degree of efficiency, for barrier–layer cells as function of energy gap, 115
Development of cadmium sulfide cells, in Germany, 136
Dewer flasks, for liquid hydrogen, 251
Direct combustion, of hydrogen, 288
Direct energy conversion, 69
 with electrochemical processes, 71
 matrix, 69
  description, 71
  examples, 71, 76
Dornier, and commercial black nickel, 103
Double wall tanks
 cost, 254
 for liquid hydrogen, 253
DSK electrode (photograph), 189
DSK principle, 188
Dye energy donors, and energy acceptors, 168

Earth's atmosphere, change in temperature, 9
Efficiency
 effect on power plant size, 1
 in ocean thermal energy conversion, 110
 of solar power plants, 152
 of sunlight conversion, and photosynthesis, 108
 in water electrolyzers, 129, 191
  as a function of current density, 192
Electric cars, 310
Electric energy
 cost, 201
 and a hydrogen economy, 33
Electrical energy, direct transmission, 199
Electrical transmission of energy, cost, 12
Electricity, from solar cells, 152
Electricity conversion, as function of grain size for cells, 127
Electrochemical energy storage, 245
Electrolysis, by ELOFLUX method, 187
Electrolytic production of hydrogen, 176
Electrolyzer plants, largest in world, 182
Electrolyzers
 new types, 183
 types (diagram), 180
 for water, 181
Electrons, gradual delivery through narrow barriers, 168
Elements, and world reserves, 29
Elliott and Turner
 and effect of lessened oil availability on future coal supplies, 51
 their prediction of coal production, 50
ELOFLUX
 electrolyzer (schematic view), 191
 water electrolyzers, 184
  water electrolyzer cell, 181
Emission, 102
 and reflection, of thermal energy collectors, 102
Energy
 carriers, and their exhaustion, 49
 collected by plants, 165
 consumption, as a function of time, 45
 expenditure, liquefaction of hydrogen, 251
 farming, with bacteria, 163
 as a function of Gross National Product, 46
 gap, and degree of efficiency in barrier-layer cells, 115
 losses, in water electrolyzers, 193
 medium
  its economics, 31
  future sources, 30
 needs of the Federal Republic of Germany, 47
 prediction of shortages, 27
 production in Germany, 48
 recovered in expansion of gases at end of pipeline, 240
 recovery plan for ailing economy, 1
 sources, and their media, 29
 storage
  in flywheels, 246
  in liquid hydrogen, 212
 suppliers, as a function of time, 44
 transmission over large distances, 199
 wastage, 71
Engine, powered by hydrogen, work of Murray and Schoeppel, 312
England, and wind energy conversion, 75
Escher, and supersonic aircraft powered by hydrogen, 318
Exhaustion of primary energy carriers, 49
Expansion of gases, and recovery of energy, 240

Fast breeder reactors, 60
  fuel consumption, 61
Federal Republic of Germany
  and coal, 48
  and primary energy consumption, 53
  supplied by solar–hydrogen economy, 24
Fermi levels in electrolytes, concept, 172
Ferredoxin, and function in photosynthesis, 161
Flame temperature, of hydrogen, 288
Flat plate, 140
  collectors, 140
Flights
  transition to hydrogen, 320
    problems, 320
Floating island, and atomic power plant, 13
Fluorescence, dyes in monolayers, 170
Flywheels, for energy storage, 246
Ford Foundation, recommendations of breeder reactors, 62
Free energy, and temperature dependence in water decomposition, 178
Fuel, and fast breeder reactors, 61
Fuel cells
  alkaline, 274
  assembly from United Technologies, 280
  and coal combustion, 17
  converted to electricity, 273
  diagram, 277
  hydrogen–oxygen, 275
  low temperature, 276
  produced in Japan, 285
  and the TARGET program, 283
  various designs, 283
Fujishima and Honda, their photoelectrochemical cell, 170
Fusion
  problems, 59
  Tokamak reactions, 55

Gaseous hydrogen, storage, 247
Gases, comparison of transmission media, 202
Gas pipelines, primary method of energy transmission, 12
Gasoline, safety, 267
Gassification, reactor for, 301
Germany
  its development of cadmium sulfide cells, 136
  its pipeline network, 206
Georgia Institute of Technology, and power tower work, 144
Grain size, efficiency of energy, 127
Gregory
  contribution to the hydrogen economy, 36
  plots of energy transmission cost, 204
Gross National Product, as a function of energy, 46
Gulf of Mexico, ideal site for ocean thermal energy conversion, 111

Hallet, work on hydrogen storage, 253
Hammond, concepts of future energy islands, 12
Haruta and Sano, their work on flameless hydrogen combustion, 293
Heat power plant, for ocean thermal gradients, 110
Heat transfer
  liquids, 104
  organic media, 103
Heliostat field, projected by Messerschmidt-Bolkow-Blohm Co., 147
High-temperature steam electrolyzer, 187
Hindenberg, 266
Hubbert, prediction of coal production, 51
Hutter, map of wind conversion energy, 80
Hydrocracking, and hydrogen, 300
Hydroelectric supplies, importance for a hydrogen economy in the Sahara Desert, 30
Hydrogen
  advantages as an energy medium, 16
  as an aircraft fuel, 317
  and air transportation, 309
  amount transported from Huelva to Karlsruhe, 222
  and automotive fuel, 309
  bound
    chemically, 255
    physically, 255
  catalytic burners, 290
  combustion
    catalytic, 287
    direct, 288
    flameless, 293
    range, 269
  comparison
    with natural gas, 31
    with other fuels, for automotive transportation, 311
  construction of electrolyzers, 179
  -containing compounds, and the storage of

Hydrogen (*cont.*)
hydrogen, 260
dangers
chemical, 268
physical, 268
distribution and transport through cylinders, 209
economy
and aircraft, 74
block diagram, 19
and chemical technology, 33
concept, its origin, 34
and liquid wastes, 33
possible structure, 18
and shipping, 35
and storage, 32
and vehicular propulsion, 34
and water, 34
fuel cells
conversion into electricity, 273
at Siemens, 279
industrial applications, 294
and internal combustion engines, 311
liquefaction (schematic diagram), 250
methods of production (tabulated), 15
-oxygen fuel cell, 275
diagram, 277
pipelines, 206
calculations, 218
capital investment, 220
high-pressure, 214
method of transport, 203
network in Germany, 207
and night operation, 227
and storage medium, 228
production
electrolytic, 125
from nighttime electricity, 35
photochemical, using monomolecular layers, 164
photoelectrochemical, 170
photolytic, 157
synfuel, 299
from water, cyclical, 15
and safety, 265, 267, 268, 270
slush, 255
storage, 243, 247
at low temperatures, 249
optimization by thermodynamic calculation, 238
thermodynamic optimization,

Hydrogen (*cont.*)
technology, 63
and its introduction, 64
recent advances, 17
and the time frame, 43
transmission, calculations of Messer–Greisheim Company, 233
transport by rail cars, 211
uses
iron ore reduction, 304
prospects for, 307
synthesis of ammonia, 298
as a transmission medium, 201
and wind energy, 107
Hydrogenation
of coal, 303
direct, 303
of oil, 300
plants in Germany, 304
Pott–Broche method, 304

Ignition of hydrogen, 288
required minimal energy, 288
Industrial use of hydrogen, 307
Inflammability of hydrogen, 288
Internal combustion engines
and their operation with hydrogen, 315
powered by hydrogen, 311
Investment cost, in hydrogen pipelines, 218
Iron ore
direct
reduction, 304
by hydrogen, 306
Isotherms
for hydrogen dissociation, 256
and pressure, 257

Joffe, contributions to thermoelectrics, 83
Justi
and distribution of oxygen, 19
first block diagram for solar hydrogen (1964), 236
originator of double skeletal catalyzer principle, 187
and porous electrode concepts, 189
and residential air conditioning by heat loss to night air, 99
Justi and Winsel, and work with Raney catalysts, 277

Karlsruhe, pipeline costs to Huelva, 221

Keeling, recordings of CO_2 in Hawaii, 9
Kerosene, comparison with liquid hydrogen, 317
Kipker, calculations of distances for oxygen pipelines, 234
Koppers–Totcek, and conduction of hydrogen, 298
Kramer, review on wind energy, 108
Krupp
  hydrogenation of iron ore, 305
  and rotary furnace process, 305
Kuhn, photoelectric conversion of molecular layers, 101

Lawaczek, suggestion of 1933, 35
Lawson
  criterion, 55
  diagram, 55
Leanness of mixture, and use of hydrogen in automobiles, 314
Levi and Zener, ocean energy conversion, 111
Liquid helium, and great expense, 44
Liquid hydrogen
  comparison with kerosene, 317
  and Dewer flask, 251
  and double wall tank, 253
  and fueling aircraft, 319
  and hydrogen storage, 213
    block diagrams, 213
  as a storage medium, 249
  at Kennedy Space Center, Cape Canaveral, FL, 254
  as a transmission medium, 214
Living standard, and population, 95
Long-distance energy transfer, through gas pipelines, 12
Long and Gregory, their paper on hydrogen economy, 36
Low-temperature heat, from a hydrogen economy, 37
Low-temperature alkaline fuel cells, 276
Lurgi, and pressure electrolyzers, 183

Magnets, superconducting, 246
Manufacture
  of cadmium sulfide, 138
    costs of CdS cells as a function of time (prediction), 133
  of polyfoils in silicon cells, 129
Map
  Europe, various shading areas, 94
Map (*cont.*)
  Huls Co., and its pipeline network, 207
  wind energy conversion, 79
Marchetti and DeBeni, and thermochemical cycles (1970), 39
Martin–Marietta, and design of solar power plants, 149
Mass production, effect on solar cell production, 139
Materials
  for energy storage, 244
  shortages, 27
  world reserves, 28
Matrix, for direct energy conversion, 69
Media, and energy sources, 29
Mercea, Grecu, and Fodor, and catalytic hydrogen conversion in space heating, 298
Messerschmidt–Bolkow–Bohn, heliostat work, 145
Metal hydrides, 257
  and storage of hydrogen, 315
  tanks, 255
  and use in cars, 315
Metallurgy and refining, and effect on hydrogen economy, 33
Methane
  and gasoline
    safety, 267
  safety, 267
Methanol synthesis, and hydrogen, 302
Microwave
  radiation
    directed, 200
    transmission, 200
  transmission of energy, 214
Monomolecular layers
  arrangement for light collection, 166
  from fatty acids, 167

Natural gas
  in pipelines, 203
  and storage, 248
  and supplies to German industry, 48
Networks, of hydrogen pipelines, 206
Ng, Gregory, and Long, their paper on transmission of hydrogen through pipes, 30
Nighttime electricity, and storage in hydrogen, 35

Nitrogen oxide emission, as a function of the leanness of mixture, 314
Nuclear energy, comparison with other primary sources, 46
Nuclear fusion, its mechanism, 54

Ocean thermal energy conversion
  diagram, 107
  in Gulf of Mexico, 111
  technical problems, 111
Ocean thermal gradients, 108
Oil, in the German energy market, 47
Oil fields and storage, 248
  for hydrogen storage, 248
OPEC
  and limitations in production, 51
  oil production through 2025, 52
Optimization of calculations
  for hydrogen pipelines, 228
  of pipe diameter, 225
Organic heat transfer media, 103
Orientation, in silicon cells, 129
  and conversion efficiency, 129
Origin of hydrogen economy concept, 34
Ostwald, suggestion of fuel cells with coal, 75
OTEC
  diagram, 107
  technical problems, 111
Oxygen
  distribution, suggestion by Justi, 19
  transmission through pipelines, 235
  -zinc cells, 74

Para-hydrogen, its effect on liquefaction costs, 250
Peltier effect, 86
Perlite, in solution layers, for liquid hydrogen, 253
Phosphoric acid fuel cells, 279
Photochemical conversion (description), 100
Photoelectrochemistry
  mechanism, 172
  production
    of hydrogen, 170
    problems, 174
    use, 170
  types of cells, 173
Photogalvanic cells (1962 suggestion), 36
  production of hydrogen from the sea, 36
Photogalvanic conversion (description), 100
Photolysis of water, 158
Photolytic production of hydrogen, 157
Photosynthesis
  diagram, 159, 161
  effects, 105
  efficiency of sunlight conversion, 101
  and ferredoxin, 161
  the Z scheme, 160
Photothermal effect, 101
Photovoltaic cells, mechanism of action, 112
Photovoltaic conversion, (description), 100
  direct energy conversion, 80
Pipe diameter, and optimization of energy costs, 224
Pipe material, to be used for hydrogen, 223
Pipelines
  diameters, for long-distance pipes, 226
  and high-pressure transmission of hydrogen, 214
  and hydrogen, 203
  length from Huelva to Karlsrhue, 220
  and natural gas, 203
  networks, and transmission of energy, 203
  system, and storage, 248
  tabulation of costs, 232
Polycrystalline cells, their structure, 125
Plants
  apparatus for collecting energy, 165
  and photosynthetic bacteria, 162
Pockles, work of 1891, 168
Pollution
  caused by CO_2, 8
  vital factor in siting of energy power plants, 3
Population, and living standard, 95
Population centers, their distance from new power plants, 6
Pores, in electrolyzer systems, 188
Porosity, and potential for water electrolysis, 190
Porous electrode
  diagram, 275
  two-layer structure, 276
Pott-Broche, their method of hydrogenation, 304
Power output, on silicon photo cells, 119
Power plants
  effect of CO_2 on choice, 24
  effect of size on efficiency, 2
  for thermal gradients in the ocean, 109
Power stations
  solar, 123
    satellite, 151

Power towers, European work, 141
Prediction
of coal production, 50
of future for CO_2, 10
Pressure
cylinders, and gaseous hydrogen, 247
decrease in hydrogen pipeline during night operation, 227
effect upon energy transportations, 106
electrolyzers, 182
of hydrides, 256
storage pipes (diagram), 231
Primary energy consumption, and Federal Republic of Germany, 53
Principle of thermal gradients in the ocean, 108
Problems
of fast breeder reactors, 60
in fusion, 59
with solar cells, 123
Production cost of electricity, as a function of degree of use, 5
Prospect for catalytic combustion of hydrogen, 294
Protective layer cells of silicon, 111

Rail cars, and transport of hydrogen, 211
Raney catalysts, 276
electron microscope photographs, 278
experimentation, 277
Range for aircraft, as function of hydrogen, 320
Rappoport, and degree of efficiency in photo cells, 115
Reaction centers, in photolysis, 161
Reactors, fast breeder, 60
Recovery of energy, from high-pressure hydrogen (calculation), 238
Reduction
of iron ore, 304, 306
by hydrogen, 305
power, in photolysis, 159
Research, length of time required before implementation of hydrogen economy, 65
Reserves, and resources, 29

Safety
aspects
experimentation with hydrogen, 270
for liquid hydrogen in aircraft, 319
Safety (*cont.*)
engineering, and physical data, 267
and gasoline, 267
of German hydrogen pipeline network, 209
and hydrogen, 265, 267
instructions, with hydrogen, 268
and methane, 267
Sahara Desert, possible site for solar energy, 30
Satellites, problems in mounting, 152
Schematic diagrams
of fuel cell distribution and production, 284
of hydrogen liquefaction, 250
Schottky cells, 126
Seebeck effect, 81
Semiconductors, use in photoelectrochemical conversion, 173
Semitrailer, and the transport of hydrogen, 211
Serephin, commercial black nickel, 102
Service life
of cadmium sulfide, 135
of cadmium sulfide cells, 136
Shell, and synthesis of ammonia, 298
Shipping, and a hydrogen economy, 74
Shirland, work on cadmium sulfide cells, 135
Shishodia, and solar energy power plant, 148
Shortages, in materials, 27
Siemens, their work on hydrogen fuel cells, 278
Silicon cells
barrier layer, 112
conversion efficiency as a function of current density, 139
diagram, 113
photo cells
characteristics, 117
intensity of radiation, 118
protective layer, 111
solar, 114
polycrystalline, 124
problems, 123
Silicon, description of manufacture, 129
Silicon grains
effective diffusion length, 126
high purity, 129
Slush hydrogen, 255
use as a transport medium, 212
Solar air conditioning (diagram), 99
Solar cells, 123
and conversion efficiency, 128
Solar chemical options, with bacteria, 163

Solar chemical production, and long-term viewpoint, 163
Solar collector
 and cavity receiver production, 150
 used in Africa, 97
Solar electric generating stations, 153
Solar electricity, competitive with that from coal?, 151
Solar energy, 99
Solar energy
 basis for use, 89
 maximum annual supply, 91
 small-scale use in solar cars, 96
 technology, ways of collecting, 140
Solar farms, in Phoenix, AZ, 148
Solar generating stations, high performance, U.S., 149
Solar heat collection, and flat collectors, 93
Solar–hydrogen economy
 cost of a major plant, 22
 example, 23
 and Federal Republic of Germany, 23
 and Justi
  detailed design of solar power plant, 21
  first block diagram, 37
 transition to method, 20
Solar light, concentration, 139
Solar power plants
 Albuquerque, NM, 148
 in the United States, 148
Solar power stations, 145
 at Almeria, Spain, 235
Solar radiation, characteristics, 92
Solar satellite stations, 153
 power stations, 151
Space heating
 by hydrogen, 294
 and water, 97
Space-produced electricity, received on earth, 153
Space stations
 criteria for future use, 154
 solar collection, 153
Spectrum, solar radiation, 92
Steam electrolysis process, 186
Steam electrolyzer, high temperature, 187
SPE water electrolysis, 186
Storage
 above-ground tanks, 247
 capacities, 250
  various means of large-scale gas storage, 250
Storage (*cont.*)
 by electrochemical means, 245
 of hydrogen, 247
  of gaseous hydrogen, 247
  in hydrogen-containing compounds, 260
  of low-temperature hydrogen, 249
  in pipelines, 217
 media, 245
 in pipeline systems, 248
  summary of calculations, 230
  by thermal processes, 243
Sulfur, removal from fuel oil and hydrogen, 299
Sunlight, distribution over the areas of the world, 21
Superconducting magnets, 246
Superconductivity
 its discovery, 200
 and nonstoichiometric compounds, 200
Superconductors, and storage of hydrogen, 213
Supersonic aircraft, powered by hydrogen, 318
Synfuel production, and hydrogen, 299
Synthesis
 of ammonia, 298
 of methanol and hydrogen, 302

Tabor, and utilization of interference fringes for thermal conversion, 102
Tanks, with metal hydrides, 255
TARGET, programmed fuel cells, 281
Technical problems with OTEC, 111
Technologies, for cadmium sulfide cells, 138
Temperature–entropy diagram, and calculation of energy recovery, 239
Temperature gradients, tropical ocean, 106
Texaco, and synthesis of ammonia, 298
Thermal energy storage, 243
Thermal gradients, oceanic, 100, 106, 108
Thermochemical processes, 193
 suggestion by Marchetti and DeBeni, 194
Thermodynamic data, and water composition, 194
Thermodynamic optimization, for hydrogen storage, 236
Thermodynamics
 of water decomposition, 177
  as a function of temperature, 178
Thermoelectric conversion, 84
 direct energy conversion, 80
 and efficiency, 85
Thermoelectricity production, 86

Thermoelectrics, 81, 83
work of Justi and Lantz, 83
Thermomagnets, 85
Thin-layer cells, production, 135
Time frame
economic introduction of space stations, 154
for a hydrogen technology, 43
introduction, 63
Tokyo Bay, and the Japanese production fuel cells, 285
Transition to extensive use of hydrogen, in industry, 308
Transmission
through directed microwave radiation, 200
of energy, 199
through hydrogen, 201
Transport capacity, under various pressures, 205
Transport of energy in gases, cost, 204
Transport of hydrogen, cost as a function of distance, 32
Transportation, air, and hydrogen, 309
Two-layer structure, porous electrodes, 276

United States, and its solar power plants, 148
United Technologies
and phosphoric acid fuel cells, 280
their work on fuel cells, 282

Vacuum condensor pressure, in power plants, 7
VARTA, their fuel cells, 38
Vehicular propulsion
in a hydrogen economy, 34
by fuel cells, 310
Velocity of hydrogen, 288

Water
electrolysis
cell, with ELOFLUX principle, 190
potential and porosity, 190
electrolyzers
and efficiency, 193
at high temperature, 185
their practical construction, 179
and a hydrogen economy, 34
photolysis, 159
photolytic decomposition, 157
splitting
conclusions, 195
contributions by Otto Warburg, 159
thermal data, 105
thermodynamic decomposition, 126
Weyss, his neglect of the original suggestions of Haldane, Stewart, and Bockris, 40
Wind energy
and the Carnot factor, 107
conversion
direct energy conversion, 77
of hydrogen, 107
in England, 78
in Europe, 78
and the work of Betz, 78
Wind generators, types, 78
Woodwell, change of CO_2 concentration, 11
World energy resources (tabulation), 44
World reserves of materials, 28

Z scheme, in photosynthesis, 160
Zedansky and Alonza, and high-pressure electrolyzer, 181
Zinc-oxygen cells, 74
Zirconia ceramics, utilization in water electrolysis, 186
Zweig, work on hydrogen for buses, 312